AF252043

The Digital and the Real World

Computational Foundations of Mathematics, Science, Technology, and Philosophy

The Digital and the Real World

Computational Foundations of Mathematics, Science, Technology, and Philosophy

Klaus Mainzer

Technical University of Munich, Germany

NEW JERSEY · LONDON · SINGAPORE · BEIJING · SHANGHAI · HONG KONG · TAIPEI · CHENNAI · TOKYO

Prof. em. Dr. Klaus Mainzer
Emeritus of Excellence
Graduate School of Computer Science
Technical University of Munich
Germany
e-mail: mainzer@tum.de

Published by

World Scientific Publishing Co. Pte. Ltd.

5 Toh Tuck Link, Singapore 596224

USA office: 27 Warren Street, Suite 401-402, Hackensack, NJ 07601

UK office: 57 Shelton Street, Covent Garden, London WC2H 9HE

Library of Congress Cataloging-in-Publication Data
Names: Mainzer, Klaus, author.
Title: The digital and the real world : computational foundations of mathematics, science,
    technology, and philosophy / by Klaus Mainzer (Technical University of Munich, Germany).
Description: New Jersey : World Scientific, 2017. | Includes bibliographical references.
Identifiers: LCCN 2017023302 | ISBN 9789813225480 (hc : alk. paper)
Subjects: LCSH: Logic, Symbolic and mathematical. | Mathematics--Philosophy.
Classification: LCC QA8.4 .M35 2017 | DDC 510.1--dc23
LC record available at https://lccn.loc.gov/2017023302

British Library Cataloguing-in-Publication Data
A catalogue record for this book is available from the British Library.

For any available supplementary material, please visit
http://www.worldscientific.com/worldscibooks/10.1142/10583#t=suppl

Desk Editors: V. Vishnu Mohan/Tan Rok Ting

Typeset by Stallion Press
Email: enquiries@stallionpress.com

Printed in Singapore

Preface

In the 21st century, digitalization is a global challenge of mankind. Even for the public, it is obvious that our world is increasingly dominated by powerful algorithms and big data. But, how computable is our world? Some people believe that successful problem solving in science, technology, and economy only depends on fast algorithms and data mining. Chances and risks are often not recognized and understood because the foundations of algorithms and information systems are not studied rigorously. Actually, they are deeply rooted in logic, mathematics, computer science, and philosophy.

Therefore, this book studies the foundations of mathematics, science, and technology from a computational point of view. We start with basics of computability theory, proof theory, and information theory. In the second step, we introduce new concepts of information and computing systems in order to overcome the gap between the digital world of computers and the analog world of physical reality and sensor experience. Digital computing relates to a discrete reality which is mathematically modeled by finite quantities like bits and integers. Analog ("real") computing assumes a continuous reality which is mathematically represented by infinite quantities like real numbers. The gap between the digital and the analog can also be found in the global information networks: The Internet of Things (IoT) consists of digital apps and supercomputers as well as trillions of analog sensors. Their mathematical modeling is a great challenge of scientific computing. The book also considers consequences for digital and analog physics, digital and analog functions of the brain, digital and real computing in economic and financial applications. The epistemic question arises whether human thinking (e.g., mathematics) is more than digital computing beyond Turing computability.

Traditionally, foundational debates on mathematics are considered as philosophical issues which are far away from the mathematical applications to solve real-world problems. However, constructive problem solving is deeply connected with the foundations of constructive logic and mathematics. A constructive solution comes with an algorithm to compute it directly. But, sometimes, mathematicians only prove indirectly that a solution exists consistently without telling us how to find the solution. If there is no constructive proof, the question arises how far away it is from being constructive. We suggest a foundational theory with degrees of constructivity, provability, and computability for practical applications. Finally, the gap between the digital and the real-world can be overcome by these degrees of constructivity, provability, and computability: How far can real-world problems be reduced to degrees of constructive proving and computing?

Obviously, this kind of foundational research has practical consequences and societal impact. In the age of big data and increasing complexity of human civilization, constructive proofs and tests of algorithms are urgently demanded to avoid a dangerous future with overwhelming computational power running out of control. Automated theorem proving which came up in the foundations of mathematics and computer science can help prevent costly and dangerous flaws in global networks. The degrees of constructivity, provability, and computability improve control, security, and reliability.

The book aims at students at graduate level, scientists and readers who are interested in foundational, interdisciplinary, and philosophical questions of mathematics, computer science, and science in general. In Munich, my colleagues, Helmut Schwichtenberg, Peter Schuster, and I initiated an international autumn school on "Proof, Computation, and Digitalization in Logic, Mathematics, Computer

Science, and Philosophy" which considers the theoretical foundations as well as their practical applications. I thank all the colleagues and postdocs of this school for their inspiration. Our Munich enterprise should offer an interdisciplinary platform of foundational discussion and research on logic, mathematics and computer science which are, without any doubt, key disciplines of modern digital civilization. It is an exciting experience that these debates are deeply involved in philosophy as well as in the societal impact of future technologies.

Klaus Mainzer
Munich

February 2017

Contents

Chapter 1

Introduction

Historically, the theory of algorithms and computability started in the beginning of the last century with foundational debates on logics, mathematics, and philosophy. Therefore, the *digital world* has its origins in logical, mathematical and philosophical theories.

Alan Turing anticipated the development of the digital general purpose computer by his famous Turing machine (Turing, 1936). Here, we have a finite state control device with a read/write head and a two-way unlimited tape consisting of an unlimited number of cells. The control device is regulated by a program which is a finite set of instructions with states of the machine, operations for printing 0, 1 and * (blank) into the cells, moving left or right one cell, and stopping the tape.

A number-theoretic function is defined to be computable if it is the input–output function of a Turing machine. Further on, a set of natural numbers is decidable if there is an effective procedure (characteristic function) that given any element of the set will decide in a finite number of steps whether or not the element is in the set. In short, for decidability of a set, its characteristic function must be Turing computable.

All number-theoretic propositions can be coded with the digits 0 and 1. Thus, Turing's theory of computability in computer science is reduced to the digital paradigm. The Turing machine is mathematically equivalent to many other effective procedures (e.g., register machines, recursive number-theoretic functions). Therefore,

Church's thesis demands that Turing-computability means (digital) computability in general.

But, there seems to be a *deep gap between digital computer science and mathematics*. Many mathematical disciplines like the calculus heavily depend on the continuous nature of real numbers. In (classical) physics, the dynamics of nature is modeled by continuous curves and differential equations. Their solutions correspond to physical events which can be predicted or explained as causal effects under certain constraints. At least in classical physics, nature is assumed to be continuous, and even quantum physics uses real and complex-valued quantities of fundamental laws of continuous differential equations (e.g., Schrödinger equation).

Our perceptions of the world seem to be continuous: Our bodily sensors receive analog signals of, e.g., electromagnetic and acoustic waves. Behind all that, there is an old and deep conflict in philosophy between the continuous and discrete characteristics of nature since the Antiqutity. Is matter continuous (Aristotle) or atomic (Democritus)? Nowadays, in technology, we distinguish bewteen digital (discrete) and analog (continuous) signal processing.

Before the digital paradigm of computer science started with the Turing machine, there was already an old tradition of *real algorithms in mathematics*. Since antiquity, undecidability problems of geometrical constructions (e.g., trisection of an angle) were discussed (Mainzer, 1980). The final answer was found in the beginning of the 19th century by Galois's result on the non-solvability by radicals of polynomial equations of degree 5 or more. These algorithms manipulate real numbers. There is a long standing tradition of decidability results in algebra and analysis leading to modern numerical analysis. We remind the reader of Newton's method for finding (approximate) zeros of polynomials in one variable. Newton's method was one of the first search algorithms in numerical analysis and scientific computing.

Nowadays, *practical algorithms* to solve equations of physics, models of climate, weather prediction, or financial mathematics mainly refer to numerical analysis. Therefore, we not only need logical–mathematical foundations of digital computing (Turing, 1936) but also foundational studies of real algorithms. They have a long tradition in mathematics from Newton, Gauss, Euler, etc., to modern numerical analysis and scientific computing.

What do real computability and decidability mean? It is amazing that, from a logical point of view, the digital world of natural numbers seems to be much more complicated than the continuous world of real analysis, although counting digits is easily done by children and the naturals numbers are a subset of the real numbers: But, according to Gödel's famous proof, arithmetic is incomplete, and a closed real field is decidable.

How far can we transfer the results of computability and decidability from digital to real mathematics? These results have consequences for the discrete (digital) and the analog (continuous and real) and with that to the modern analog world of sensor technology and human experience by analog sensors.

Computability of algorithms in computer science is also deeply linked with the *provability of theorems in mathematics*. Since antiquity, truth of mathematical theorems was verified by logical proofs in axiomatic systems. Logical-axiomatic proofs became the paradigm in science and philosophy (Hilbert, 1918). In the 20th century, the question arose whether the true propositions of a theory can be described completely, correctly, and consistently in formal systems. Famous logicians, mathematicians, and philosophers of the 20th century (e.g., Hilbert, Gödel, and Turing) demonstrated the possibilities and limitations of formalization.

It is well known that mathematical proofs sometimes only guarantee the existence of a solution without providing an effective algorithm and constructive problem-solving procedure. The reason is that mathematics as it is normally practised is based on the law of excluded middle: Either a statement is true or its negation is true. According to this law, it is sufficient to show indirectly that the assumption of non-existence of a solution is false (Mainzer, 1970). But, for practical applications, it is, of course, a great disadvantage that indirect proofs do not tell us how to find a constructive solution step by step (Bishop, 1967).

From a digital point of view, constructive proofs should be realized by digital systems. If mathematics is restricted to computable functions of natural numbers, Turing machines will do the job. But, in higher mathematics and physics (e.g., functional analysis), we have to consider functionals and spaces of higher types. Therefore, a general concept of a digital information system is necessary to compute finite approximations of functionals of higher types. Contrary to the machine orientation of computational mathematics, intuitionistic mathematics (INT) is rooted in the philosophy of human creativity. Mathematics is understood as a human activity of constructing and proving step by step. In the rigorous sense of Brouwer's intuitionism, we even get a concept of real continuum and infinity which differs from ordinary understanding of classical mathematics. In the foundational research programs of proof mining and reverse mathematics, we offer degrees of constructivity and provability to classify problem solving and proving for different mathematical applications.

Proof mining has the aim to obtain the missing information in an incomplete theorem and to extract effective procedures by a purely logical analysis of mathematical proofs and principles (Kohlenbach, 2008). Sometimes, it is even sufficient to find bounds for search processes of problem solving. At least, proof mining tells us how far away a proof is from being constructive. The research program of proof mining goes back to Georg Kreisel's proof-theoretic research. He illustrates the extraction of effective procedures from proofs as "unwinding proofs" (Feferman, 1996).

From a theoretical point of view, proof mining is an important link between logic, mathematics, and computer science. Logic is no longer only a formal activity besides and separated from mathematics. Actually, metatheorems of proof mining deliver *logical tools to solve mathematical problems* more effectively and to obtain new mathematical information: Proofs are more than verification of theorems!

From a practical point of view, proof theory also has practical consequences in applied computer science. In this case, instead of formal theories and proofs, we consider formal models and computer programs of processes, e.g., in industry. Formal derivations of formula correspond to, e.g., steps of industrial production. In order to avoid additional costs of mistakes and biases, we should test and prove that the *computer programs* are correct and sound before applying them. *Automated theorem proving* is applied to integrated designs and verification. Software and hardware designs must be verified to prevent costly and life-threatening errors. Dangerous examples in the past were flaws in the microchips of space rockets and medical equipments. Network security of the Internet is a challenge for banks, governments, and commercial organizations. In the age of Big Data and increasing complexity of our living conditions, rigorous proofs and tests are urgently demanded to avoid a dangerous future with overwhelming computational power running out of control.

But, in real mathematics, proofs cannot all be reduced to effective procedures. Actually, there are *degrees of constructivity, computability, and provability*. How strong must a theory be in order to prove a certain theorem? Therefore, we try to determine which axioms are required to prove a theorem. How constructive and computational are these assumptions?

Instead of going forward from axioms to theorems in usual proofs, we prove in the reverse way (backward) from a given theorem to the assumed axioms. Therefore, this research program is called *reverse mathematics* (Friedman, 1975). If an axiom system S proves a theorem T and theorem T together with an axiom system S' (the reversal) prove axiom system S, then S is called equivalent to theorem T over S'. In order to determine the degrees of constructivity and computability, one tries to characterize mathematical theorems by equivalent subsystems of arithmetic. The formulas of these arithmetic subsystems can be distinguished by different degrees of complexity.

In (second-order) arithmetic, all objects are represented as natural numbers or sets of natural numbers. Therefore, proofs about real numbers must use Cauchy sequences of rational numbers which can be represented as sets of natural numbers. In reverse mathematics, theorems and principles of real mathematics are characterized by equivalent subsystems of arithmetic with different degrees of constructivity and computability (Ishihara, 2005). Some of them are computable in Turing's sense, but others are definitely not and need stronger tools beyond Turing computability. Therefore, it is proved that only parts of real mathematics can be reduced to the digital paradigm. Obviously, reverse mathematics and proof mining are important research programs to clarify and deepen the connections between mathematics, computer science, and logic in a rigorous way.

Another important bridge between logic, mathematics, and computer science is type theory. Types of data (resp., terms) are distinguished in computer programs of computer science as well as in formal systems of mathematics, in order to avoid software bugs (resp., logical paradoxes). Martin-Löf's *intuitionistic type theory* offers both a philosophical foundation of constructive mathematics as well as a proof assistant in computer science (e.g., Coq). Recently, homotopy type theory extends formalization from set-theoretical objects up to categories related to more and more large cardinals. *Homotopy type theory* (HoTT) tried to develop a *universal* ("*univalent*") *foundation of mathematics* as well as computer languages with respect to proof assistants for advanced mathematical proofs. The question arises how far univalent foundations of mathematics can be constructive.

Philosophically, constructive foundations of mathematics are deeply rooted in the epistemic tradition of efficient reasoning with the *principle of parsimony* in explanations, proofs, and theories. It was the medieval logician and philosopher William of Ockham (1285–1347) who first demanded that one should prefer explanations and proofs with the fewest number of assumed abstract concepts and principles. The reason is that, according to Ockham, universals (e.g., mathematical sets) are only abstractions from individuals (the elements of a set) without real existence. Ockham's principle of parsimony later became popular as "Ockham's razor" reducing abstract principles as far as possible.

The question arises how far reduction is possible without loss of essential information. In order to analyze real computation in mathematics, we need an extension of *digital computability beyond Turing computability. Real computing machines* can be considered idealized analog computers accepting real numbers as given infinite entities (Blum *et al.*, 1989). They are not only theoretically interesting but also practically inspiring with respect to their applications in scientific computing and technology (e.g., sensor technology).

These aspects are also exciting for *computational neuroscience* and *cognitive science*. Appropriate neural networks or cellar automata are computationally equivalent to Turing machines. Further on, we can distinguish a hierarchy of automata and machines which can realize formal languages with increasing complexity. Examples are finite automata accepting regular languages or Turing machines accepting Chomsky grammars of natural languages. Actually, (recurrent) neural networks with rational numbers as synaptic weights are computationally equivalent to Turing machines accepting recursive languages like Chomsky grammars. It can be proven that (recurrent) neural networks with non-computable real weights can even operate on non-recursive languages (Siegelmann, 1994). Human brains also operate on non-recursive natural languages. If analog neural networks are considered as models of human brains, they seem to be more powerful than digital Turing machines.

The foundational debate on the digital and analog, discrete and continuous also has deep consequences for *physics*. In classical physics, the real numbers are assumed to correspond to continuous states of physical reality. For example, electromagnetic or gravitational fields are mathematically modeled by continuous manifolds. Thus, "real computing" (i.e., computing with real numbers) seems to model effective procedures in a continuous reality. Fluid dynamics illustrates this paradigm with continuous differential equations.

But, in modern quantum physics, a "coarse grained" reality seems to be more appropriate. Quantum systems are characterized by discrete quantum states which can be defined as quantum bits. Instead of classical information systems, following the digital concept of a Turing machine, quantum information systems with quantum algorithms and quantum bits open new avenues to *digital physics*. *Information* and *information systems* are no longer fundamental categories only of information theory, computer science, and logic but also of *physics* (Mainzer, 2016b).

But, is it possible to reduce the real world to a quantum computer as an extended concept of a universal quantum Turing machine? "It from bit" proclaimed physicist John A. Wheeler (1990). On the other side, fundamental symmetries of physics (e.g., Lorentz symmetry and electroweak symmetry) are continuous (Audretsch and Mainzer, 1996; Mainzer, 2005). Einstein's space–time is also continuous. Are they only abstractions (in the sense of Ockham) and approximations to a discrete reality?

Some authors proclaim a discrete cosmic beginning with an initial quantum system (e.g., quantum vacuum) evolving in an expanding macroscopic universe with increasing complexity which can be approximately modeled by continuous mathematics. In this case, physical reality is discrete and continuous mathematics with real numbers only a (ingenious) human invention. But physical laws including quantum theory (e.g., Hilbert space) are infused with real numbers and the mathematics of the continuum. Thus, from an extreme Platonic point of view, we could also assume a layer much deeper than the discrete quantum world — the universe of mathematical structures themselves as primary reality with infinity and continuity.

Anyway, besides all ontological speculations, mathematics is fundamental for science and cannot be reduced to the digits and the discrete nature of computer science. We should distinguish *degrees of constructivity, computability, and provability* in the rigorous sense of *real computing, proof mining*, and *reverse and univalent mathematics*.

In this case, Stephen Wolfram's vague concept of computational equivalence can be made precise and corrected (Wolfram, 2002). He demanded that the natural processes of atomic, molecular, and cellular systems should be considered as "computationally equivalent" procedures of Turing machines, cellular automata, and neural networks (in the sense of a physically extended Church's thesis). This idea already came up with John von Neumann's and Konrad Zuse's cellular automata (Zuse, 1969). At least, the mathematical models and theories of these natural systems can be distinguished by different degrees of complexity. Therefore, we need the corresponding mathematical theories and models with their equations and laws in order to derive reliable predictions, classifications, and explanations. Quasi-empirical computer experiments in the sense of Wolfram, even with powerful computers and algorithms, are not sufficient.

The exponentially increasing power of supercomputers and global networks is overwhelming. Success and efficiency in science and economy seem to only depend on fast algorithms and huge databases. In economy, they rapidly predict future trends and profiles of products and customers. In science, they recognize correlations and patterns in huge collections of data (e.g., elementary particle colliders in high energy physics, machine learning algorithms in molecular biology). Science (like economy) seems to be more and more driven by fast algorithms and a huge amount of data: *Big Data −The end of theory*? asked Chris Anderson, an influential publisher of digital media.

Wolfram even proclaimed a "new kind of science", in which *computer experiments* would replace mathematical proofs and theories. Wolfram had simulated extensive pattern formations of cellular automata on high performance computers, discovered some remarkable correlations, and classified the patterns based on his observations. But one can prove that only the fundamental mathematical laws of cellular automata allow accurate forecasting and classification of patterns (Mainzer and Chua, 2011).

This argument can be extended to the emergence of patterns in nature, taking in physics, chemistry, biology, and brain research (Mainzer and Chua, 2013). Here too, we found that it was only when the basic equations were known that accurate declarations and forecasts could be made on the emergence of structure and patterns. What we can generally say about science is that theory is often the best way to solve a problem. How will a mountain of data help me if I don't know what I am looking for?

Correlation cannot replace causation. Causation is explained by causal laws which are mathematically represented by equations of dynamical systems. Quasi-empirical computer experiments with Big Data may be helpful for a rough orientation. But, we need a deep analysis of the computational foundations in mathematics, science, technology and philosophy for reliable tools of problem solving in a digital world. In the end, we need *well-founded mathematical theories* with different *degrees of constructivity, computability, and provability.*

This book starts with "Basics of Computability" (Chapter 2) from Turing's theory of computability to decidable and undecidable problems, followed by "Hierarchies of Computability" (Chapter 3) with computational degrees and complexity of problem solving. Chapter 4 explains "Constructive Proof Theory" with different degrees of constructive proofs. Chapters 2–4 consider computability and provability on the basis of elementary number theory. In order to prepare proof mining in higher mathematics, Turing-computable functions are no longer sufficient. What do computable functionals of higher type (e.g., functions or sets of number-theoretic functions) mean? Chapter 5 "Computational Mathematics and Digital Information Systems" introduces a general concept of information system beyond Turing computability. The question arises how far the corresponding functionals can be approximated by constructive procedures.

Chapter 6 considers the foundations of "Intuitionistic Mathematics and Human Creativity". INT is not only interesting because of its rigorous ban of the law of excluded middle. Brouwer's philosophical understanding of mathematical constructing and proving leads to a different concept of real continuum and infinity. His fan theorem and bar theorem are consequences of the intuitionistic philosophy. In the past, intuitionism and classical mathematics often fought against one another as different ideological schools. But, independent of ideological points of view, these principles can be considered as additional axioms which may be accepted or not like all axioms or hypotheses in science. Anyway, independent of their philosophical meaning, we can prove mathematical implications and equivalent theorems of these principles in order to classify mathematical theorems and theories according to their degrees of constructivity and provability.

A main motivation of this book is to overcome the gaps between logic, mathematics, and computer science. Nowadays, logic is often considered as a discipline of only theoretical and philosophical interest with no practical relevance for ordinary mathematics and its applications. In Chapters 7 and 8, we study two foundational research programs bridging the gap between theory and application, philosophy and problem solving.

Chapter 7 is dedicated to "Proof Mining bridging Logic, Mathematics, and Computer Science". Starting with extractions of effective information from proofs in elementary number theory, proof mining is extended to, e.g., numerical analysis and functional analysis which are important for practical applications. The main idea is that proofs can be characterized by more or less constructive and computable functionals. Actually, these functionals can be considered as elements of information systems with different degrees of complexity below or beyond Turing computability. Historically, Gödel started with the class of primitive recursive functionals which was extended in his Dialectica interpretation. Later on, a variety of similar interpretations were studied to analyze proofs in different mathematical theories. Sometimes, functional interpretations make the extraction of effective computer programs possible to realize proofs automatically. There are interactive proof assistants such as Coq (Bertot and Castéran, 2004), HOL, Isabelle (Nipkow *et al.*, 2002), or MINLOG (Schwichtenberg, 2006). We will consider some applications of MINLOG because of its natural understanding of constructive logic. In general, automated theorem proving has deep technical and societal impact in order to prevent errors and flaws in software and hardware design, in the Internet and global communication (cf. Chapter 16). This is an important breakthrough to link proof theory with computer science and mathematics.

The link between mathematics, proof theory, and computer science is also supported by the research program of "Reverse Mathematics bridging Logic, Mathematics, and Computer Science", which is considered in Chapter 8. Reverse mathematics allows one to determine the proof-theoretic strength, degree of computability and complexity of theorems by classifying them with respect to equivalent theories and proofs. Many theorems of classical mathematics can be classified by subsystems of second-order arithmetic $\mathbb{Z}_2$ with variables of natural numbers and variables of sets and functions of natural numbers.

Chapter 9 considers the development "From Intuitionistic to Homotypy Type Theory — Bridging Logic, Mathematics, and Computer Science". Actually, types of data resp. terms are used in programming languages as well as in formal systems of mathematics in order to avoid software failures resp. logical paradoxes. An important evolution taking place in current mathematics is the transition into a new era of automated tools for proof construction and verification which are known as proof assistants. Martin-Löf intuitionistic type theory was an important step to the most powerful proof assistants (such as Coq). An exciting development is the recent *univalent foundations project* of the Institute for Advanced Study (Princeton) which provides new semantics for type theories using notions from geometry and topology up to abstract mathematical categories. The key question is how far constructive proof procedures have the effect of extending the validity of mathematical reasoning to the widest possible context.

Chapter 10, "Real Computability and Real Analysis", transfers the digital concepts of computability and decidability to the continuous world of real numbers. Decision machines and search procedures are introduced over real and complex numbers. A universal machine (generalizing Turing's universal machine in the digital world) can be defined over a mathematical ring (and with that over real and complex numbers). On this basis, decidability and undecidability of several theoretical and practical mathematical problems (e.g., Mandelbrot set, Newton's method) can rigorously be proved.

Like in the digital world, we can introduce a "Complexity Theory of Real Computing" (Chapter 11). Polynomial time reductions as well as the class of NP problems are studied over a general mathematical ring (and with that on real and complex numbers). Real computability can be extended in a polynomial hierarchy for unrestricted machines over a mathematical field.

Chapter 12 discusses the consequences of "Real Computing and Neural Networks". Neural networks are models of natural brains in evolution with different degrees of complexity. Mathematically, they are equivalent to computing machines with different degrees of complexity from finite automata to Turing machines. Otherwise, the simulation of neural networks on digital computers would not be possible. Computing machines can understand formal languages according to their degree of computability from simple regular languages (finite automata) to Chomsky grammars of natural languages (Turing machines). Analog neural networks with real synaptic weights are beyond digital Turing computability. They correspond to real computing. Actually, they can operate on non-computable languages like human brains.

Brains and computers are examples of information processing machines. What does information mean? Chapter 13 distinguishes the "Complexity of Algorithmic Information". Chaitin's concept of algorithmic information is closely connected with Gödel's incompleteness and Turing's halting problem of Turing machines (Chaitin, 2007). Laws and theories can be understood as algorithmic information compression. But Chaitin's algorithmic information is reduced to the digital world.

If we consider dynamical systems in the "real" world of physics, chemistry or biology, we need concepts of continuous information flow which are explained in Chapter 14, "Complexity of Information Dynamics". Basic information measures from Shannon's information entropy to Kolmogorov–Sinai entropy are distinguished. How can continuous dynamics be related to discrete symbolic dynamics?

Chapter 15 "Digital and Real Physics," considers the foundations of digital physics. In this case, the universe itself is assumed to be a gigantic information system with quantum bits as elementary states. Deutsch (1985) discussed the quantum versions of Turing computability and Church's thesis. They are interesting for quantum computers, but still reduced to the digital paradigm. Obviously, modeling the "real" universe also needs real computing. Are continuous models only useful approximations of an actually discrete reality?

With Chapter 16, the book ends with "Digital and Real Computing in the Social World". The chapter starts with the question why it is so difficult to model the social world. Economics and financial mathematics are highly mathematized, but they miss the success of explanation and predictions of mathematical natural sciences. We consider the degrees of constructivity, provability, and computability of fundamental economic and financial theorems in the sense of reverse mathematics. Ockham's principle of parsimony is actually a rule of economic efficiency in theory building: The degree of theoretical abstractions corresponds to the "price" we are willing to pay for a reliable tool of problem solving.

Scientific modeling of complex social systems is extended to global information and computer networks of Internet of Things (IoT) with Big Data. These complex networks are digital with their apps and supercomputers and also analog with trillions of sensors. Their mathematical modeling is a great challenge of scientific computing. It is also deeply connected with real computing, proof mining and reverse mathematics. Chapter 17 is with a philosophical outlook and a plea for more foundational research in an age of exponentially growing technologies.

Chapter 2

Basics of Computability

The early historical roots of computer science can be found in age of classical mechanics. The mechanization of thoughts begins with the invention of mechanical devices for performing elementary arithmetic operations automatically. A mechanical calculation machine executes serial instructions step by step. In general, the traditional design of a mechanical calculation machine contains the following devices.

First, there is an input mechanism by which a number is entered into the machine. A selector mechanism selects and provides the mechanical motion to cause the addition or subtraction of values on the register mechanism. The register mechanism is necessary to indicate the value of a number stored within the machine, technically realized by a series of wheels or disks. If a carry is generated because one of the digits in the result register advances from 9 to 0, then that carry must be propagated by a carry mechanism to the next digit or even across the entire result register. A control mechanism ensures that all gears are properly positioned at the end of each addition cycle to avoid false results or jamming the machine. An erasing mechanism has to reset the register mechanism to store a value of zero.

Wilhelm Schickard (1592–1635), Professor of Hebrew, oriental languages, mathematics, astronomy, and geography, is presumed to be the first inventor of a mechanical calculating machine for the first four rules of arithmetic. The adding and subtracting part of his machine is realized by a gear drive with an automatic carry mechanism. The multiplication and division mechanism is based on Napier's multiplication tables. Blaise Pascal (1623–1662), the brilliant French mathematician and philosopher, invented an adding and subtracting machine with a sophisticated carry mechanism which in principle is still realized in our hodometers of today (Williams, 1985).

But it was Leibniz's mechanical calculating machine for the first four rules of arithmetic which contained each of the mechanical devices from the input, selector, and register mechanism to the carry, control, and erasing mechanism. The Leibniz's machine became the prototype of a hand calculating machine. If we abstract from the technical details and particular mechanical constructions of Leibniz's machine, then we get a model of an ideal calculating machine.

Definition of Leibniz's mechanical calculating machine. Figure 1 is a scheme of this ideal machine with a crank C and three number stores SM, TM, RM. Natural numbers can be entered in the set-up (input) mechanism SM by the set-up handles SH. If crank C is turned to the right, then the contents of SM are added to the contents of the result mechanism RM (Register Machine), and the contents of the turning mechanism TM (Turing Machine) are raised by 1. A turn to the left with crank C subtracts the contents of SM from the contents of RM and diminishes the contents of TM by 1.

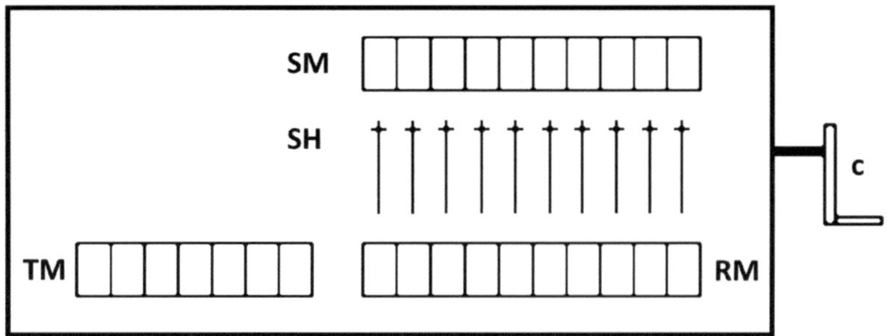

Fig. 1. Hand calculating machine.

Addition means the following. At the beginning of the calculation, the erasing procedure is implemented by setting TM and RM to zero. Then the first number is set up in SM by SH. A turn to the right of crank C transports this number into RM. In other words, the number is added to the zero in RM. Now, the second number is set up in the SM and added to the contents of RM by a turn to the right. The sum of both numbers can be read in the RM. After turning the crank twice to the right, the TM shows 2. Multiplication only means a repeated addition of the same number. The product $b \cdot a$ results from adding the number a to itself b times.

Leibniz even designed a mechanical calculating machine for the binary number system with only two digits 0 and 1, which he discovered some years earlier (von Mackensen, 1974). He described a mechanism for translating a decimal number into the corresponding binary number and vice versa. As modern electronic computers only have two states 1 (electronic impulse) and 0 (no electronic impulse), Leibniz truly became one of the pioneers of computer science.

Leibniz's historical machines suffered from many technical problems because the materials and technical skills then available were not up to the demands. Nevertheless, his design is part of a general research program (Leibniz, 1998).

Leibniz's mathesis universalis

Leibniz's *mathesis universalis* (Scholz, 1961) intended to simulate human thinking by calculation procedures ("algorithms") and to implement them on mechanical calculating machines. Leibniz proclaimed two basic subdisciplines:

- An *ars iudicandi* should allow every scientific problem to be decided by an appropriate arithmetic algorithm after its codification into numeric symbols.
- An *ars inveniendi* should allow scientists to seek and enumerate possible solutions of scientific problems.

Leibniz's *mathesis universalis* already seems to foreshadow the famous Hilbert program in our century with its demands for formalization and axiomatization of mathematical knowledge. Actually, Leibniz developed some procedures to formalize and codify languages. He was deeply convinced that there are universal algorithms to decide all problems in the world by mechanical devices.

In the 19th century, it was the English mathematician and economist Charles Babbage who not only constructed the first program-controlled calculation machine (the "analytical engine") but also studied its economic and social consequences (Bromley, 1982). A forerunner of his famous book *On the Economy of Machinery and Manufactures* (1841) was Adam Smith's idea of economic laws, which paralleled Newton's mechanical laws. In his book *The Wealth of Nations*, Smith described the industrial production of pins as an algorithmic procedure and anticipated Henry Ford's idea of program-controlled mass production in industry.

The modern formal logic of Frege and Russell and the mathematical proof theory of Hilbert and Gödel have been mainly influenced by Leibniz's program of *mathesis universalis*. The hand calculating machine which was abstracted from the Leibniz's machine can easily be generalized to Marvin Minsky's so-called register machine (Minsky, 1961; Sheperdson and Sturgis, 1963). It allows the general concept of computability to be defined in modern computer science.

Definition of a register machine. A hand calculating machine had only two registers TM and RM, and only rather small natural numbers can be input. An ideal register machine has finite number of registers which can store any finite number of a desired quantity. The registers are denoted by natural numbers $i = 1, 2, 3, \ldots$. The contents of register i are denoted by $\langle i \rangle$. As an example, the device $\langle 4 \rangle := 1$ means that the content of the register with number 4 is 1. The register is empty if it has the content 0.

In the hand calculating machine, an addition or subtraction was realized only for the two registers $\langle SM \rangle$ and $\langle RM \rangle$, with $\langle SM \rangle + \langle RM \rangle$ or $\langle RM \rangle - \langle SM \rangle$ going into the register RM. In a register machine, the result of subtraction $\langle i \rangle - \langle j \rangle$ should be 0 if $\langle j \rangle$ is greater than $\langle i \rangle$. This modified subtraction is denoted by $\langle i \rangle \mathbin{\dot-} \langle j \rangle$. In general, the program of an ideal register machine is defined using the following elementary procedures as building blocks:

(1) Add 1 to $\langle i \rangle$ and put the result into register i, in short $\langle i \rangle := \langle i \rangle + 1$.
(2) Subtract 1 from $\langle i \rangle$ and put the result into register i, in short: $\langle i \rangle := \langle i \rangle \mathbin{\dot-} 1$.

These two elementary procedures can be composed using the following concepts:

(3) If P and Q are well-defined programs, then the chain $P \to Q$ is a well-defined program. $P \to Q$ means that a machine has to execute program Q after program P.
(4) The iteration of a program, which is necessary for multiplication, for instance, as iterated addition is controlled by the question of whether a certain register is empty.

A diagram illustrates this feedback:

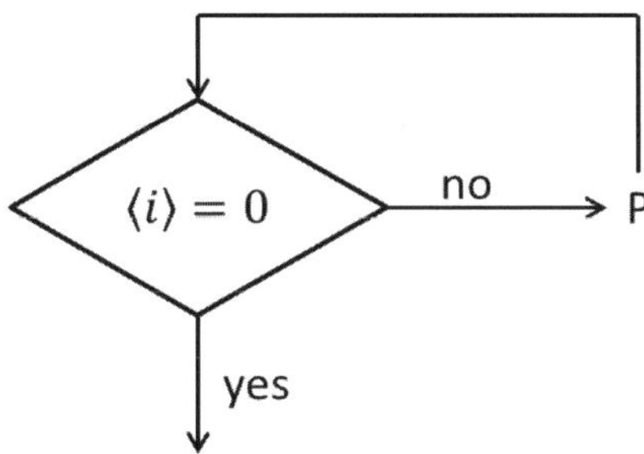

If P is a well-defined program, then execute P until the content of the register with number i is zero.

Example of a register machine. Each elementary operation (1) and (2) of a program is counted as a step of computation. A simple example is the following addition program:

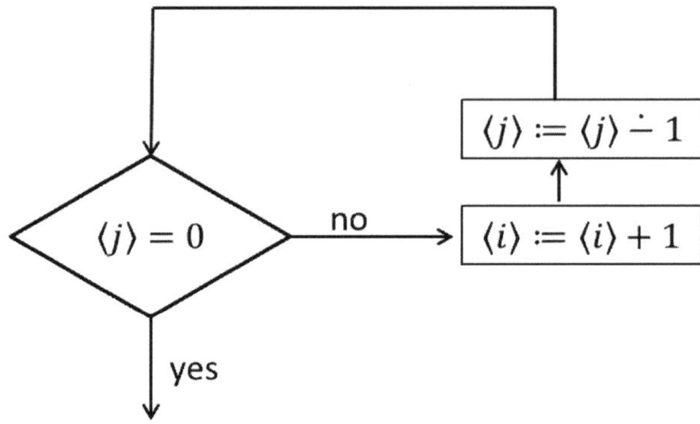

Each state of the machine is illustrated by the following matrix, which incrementally adds the content y of register $\langle j \rangle$ to the content x of register $\langle i \rangle$ and simultaneously decrements the content of $\langle j \rangle$ to zero. The result $x + y$ of the addition is shown in register $\langle j \rangle$:

$$
\begin{array}{cc}
\langle i \rangle & \langle j \rangle \\
x & y \\
x+1 & y \mathbin{\dot-} 1 \\
\vdots & \vdots \\
x+y & y \mathbin{\dot-} y
\end{array}
$$

Definition of register machine computability. A register machine (RM) with program F is defined to compute a function f with n arguments if for arbitrary $x_1, \ldots, x_n$ in the register $1, \ldots, n$ (and zero in all other ones), the program F is executed and stops after a finite number of steps with the arguments of the function in the register $1, \ldots, n$ and the function value $f(x_1, \ldots, x_n)$ in register $n + 1$.

The program

$$\langle 1 \rangle := x_1; \ldots; \langle n \rangle := x_n$$
$$\downarrow$$
$$F$$
$$\downarrow$$
$$\langle n + 1 \rangle := f(x_1, \ldots, x_n)$$

works according to a corresponding matrix. A function f is called computable by an RM (RM-computable) if there is a program F computing f.

The number of steps which a certain program F needs to compute a function f is determined by the program and depends on the arguments of the function. The complexity of program F is measured by a function $s_F(x_1, \ldots, x_n)$ computing the steps of computation according to program F. Sometimes, $s_F(x_1, \ldots, x_n)$ is also said to measure the computational time of program F. For example, the matrix of the addition program for $x + y$ shows that y elementary steps of adding 1 and y elementary steps of subtracting 1 are necessary. Thus $s_F(x, y) = 2y$.

Definition of RM complexity. As an RM-computable function f may be computed by several programs, a function g is called the step counting function of f if there is a program F to compute f with $g(x_1, \ldots, x_n) = s_F(x_1, \ldots, x_n)$ for all arguments $x_1, \ldots, x_n$. The complexity of a function is defined as the complexity of the best program computing the function with the least number of steps.

Obviously, Minsky's register machine is an intuitive generalization of a hand calculating machine *à la* Leibniz. But, historically, some other equivalent formulations of machines were at first introduced independently by Alan Turing (1936) and Emil Post (1936) in 1936.

Definition of a Turing machine. A Turing machine (Fig. 2) can carry out any effective procedure provided it is correctly programmed. It consists of

(a) a control box in which a finite program is placed,
(b) a potentially infinite tape, divided lengthwise into squares,
(c) a device for scanning, or printing on one square of the tape at a time, and for moving along the tape or stopping, all under the command of the control box.

If the symbols used by a Turing machine are restricted to a stroke / and a blank *, then an RM-computable function can be

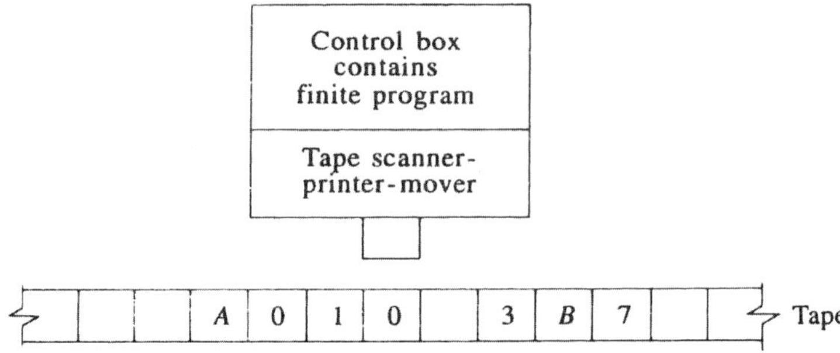

Fig. 2. Turing machine.

proved to be computable by a Turing machine and vice versa. We
must remember that every natural number x can be represented by a
sequence of x strokes (for instance 3 by $///$), each stroke on a square
of the Turing tape. The blank * is used to denote that the square is
empty (of the corresponding number is zero). In particular, a blank
is necessary to separate sequences of strokes representing numbers.
Thus, a Turing machine computing a function f with arguments
$x_1, \ldots, x_n$ starts with tape $\cdots * x_1 * x_2 * \cdots * x_n * \cdots$ and stops
with $\cdots * x_1 * x_2 * \cdots * x_n * f(x_1, \ldots, x_n) * \cdots$ on the tape.

From a logical point of view, a general purpose computer —
as constructed by associates of John von Neumann in America
and independently by Konrad Zuse in Germany — is a technical
realization of a universal Turing machine which can simulate any
kind of Turing program (Herken, 1995). Analogously, we can define
a universal register machine which can execute any kind of register
program. Actually, the general design of a von Neumann computer
consists of a central processor (program controller), a memory, an
arithmetic unit, and input–output devices. It operates step by step
in a largely serial fashion. A present-day computer *à la* von Neumann
is really a generalized Turing machine.

The efficiency of a Turing machine can be increased by the introduction of several tapes, which are not necessarily one-dimensional,
each acted on by one or more heads, but reporting back to a
single control box which coordinates all the activities of the machine
(Fig. 3) (Arbib, 1987, p. 131). Thus, every computation of such a

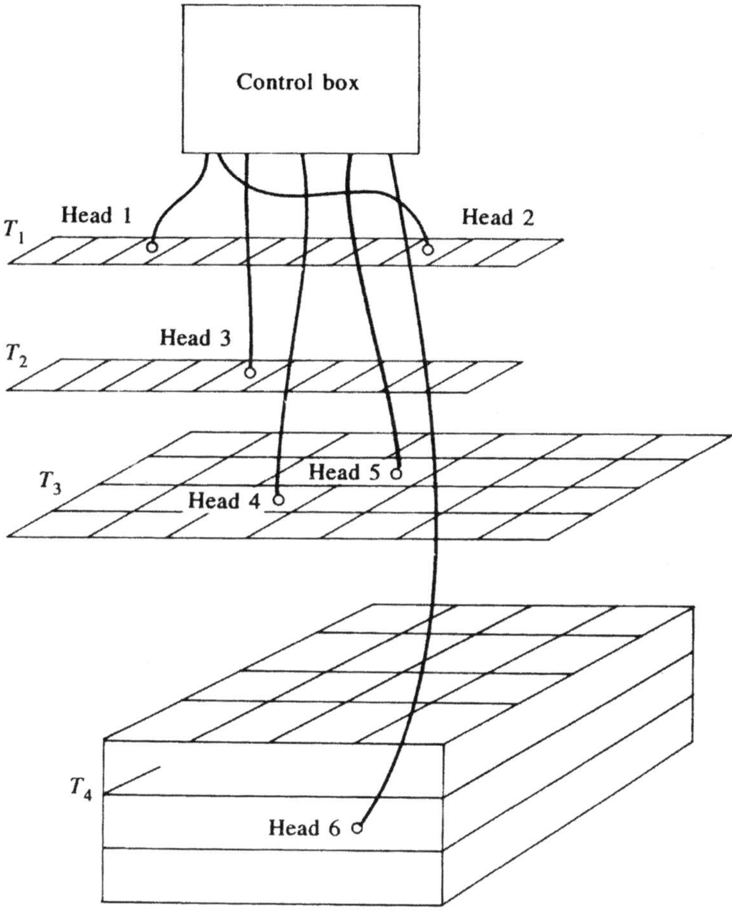

Fig. 3. Turing machine with several tapes.

more effective machine can be done by an ordinary Turing machine. Concerning the complex system approach, even a Turing machine with several multidimensional tapes remains a sequential program-controlled computer, differing essentially from self-organizing systems like neural networks.

Besides Turing and register machines, there are many other mathematically equivalent procedures for defining computable functions. Recursive functions are defined by unbounded search, procedures of functional substitution and iteration, beginning with some elementary functions, for instance, substitution and iteration, beginning with some elementary functions, for instance, the successor function $n(x) = x + 1$, which are obviously computable. All these definitions of computability by Turing machines, register machines, recursive functions, etc., can be proved to be mathematically equivalent. Obviously, each of these precise concepts defines a procedure which is intuitively effective. Thus, Alonzo Church postulated his thesis that the informal intuitive notion of an effective procedure is identical with one of these equivalent precise concepts, such as that of a Turing machine.

Church's thesis. *Every computational procedure (algorithm) can be calculated by a Turing machine.*

Church's thesis cannot be proved, of course, because mathematically precise concepts are compared with an informal intuitive notion. Nevertheless, the mathematical equivalence of several precise concepts of computability which are intuitively effective confirms Church's thesis. Consequently, we can speak about computability, effectiveness, and computable functions without referring to particular effective procedures ("algorithms") like Turing machines, register machines, recursive functions, etc. According to Church's thesis, we may in particular say that every computational procedure (algorithm) can be calculated by a Turing machine. So every recursive function, as a kind of machine program, can be calculated by a general purpose computer (Feferman, 2006).

To be more precise, let us consider a class of functions which was accepted as intuitively computable from the very beginning (Kleene, 1974; Shoenfield, 1967).

Definition of primitive recursive functions. The class $\mathcal{F}_{pr}$ of primitive recursive functions is defined by the following conditions:

(1) Z, S, p_i^n belong to $\mathcal{F}_{pr}$:

> Z is the zero function with $Z(x) = 0$,
> S is the successor function with $S(x) = x + 1$,
> p_i^n is a projection function with $p_i^n(x_0, \ldots, x_n) = x_i$ $(0 \leq i \leq n)$.

(2) $\mathcal{F}_{pr}$ is closed under composition:

If the functions f, g_j belong to $\mathcal{F}_{pr}$ with $f : \mathbb{N}^k \to \mathbb{N}$ and $g_j : \mathbb{N}^n \to \mathbb{N}$ $(1 \le j \le k)$, then there is a function h from $\mathcal{F}_{pr}$ satisfying

$$h(\vec{x}) = f(g_1(\vec{x}), \ldots, g_k(\vec{x})) \quad (\text{with } \vec{x} = x_1, \ldots, x_n).$$

(3) $\mathcal{F}_{pr}$ is closed under recursion:

For functions $f : \mathbb{N}^n \to \mathbb{N}$ and $g : \mathbb{N}^{n+2} \to \mathbb{N}$, there is a function h from $\mathcal{F}_{pr}$ such that

$$h(0, \vec{x}) = f(\vec{x}),$$
$$h(S(y), \vec{x}) = g(h(y, \vec{x}), y, \vec{x}).$$

(4) $\mathcal{F}_{pr}$ is the least class satisfying conditions (1)–(3).

The class of primitive recursive functions can be extended by an intuitively computable search procedure for minimal numbers satisfying certain computable conditions.

Definition of recursive functions. The class $\mathcal{F}_r$ of total recursive functions is the least class of functions satisfying the conditions (1)–(3) of $\mathcal{F}_{pr}$ (with $\mathcal{F}_r$ replacing $\mathcal{F}_{pr}$) and the following condition:

(5) $\mathcal{F}_r$ is closed under the application of the μ-operator:

$f(\vec{y}) = \mu x(g(x, \vec{y}) = 0)$, i.e., the least x satisfying $g(x, \vec{y}) = 0$ with g from $\mathcal{F}_r$ if such an x exists, or formally $\forall y \exists x(x, \vec{y}) = 0$.

Sometimes it is not clear that functions are totally defined. $f(x) \simeq g(x)$ means that function $f(x)$ is defined iff function $g(x)$ is defined, and if they are defined, then their values are equal. Now, we can extend the class of total recursive functions to the class of partial recursive functions.

Definition of partial recursive functions. The class $\mathcal{F}'_r$ of partial recursive functions is defined by the following conditions:

(1) $\mathcal{F}_r$ is a subset of $\mathcal{F}'_r$.

(2) $\mathcal{F}'_r$ is closed under composition:
If the functions h, g_j belong to $\mathcal{F}'_r$ with $h : \mathbb{N}^k \to \mathbb{N}$ and $g_j : \mathbb{N}^n \to \mathbb{N}(1 \leq j \leq k)$, then there is a function f of $\mathcal{F}'_r$ satisfying

$$f(\vec{x}) \simeq h(g_1(\vec{x}), \ldots, g_k(\vec{x})).$$

(3) $\mathcal{F}'_r$ is closed under the application of the μ-operator:
$f(\vec{y}) \simeq \mu x(g(x, \vec{y}) = 0)$, i.e., the least x satisfying $g(x, \vec{y}) = 0$ with g from $\mathcal{F}_r$ if such an x exists, or formally $\forall y \exists x(x, \vec{y}) = 0$.

It turns out that all partial recursive functions are computable in the sense of Church's thesis. Now, we are able to define effective procedures of decision and enumerability, which were already demanded by Leibniz's program of a *mathesis universalis* (Davis, 1958). The characteristic function χ_M of a subset M of natural numbers is defined by $\chi_M(x) = 1$ if x is an element of M, and as $\chi_M(x) = 0$ otherwise.

Definition. A set M is defined as effectively decidable if its characteristic function saying whether or not a number belongs to M is effectively computable.

Programs and computations consist of lists of symbols which can be coded by natural numbers. Then, we can define a decidable predicate $T(x, \vec{y}, z)$ (Kleene's T-predicate) with the meaning that "z codes a (terminating) computation according to program x for arguments $\vec{y}$". The total computable result-extracting function U extracts the result from the code for a terminating computation. It can be proven that each computable function f with code x of its computer program can be represented by the following form.

Kleene's normal form.

$$f(\vec{y}) \simeq U(\mu z \chi_T(x, \vec{y}, z) = 1)$$
$$\simeq U(\mu z T(x, \vec{y}, z)).$$

The partially defined expression $[x](\vec{y})$ denotes the result of applying program x to the arguments of function f. Actually, x is the code number of a machine program which is also called "machine number" of the computable function f. Obviously, $[x](\vec{y})$ is defined if there is a (terminating) computation z according to program x for input arguments $\vec{y}$:

$$[x](\vec{y}) \text{ is defined } \leftrightarrow \exists z T(x, \vec{y}, z).$$

A set M is defined as effectively enumerable if there exists an effective (computable) procedure f for generating its elements, one after another (formally $f(1) = x_1, f(2) = x_2, \ldots$ for all elements $x_1, x_2, \ldots$ from M). The definition of recursive enumerability can be generalized from sets to predicates.

Definition of recursive enumerability. A predicate P is recursively enumerable, if P is empty or there is a recursive function f with $P(x) \leftrightarrow \exists y f(y) = x$ for all x.

Recursive enumerability can be interpreted as a formal definition of Leibniz's *ars inveniendi*. It can be proven (Hermes, 1978) that for all recursively enumerable predicates P there is a recursively decidable predicate Q.

Theorem. $P(x) \leftrightarrow \exists y\, Q(x, y)$ *for all* x.

Proof. ("$\rightarrow$") Let P be recursively enumerable. If P is empty, then P is also recursively decidable. We define $Q(x, y): \leftrightarrow P(x) \wedge y = y$.

It follows by definition

$$P(x) \leftrightarrow \exists y (P(x) \wedge y = y)$$
$$\leftrightarrow \exists y Q(x, y) \text{ for all } x.$$

If P is not empty, then there is by definition a recursive function f with

$$P(x) \leftrightarrow \exists y f(y) = x$$
$$\leftrightarrow \exists y Q(x, y) \text{ with a recursive predicate } Q(x, y) : \leftrightarrow f(y) = x.$$

("$\leftarrow$") We assume that P can be represented as

$$P(x) \leftrightarrow \exists y\, Q(x,y) \text{ with a recursively decidable predicate } Q.$$

We can assume that P is not empty, otherwise the statement is trivial. Now, we assume a number $\bar{x}$ with $P(\bar{x})$ and define a recursive function:

$$f(y) = \begin{cases} \sigma_{2,1}(y) & \text{if } Q(\sigma_{2,1}(y), \sigma_{2,2}(y)), \\ \bar{x} & \text{otherwise} \end{cases}$$

with primitive recursive encoding function

$$\sigma_2(x,y) := 2^x(2y+1)\dot{-}1$$

and inverse functions

$$\sigma_{2,1}(z) = \exp(0, z+1),$$

$$\sigma_{2,2}(z) = \frac{\frac{z+1}{2^{\exp(0,z+1)}}\dot{-}1}{2}.$$

Obviously, we get

$$\sigma_{2,1}(\sigma_2(x,y)) = x,$$

$$\sigma_{2,2}(\sigma_2(x,y)) = y,$$

$$\sigma_2(\sigma_{2,1}(z), \sigma_{2,2}(z)) = z.$$

In the next step, we prove that f enumerates the elements of P according to the definition of recursive enumerability, i.e., $P(x) \leftrightarrow \exists y f(y) = x$.

("$\rightarrow$") In case of $P(x)$, there is a z with $Q(x,z)$ according to the assumed representation of P. In this case, we define $y = \sigma_2(x,z)$. It follows $Q(\sigma_{2,1}(y), \sigma_{2,2}(y))$. By definition of f, we get $f(y) = \sigma_{2,1}(y) = x$.

("$\leftarrow$") Now, we assume that there is a y with $f(y) = x$. We must prove $P(x)$. According to the definition of f, we have to distinguish the following two cases:

(1) $Q(\sigma_{2,1}(y), \sigma_{2,2}(y))$ is true: In this case, $f(y) = \sigma_{2,1}(y)$. It follows $Q(f(y), \sigma_{2,2}(y))$, i.e.,

$Q(x, \sigma_{2,2}(y))$. Therefore, there is a z with $Q(x, z)$.

(2) $Q(\sigma_{2,1}(y), \sigma_{2,2}(y))$ is not true: According to our definition of f, it is $f(y) = \bar{x}$. According to the assumption of "$\leftarrow$", it is $x = \bar{x}$. It follows $P(x)$ because of $P(\bar{x})$. $\square$

The last theorem and Kleene's normal form can now be used to characterize recursive enumerable predicates (Rogers, 1967).

Kleene's enumeration of recursively enumerable predicates.
Let P be a recursively enumerable predicate

$$P(y) \leftrightarrow \exists z\, Q(y, z)$$

with Q recursive for all y. Define a recursive partial function $f(y) \simeq \mu z Q(y, z)$ with machine number x. Then

$$P(y) \leftrightarrow [x](y) \text{ is defined } \leftrightarrow \exists z\, T(x, y, z).$$

The number x is called the recursive enumerable (RE) index of the recursive enumerable predicate P. Intuitively spoken, Kleene's T-predicate enumerates all recursive enumerable predicates by their RE indices (Soare, 1987).

Post's theorem of decidability and enumerability

Decidability can be characterized by enumerability: A set or predicate is recursively decidable if the set or predicate and its complementary set or predicate are recursively enumerable (Post's theorem). The complementary set $\bar{M}$ of M contains all elements which do not belong to M. If M and $\bar{M}$ are recursively enumerable, then we can enumerate their elements step by step in order to decide if a given number does belong to M or not. Thus, M is recursively decidable. By definition, it follows that every recursively decidable set is recursively enumerable. But there are recursively enumerable sets which are not decidable. These are the first hints that there are limits to Leibniz's original program of a *mathesis universalis*, based on the belief in universal decision procedures.

At this point, Turing's famous halting problem comes in: Is there a universal decision procedure to determine whether an arbitrary computer program stops after finite steps for an arbitrary input? Turing proved that the halting problem is in principle unsolvable. Then Gödel's incompleteness (Gödel, 1931) is only a corollary of Turing's proof (Chaitin, 1998).

Turing started his proof with the question whether real numbers are computable. A real number like $\pi = 3.1415926\ldots$ has an infinite number of digits that seem to be randomly distributed behind the decimal point. Nevertheless, there are simple finite programs for calculating the digits step by step with increasing precision of π. In this sense, π is called a computable real number. In a first step, Turing constructed an uncomputable real number.

Definition of an uncomputable number. According to Kleene's normal form, a machine program can be coded by a machine number. Imagine a list of all possible computer programs that are ordered according to their increasing machine numbers $p_1, p_2, p_2, \ldots$. If a program computes a real number with an infinite number of digits behind the decimal point (e.g., π), then they should be written down behind the corresponding program number. (The number before the decimal point is neglected.) Otherwise, there is a blank line in the list:

$$p_1 \; - \; .d_{\underline{11}}d_{12}d_{13}d_{14}d_{15}d_{16}d_{17}\ldots$$

$$p_2 \; - \; .d_{21}d_{\underline{22}}d_{23}d_{24}d_{25}d_{26}d_{27}\ldots$$

$$p_3 \; - \; .d_{31}d_{32}d_{\underline{33}}d_{34}d_{35}d_{36}d_{37}\ldots$$

$$p_4$$

$$p_5 \; - \; .d_{51}d_{52}d_{53}d_{54}d_{\underline{55}}d_{56}d_{57}\ldots$$

$$\vdots$$

Following Cantor's diagonal procedure, Turing changed the underlined digits on the diagonal of the list and put these changed digits (marked with $\neq$) together into a new number with a decimal point in front:

$$-. \neq d_{11} \neq d_{12} \neq d_{13} \neq d_{14} \neq d_{15} \neq d_{16} \neq d_{17} \cdots .$$

This new number cannot be in the list because it differs from the first digit of the first number behind p_1, the second digit of the second number behind p_2, etc. Therefore, it is an uncomputable real number. With this number, Turing got the unsolvability of the halting problem.

Unsovability of the halting problem

If we could solve the halting problem, then we could decide if the nth computer program ever puts out an nth digit behind the decimal point. In this case, we could actually carry out Cantor's diagonal procedure and compute a real number, which, by its definition, has to differ from any computable real number.

The unsolvability of the halting problem refutes Hilbert's Entscheidungsproblem. If there is a complete formal axiomatic system from which all mathematical truth follows, then it would give us a procedure to decide if a computer program will ever halt. We just run through all the possible proofs until we either find a proof that the program halts, or we find a proof that it never halts. So if Hilbert's finite set of axioms from which all the mathematical truth should follow were possible, then by running through all possible proofs while checking which ones are correct, we would be able to decide if a computer program halts. That is impossible using Turing's proof.

The unsolvability of Turing's halting problem is not only fundamental for computability theory, but it also has deep consequences for mathematical foundations. An example is Hilbert's 10th problem which is also proved to be in principle unsolvable due to Turing's halting problem. In 1900, David Hilbert asked for an algorithm which will decide whether a so-called diophantine equation has a solution (Hilbert, 1900). This was the 10th problem of his famous list of 23 mathematical problems which were still unsolved at the beginning of the 20th century. Algebraic equations, which involve only multiplication, addition, and exponentiation of whole numbers, were named after the third-century Greek mathematician Diophantos of Alexandria.

Unsovability of Hilbert's 10th problem

In 1970, J. V. Matiyasevich from the Steklov Institute of Mathematics in former Leningrad (St. Petersburg) proved that Hilbert's 10th problem is equivalent to Turing's halting problem and, consequently, not decidable.

Matiyasevich used results of Davis *et al.* (1961). According to Lagrange's representation of natural numbers as sum of four quadratic whole numbers, Hilbert's 10th problem can be reduced to the existence of solutions in natural numbers.

A predicate D is called diophantine if it is definable *by* predicates $x + y = z$, $x \cdot y = z$, $x^y = z$, and logical operations $\vee$ (or), $\wedge$ (and), and $\exists$ (existence quantifier):

$$D(x_1, \ldots, x_n)$$
$$\leftrightarrow \exists y_1, \ldots, y_r \quad P(x_1, \ldots, x_n, \quad y_1, \ldots, y_r) \text{ with } P \text{ recursive}$$
$$\leftrightarrow \exists y_1, \ldots, y_r \quad \chi_p(x_1, \ldots, x_n, \quad y_1, \ldots, y_r) = 1$$

with computable characteristic function χ_p as polynom.

According to our theorem of recursive enumerability, it follows that every diophantine predicate is recursively enumerable. Vice versa, it can be proven that every enumerable predicate is diophantine. Matiyasevich used the Fibonacci sequence to define an appropriate diophantine predicate. The halting problem can be represented by an enumerable, but not decidable predicate. Therefore, the corresponding diophantine predicate is also not decidable.

Chapter 3

Hierarchies of Computability

According to Church's thesis, Turing computability is a representative definition of computability in general. In the following, we want to consider problems with degrees of complexity below and beyond this limit. Below this limit, there are many practical problems concerning certain limitations on how much the speed of an algorithm can be increased. Especially among mathematical problems, there are some classes of problems that are intrinsically more difficult to solve algorithmically than others. Thus, there are degrees of computability for Turing machines which are made precise in complexity theory in computer science.

Complexity classes of problems (or corresponding functions) can be characterized by complexity degrees, which give the order of functions describing the computational time (or number of elementary computational steps) of algorithms (or computational programs) depending on the length of their inputs. The lengths of inputs may be measured by the number of decimal digits. According to the machine language of a computer, it is convenient to represent decimal numbers in their binary codes with only binary numbers 0 and 1 and to define their length by the number of binary digits. For instance, 3 has the binary code $11 = 1 \cdot 2^1 + 1 \cdot 2^0$ with the length 2.

Definition of linear computational time. A function f has linear computational time if the computational time of f is not greater than $c \cdot n$ for all inputs with length n and a constant c.

Example of linear computational time. The addition of two (binary) numbers has obviously only linear computational time. For instance, the task $3 + 7 = 10$ corresponds to the binary calculation

$$011$$
$$\underline{111}$$
$$1010$$

which needs five elementary computational steps of adding two binary digits (including carrying). We remind the reader that the elementary steps of adding binary digits are $0 + 0 = 0$, $0 + 1 = 1$, $1 + 0 = 1$, $1 + 1 = 0$ carry 1. It is convenient to assume that the two numbers which should be added have equal length. Otherwise, we simply start the shorter one with a series of zeros, for instance 111 and 011 instead of 11. In general, if the length of the particular pair of numbers which should be added is n, the length of a number is $\frac{n}{2}$, and thus, we need no more than $\frac{n}{2} + \frac{n}{2} = n$ elementary steps of computation including carrying.

Definition of quadratic computational time. A function f has quadratic computational time if the computational time of f is not greater than $c \cdot n^2$ for all inputs with length n and a constant c.

Example of quadratic computational time. A simple example of quadratic computational time is the multiplication of two (binary) numbers. For instance, the task $7 \cdot 3 = 21$ corresponds to the binary calculation:

$$111 \cdot 011$$
$$000$$
$$111$$
$$\underline{111}$$
$$10101$$

According to former conventions, we have $n = 6$. The number of elementary binary multiplications is $\frac{n}{2} \cdot \frac{n}{2} = \frac{n^2}{4}$. Including carrying, the number of elementary binary additions is $\frac{n}{2} \cdot \frac{n}{2} - \frac{n}{2} = \frac{n^2}{4} - \frac{n}{2}$. In all, we get $\frac{n^2}{4} + \frac{n^2}{4} - \frac{n}{2} = \frac{n^2}{2} - \frac{n}{2}$, which is smaller than $\frac{n^2}{2}$.

Definition of polynomial and exponential computational time.

- A function f has polynomial computational time if the computational time of f is not greater than $c \cdot n^k$, which is assumed to be the leading term of a polynomial $p(n)$.
- A function f has exponential computational time if the computational time of f is not greater than $c \cdot 2^{p(n)}$.
- Many practical and theoretical problems belong to the complexity class P of all functions which can be computed by a deterministic Turing machine in polynomial time.

In the history of mathematics, there have been some nice problems of graph theory to illustrate the basic concepts of complexity theory (Grötschel *et al.*, 1988).

Complexity of Euler's Königsberg river problem

In 1736, the famous mathematician Leonard Euler (1707–1783) solved one of the first problems of graph theory. In the city of Königsberg, the capital of former Eastern Prussia, the so-called old and new river Pregel are joined in the river Pregel. In the 18th century, there were seven bridges connecting the southern s, northern n, and eastern e regions with the island i (Fig. 4a). Is there a route which crosses each bridge only once and returns to the starting point?

Euler reduced the problem to graph theory. The regions n, s, l, e are replaced by vertices of a graph, and the bridges between two regions by edges between the corresponding vertices (Fig. 4b).

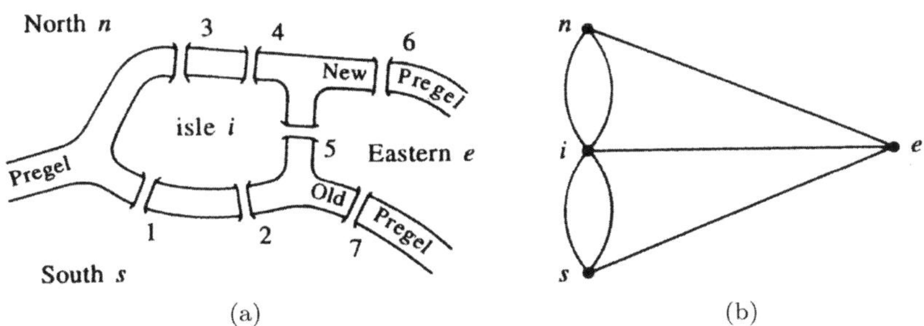

Fig. 4. Euler's Königsberg river problem.

In the language of graph theory, Euler's problem is whether for every vertex there is a route (an "Euler circuit") passing each edge exactly once, returning finally to the starting point. For arbitrary graphs, Euler proved that an Euler circuit exists if and only if each vertex has an even number of edges (the "Euler condition"). As the graph of Fig. 4b does not satisfy this condition, there cannot be a solution of Euler's problem in this case. In general, there is an algorithm testing an arbitrary graph by Euler's condition if it is an Euler circuit. The input of the algorithm consists of the set V of all vertices $1, \ldots, n$ and the set E of all edges, which is a subset of the set with all pairs of vertices. The computational time of this algorithm depends linearly on the size of the graph, which is defined by the sum of the numbers of vertices and edges.

Complexity of Hamilton's problem

In 1859, the mathematician William Hamilton (1805–1865) introduced a rather similar problem that is much more complicated than Euler's problem (Mainzer, 2007a, pp. 189–190). Hamilton considered an arbitrary graph, which means nothing else than a finite collection of vertices, a certain number of pairs of which are connected together by edges. Hamilton's problem is whether there is a closed circuit (a "Hamilton's circuit") passing each vertex (not each edge as in Euler's problem) exactly once. Figure 5 shows a graph with a Hamilton circuit passing the vertices in the order of numbering.

However, unlike the case of Euler's problem, we do not know any condition which exactly characterizes whether a graph contains a Hamilton circuit or not. We only can define an algorithm testing whether an arbitrary graph contains a Hamilton circuit or not. The algorithm tests all permutations of vertices to see if they form a Hamiltonian circuit. As there are $n!$ different permutations of n vertices, the algorithm does not need more than $c \cdot n!$ steps with a constant c to find a solution. It can easily be proved that an order of $n!$ corresponds to an order of n^n. Consequently, an algorithm for the Hamilton problem needs exponential computational time, while the Euler problem can be solved algorithmically in linear computational

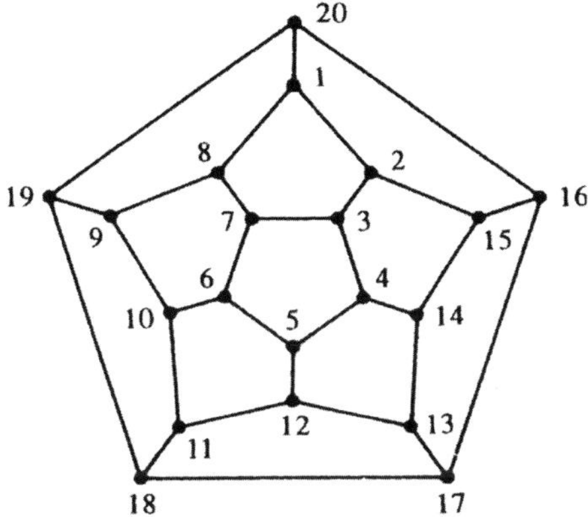

Fig. 5. Hamilton's problem.

time. Thus, Hamilton's problem cannot practically be solved by a computer even for small numbers n.

The main reason for a high computational time may be a large number of single subcases which must be tested by a deterministic computer which is allowed to choose a computational procedure at random among a finite number of possible ones instead of performing them step by step in a serial way. Let us consider Hamilton's problem again. An input graph may have n vertices $v_1, \ldots, v_n$. A non-deterministic algorithm chooses a certain order $v_{i_1}, \ldots, v_{i_n}$ of vertices in a non-deterministic, random way. Then the algorithm tests whether this order forms a Hamiltonian circuit. The question is whether for all number $j(j = 1, \ldots, n-1)$ the successive vertices v_{i_j} and $v_{i_{j+1}}$ and the beginning and starting vertices v_{i_n} and v_{i_1} are connected by an edge. The computational time of this non-deterministic algorithm depends linearly on the size of the graph.

In general, NP means the complexity class of functions which can be computed by a non-deterministic Turing machine in polynomial time. Hamilton's problem is an example of an NP-problem. Another NP-problem is the "travelling salesman problem", which is rather like Hamilton's problem except that the various edges have numbers attached to them. One seeks that Hamilton circuit for which the sum of the numbers or more intuitively the distance travelled by the salesman is a minimum.

Can NP-problems be reduced to P-problems?

By definition every P-problem is an NP-problem. But it is a crucial question of complexity theory whether $P = NP$ or, in other words, whether problems which are solved by non-deterministic computers in polynomial time can also be solved by a deterministic computer in polynomial time.

Hamilton's problem and the travelling salesman problem are examples of so-called NP-complete problems.

Complexity of NP-complete problems

A problem is called an NP-complete problem if any other NP-complete problem can be converted into it in polynomial time.

Consequently, if an NP-complete problem is actually proved to be a P-problem (if for instance a deterministic algorithm can be constructed to solve Hamilton's problem in polynomial time), then it would follow that all NP-problems are actually in P. Otherwise if $P \neq NP$ then no NP-complete problem can be solved with a deterministic algorithm in polynomial time.

Obviously, complexity theory delivers degrees for the algorithmic power of Turing machines or Turing-type computers. The theory has practical consequences for scientific and industrial applications. But does it imply limitations for mathematics and the human mind? The fundamental questions of complexity theory (for example $N = NP$ or $N \neq NP$) refer to the measurement of the speed, computational time, storage capacity, and so on, of algorithms. What about the complexity of mathematical problems beyond Turing computability?

Enumerability is only the first step on a ladder with increasing computational complexity. By adding unrestricted quantifiers to recursive predicates, degrees of computability can be extended beyond Turing computability (Hinman, 1978; Kleene, 1955a; Shoenfield, 1967):

Definition of arithmetical hierarchy. A predicate P is arithmetical iff it has an explicit definition

$$(*)\quad P(x) \leftrightarrow Q_1 \ldots Q_n\, R(x, x_1, \ldots, x_n)$$

for all numbers x with R recursive and each number quantifier Q_i for either $\exists x_i$ or $\forall x_i$ with number variables x_i.

Two quantifiers are of the same kind if they are both existential or both universal. Two adjacent quantifiers of the same kind can be replaced by a single quantifier (contraction of quantifiers).

A predicate is $\Sigma_n^0 (\Pi_n^0)$ for $n \geq 1$ iff it has an explicit definition $(*)$ with no two adjacent quantifiers of the same kind and the first quantifier existential (universal), e.g., for all x,

$$\Sigma_2^0 : P(x) \leftrightarrow \exists x_1 \forall x_2 R(x, x_1 x_2),$$

$$\Pi_2^0 : P(x) \leftrightarrow \forall x_1 \exists x_2 R(x, x_1 x_2).$$

Every arithmetical predicate is Σ_n^0 or Π_n^0 for some $n \geq 1$. This classification of arithmetical predicates is called the arithmetical hierarchy.

A predicate is Δ_n^0 iff it is both Σ_n^0 and Π_n^0.

Every recursively enumerable predicate P can be represented by $P(x) \leftrightarrow \exists y\, Q(x, y)$ for all x with a recursive predicate Q. Therefore, Σ_1^0 is the class of recursively enumerable predicates. Recursive predicates are closed with respect to logical connectives, e.g., $\neg$ (negation), $\vee$ (disjunction: "or"), and $\wedge$ (conjunction: "and"). In classical logic, $\exists y\, Q(x, y)$ is equivalent with $\neg \forall y \neg Q(x, y)$. It follows that P is Δ_1^0 iff both P and $\neg P$ are enumerable because of double negation. Thus, because of Post's theorem, Δ_1^0 is the class of the recursive (Turing-computable) predicates.

There are general properties of the arithmetical hierarchy:

(1) A recursive predicate is Σ_n^0 and Π_n^0 for all n.
(2) $\Sigma_m^0 \subseteq \Sigma_n^0 (\Pi_n^0)$, $\Pi_m^0 \subseteq \Sigma_n^0 (\Pi_n^0)$ for all $n > m$: If predicate P is Σ_m^0 or Π_m^0, then P is Σ_n^0 and Π_n^0.
(3) If P and Q are $\Sigma_n^0 (\Pi_n^0)$, then $P \vee Q$ and $P \wedge Q$ are $\Sigma_n^0 (\Pi_n^0)$.
(4) If P is $\Sigma_n^0 (\Pi_n^0)$, then $\neg P$ is $\Pi_n^0 (\Sigma_n^0)$.

Therefore, the arithmetical hierarchy can be illustrated in the following way:

$$
\begin{array}{ccc}
\vdots & \vdots & \vdots \\
\Sigma^0_{n+1} & \Delta^0_{n+1} & \Pi^0_{n+1} \\
\Sigma^0_n & \Delta^0_n & \Pi^0_n \\
\vdots & \vdots & \vdots \\
\Sigma^0_2 & \Delta^0_2 & \Pi^0_2 \\
\Sigma^0_1 & \Delta^0_1 & \Pi^0_1
\end{array}
$$

Kleene's normal form of recursive functions allows to enumerate all recursive functions and recursively enumerable predicates (Chapter 2). It can be generalized for the arithmetical hierarchy:

Kleene's arithmetical enumeration theorem. *For each $n \geq 1$, there is a Σ^0_n (Π^0_n) predicate which enumerates the class of all Σ^0_n (Π^0_n) predicates.*

Further on, it can be proven that the inclusions between the Σ^0_n and Π^0_n predicates, i.e., $\Sigma^0_m \subseteq \Sigma^0_n$ (Π^0_n), $\Pi^0_m \subseteq \Sigma^0_n$ (Π^0_n) for all $n > m$, are the only ones of their type:

Kleene's arithmetical hierarchy theorem. *For each $n \geq 1$, there is a Σ^0_n predicate which is not Π^0_n and hence not Σ^0_m or Π^0_m for any $m > n$. Then $\neg P$ is Π^0_n but not Σ^0_n and hence not Σ^0_m or Π^0_m for any $m > n$.*

Relative computability and oracle machines

Computations can depend on external agencies supplying answers to questions about a set (or property) M by an unknown procedure. According to Turing, the external agency of an unknown procedure is called an oracle (Turing, 1939). In Fig. 6, the computation of a function value $f(x)$ depends on an oracle deciding if certain values satisfy a property M (Rogers, 1967). Most of our calculations in

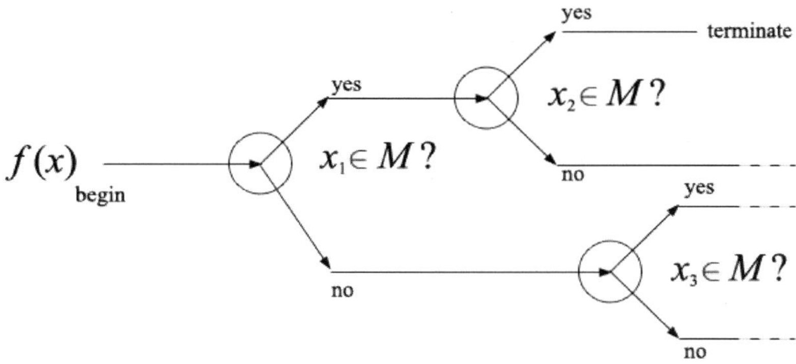

Fig. 6. Relative computability with an oracle.

everyday life depend on assumptions we rely on without knowing a final confirmation. Thus, oracles lead to an important extension of relativized computation.

A function f is computable relative to M iff the program of f is extended with an oracle M. An oracle can also be a machine operation (total function) φ supplying values $\varphi(x)$ by an unknown procedure. Random oracles depend on random procedures.

Definition of relative computability.

- A function f is recursive in φ iff f is computable relative to φ.
- A predicate P is recursive (decidable) in φ iff its characteristic function χ_P is computable relative to φ.
- P is Σ_n^0 (Π_n^0, Δ_n^0) in φ iff P is defined by Σ_n^0 (Π_n^0, Δ_n^0) quantifiers and a predicate R recursive in φ.

Relative computability with respect to a function φ can be generalized to a set Φ of functions $\varphi_0, \varphi_1, \ldots$ We say that set Φ is recursive in set Ψ with functions $\psi_0, \psi_1, \ldots$ if every member of Φ is recursive in Ψ. If Ψ is empty, then the functions recursive in Ψ are just the recursive functions. Therefore, relativized Turing computability includes Turing computability as a special case. Obviously, relativized computability satisfies the property of transitivity: If Φ is recursive in Ψ, then every function recursive in Φ is also recursive in Ψ (Shoenfield, 1967).

The members of Φ can be ordered as a finite sequence and contracted by a function $\Phi^*(x) := \langle \varphi_0(x), \ldots, \varphi_{n-1}(x) \rangle$ with $\varphi_i(x) = (\Phi^*(x))_i$ and $0 \le i \le n-1$. The contraction function Φ^* is recursive in $\varphi_0, \ldots, \varphi_{n-1}$ and each φ_i is recursive in Φ^*. With this definition in mind, we can replace relativized recursiveness (enumerability) of partial functions (predicates) by the recursiveness (enumerability) of a partial functional (relation). A functional also has functions as arguments.

Replacement lemma.

- *A partial function f is recursive in a finite sequence Φ iff there is a recursive partial functional f' with $f(x) \simeq f'(\Phi^*, x)$ for all x.*
- *A predicate P is recursively enumerable in a finite sequence Φ iff there is a recursively enumerable relation P' with $P(x) \leftrightarrow P'(\Phi^*, x)$ for all x.*

With the replacement lemma, we can prove some remarkable properties of relativized computability:

- *A predicate P is Σ^0_{n+1} iff it is recursively enumerable in the set of Π^0_n predicates.*
- *A predicate is Δ^0_{n+1} iff it is recursive in the set of Π^0_n predicates.*

By the replacement lemma, Kleene's normal form of recursive functions and recursively enumerable predicates (Chapter 2) can also be extended for relativized "machine numbers" for oracles:

Definition of relativized machine numbers. A number p is called machine number of function f relativized to oracle set Φ iff

$$[p]^{\Phi}(x) := f(x) \simeq f'(\Phi^*, x) \simeq [p](\Phi^*, x)$$

with recursive partial functional f' for all x.

A number p is called recursively enumerable (RE) index of the recursively enumerable predicate P relativized to oracle set Φ iff

$$P(y) \leftrightarrow [x](y) \text{ is defined } \leftrightarrow \exists z\, T(p, \Phi^*, y, z).$$

Intuitively spoken, Kleene's relativized T-predicate enumerates all relativized recursive enumerable predicates by their relativized RE indices.

With relativized computability, we can determine degrees of computational solvability in problem solving. How far can problem solving be reduced to Turing computability?

Definition of Turing reducibility. A predicate A is Turing reducible to a predicate B (notation: $A \leq_T B$) iff A is recursive in B.

As mentioned before, the relation $\leq_T$ is transitive, further on symmetric and reflexive. Thus, $\leq_T$ leads to an equivalence relation $\equiv_T$ postulating that two predicates are equivalent iff each is recursive in the other, i.e.,

$$A \equiv_T B \text{ iff } A \leq_T B \text{ and } B \leq_T A.$$

Intuitively speaking, two predicates are equivalent if they are equally difficult to decide. Turing reducibility can also be applied to relativized computability of functions. The corresponding equivalence classes are called Turing degrees of computability. They are denoted by $\mathbf{A}, \mathbf{B}, \ldots$ with predicates $A, B, \ldots$ as representatives of the corresponding equivalence classes.

Partial ordering of Turing degrees

$\mathbf{A} \leq \mathbf{B}$ means that a predicate of Turing degree $\mathbf{A}$ is recursive in a predicate of Turing degree $\mathbf{B}$. Then, by transitivity, any predicate of Turing degree $\mathbf{A}$ is recursive in any predicate of Turing degree $\mathbf{B}$. According to the transitivity property, one can easily verify that Turing reducibility $\leq_T$ satisfies the properties of a partial ordering:

$$\mathbf{A} \leq \mathbf{A}$$

$$\mathbf{A} \leq \mathbf{B} \wedge \mathbf{B} \leq \mathbf{A} \rightarrow \mathbf{A} = \mathbf{B}$$

$$\mathbf{A} \leq \mathbf{B} \wedge \mathbf{B} \leq \mathbf{C} \rightarrow \mathbf{A} \leq \mathbf{C}$$

$\mathbf{A} < \mathbf{B}$ means that $\mathbf{A} \leq \mathbf{B}$ and $\mathbf{A} \neq \mathbf{B}$. The set of all recursive predicates is obviously a Turing degree which is denoted by $\mathbf{0}$ as smallest degree.

A Turing degree $\mathbf{A}$ is said to be recursively enumerable in Turing degree $\mathbf{B}$ if a predicate of $\mathbf{A}$ is recursively enumerable in a predicate of $\mathbf{B}$. Then by transitivity that predicate of $\mathbf{A}$ is recursively enumerable in every predicate of $\mathbf{B}$. Obviously, a degree $\mathbf{A}$ is recursively enumerable iff it is recursively enumerable in $\mathbf{0}$, that means, if it contains a recursively enumerable predicate. If $\mathbf{A}$ is recursively enumerable in $\mathbf{B}$ and $\mathbf{B} \leq \mathbf{C}$, then $\mathbf{A}$ is recursively enumerable in $\mathbf{C}$ by transitivity. Thus, a recursively enumerable degree is recursively enumerable in every degree.

Theorem. *Among the degrees recursively enumerable in* $\mathbf{A}$, *there is a largest one.*

Proof. We define a degree $\mathbf{B}$ which is the largest degree recursively enumerable in $\mathbf{A}$: Let A be a predicate in $\mathbf{A}$ and $\mathbf{B}$ be defined as

$$B(x) :\leftrightarrow \exists z T((x)_0, \chi_A, (x)_1, z)$$

for all x. The degree $\mathbf{B}$ of B is recursively enumerable in $\mathbf{A}$. Why is $\mathbf{B}$ the largest such degree?

Let $\mathbf{C}$ be recursively enumerable in $\mathbf{A}$. Then $\mathbf{C}$ contains a predicate C which is recursively enumerable in A. By the relativized enumeration theorem, there is a RE index p with

$$C(x) : \leftrightarrow \exists z\, T(p, \chi_A, x, z)$$

$$\leftrightarrow B(\langle p, x \rangle)$$

for all x. Obviously C is recursive in B. Therefore, it follows $\mathbf{C} \leq \mathbf{B}$.
$$\square$$

Definition of Turing jump. The largest degree which is recursively enumerable in $\mathbf{A}$ is called the (Turing) jump $\mathbf{A}'$ of $\mathbf{A}$.

The largest recursively enumerable degree is $\mathbf{0}'$. If $\mathbf{A} \leq \mathbf{B}$, then $\mathbf{A}'$ is recursively enumerable in $\mathbf{A}$ and hence in $\mathbf{B}$. Therefore, $\mathbf{A}' \leq \mathbf{B}'$. With Turing jump, we can prove:

Proposition. *There is no largest degree.*

Proof. We know by the relativized arithmetical hierarchy theorem that every set recursive in A is recursively enumerable in A but not vice versa. It follows that $\mathbf{A} < \mathbf{A}'$. So there is no largest degree. $\square$

Finally, two questions arise:

(a) Is the set of degrees linearly ordered?
(b) Are there recursively enumerable degrees other than $\mathbf{0}$ and $\mathbf{0}'$?

The following theorem which we mention without proof (Shoenfield, 1967, p. 171) answers both questions:

Friedberg–Muchnik theorem. *There are recursively enumerable sets A and B such that A is not recursive in B and B is not recursive in A.*

(a) *Let A and B be the recursively enumerable predicates of this theorem and $\mathbf{A}$ and $\mathbf{B}$ their degrees. It follows that $\mathbf{A} \leq \mathbf{B}$ and $\mathbf{B} \leq \mathbf{A}$ are both false. Therefore, the set of degrees is not linearly ordered.*
(b) *Concerning the second question, we note that $\mathbf{0} \leq \mathbf{C} \leq \mathbf{0}'$ for every recursively enumerable degree $\mathbf{C}$. Consequently, $\mathbf{A}$ and $\mathbf{B}$ (with A and B recursively enumerable predicates of the Friedberg–Muchnik theorem) are recursively enumerable degrees different from $\mathbf{0}$ and $\mathbf{0}'$.*

The complexity degrees of arithmetical hierarchy can be generalized from arithmetic to analysis. In this case, the quantifiers are also referred to as function variables (Kleene, 1958, 1959):

Definition of analytical hierarchy. A predicate P is analytical iff it has an explicit definition

$$(*) \quad P(x) \leftrightarrow Q_1 \ldots Q_n R(x, x_1, \ldots, x_k, \alpha_1, \ldots, \alpha_m)$$

for all numbers x with R recursive and each quantifier Q_i for either $\exists x_i$ or $\forall x_i$ with number variables x_i or for either $\exists \alpha_i$ or $\forall \alpha_i$ with function variables α_i.

Two quantifiers are of the same kind if they are both existential or both universal. They are of the same type if they are both on number variables or function variables. There is a (recursive) procedure to bring the prefix of ($*$) to a normalized form:

(1) A number quantifier is replaced by a function quantifier of the same kind:

$$\exists x\, P(x) \leftrightarrow \exists \alpha\, P(\alpha(0))$$

$$\forall x\, P(x) \leftrightarrow \forall \alpha\, P(\alpha(0)).$$

(2) Two adjacent quantifiers of the same kind and type are replaced by a single quantifier of the same kind and type.
(3) If a function quantifier immediately follows a number quantifier, it is brought to the front of that number quantifier.

The prefix $Q_1 \ldots Q_n$ is normalized if $Q_1 \ldots Q_{n-1}$ ($n \geq 2$) are function quantifiers and Q_n is a number quantifier and the quantifiers alternate in kind.

A predicate is $\Sigma_n^1 (\Pi_n^1)$ for $n \geq 1$ iff it has an explicit definition ($*$) in which the prefix is normalized, the number of function quantifiers is n, and the first quantifier is existential (universal).

Every analytical predicate is either arithmetical or Σ_n^1 or Π_n^1 for some $n \geq 1$. This classification of analytical predicates is called the analytical hierarchy.

A predicate is Δ_n^1 iff it is both Σ_n^1 and Π_n^1.

There are general properties of the analytical hierarchy analog to the arithmetical hierarchy:

(1) An arithmetic predicate is Σ_n^1 and Π_n^1 for all n.
(2) $\Sigma_m^1 \subseteq \Sigma_n^1 (\Pi_n^1)$, $\Pi_m^1 \subseteq \Sigma_n^1 (\Pi_n^1)$ for all $n > m$: If predicate P is Σ_m^1 or Π_m^1, then P is Σ_n^1 and Π_n^1.
(3) If P and Q are $\Sigma_n^1 (\Pi_n^1)$, then $P \vee Q$ and $P \wedge Q$ are $\Sigma_n^1 (\Pi_n^1)$.
(4) If P is $\Sigma_n^1 (\Pi_n^1)$, then $\neg P$ is $\Pi_n^1 (\Sigma_n^1)$.

Therefore, the analytical hierarchy can be illustrated in the following way:

$$
\begin{array}{ccc}
\vdots & \vdots & \vdots \\
\Sigma^1_{n+1} & \Delta^1_{n+1} & \Pi^1_{n+1} \\
\Sigma^1_n & \Delta^1_n & \Pi^1_n \\
\vdots & \vdots & \vdots \\
\Sigma^1_2 & \Delta^1_2 & \Pi^1_2 \\
\Sigma^1_1 & \Delta^1_1 & \Pi^1_1
\end{array}
$$

It follows with (4) that a predicate P is Δ^1_n iff both P and $\neg P$ are Σ^1_n and iff both P and $\neg P$ are Π^1_n. In analogy to the arithmetical hierarchy, there is an analytical enumeration theorem and analytical hierarchy theorem.

The analytical hierarchy can also be relativized, and corresponding properties can be applied to the relativized case:

Definition of relativized analytical hierarchy. A predicate P is analytical in Φ iff it has an explicit definition

$$(*) \quad P(x) \leftrightarrow Q_1 \ldots Q_n R(x, x_1, \ldots, x_k, \alpha_1, \ldots, \alpha_m)$$

for all numbers x with R recursive in Φ and each quantifier Q_i for either $\exists x_i$ or $\forall x_i$ with number variables x_i or for either $\exists \alpha_i$ or $\forall \alpha_i$ with function variables α_i.

A predicate is $\Sigma^1_n (\Pi^1_n)$ in Φ for $n \geq 1$ iff it has an explicit definition $(*)$ in which the prefix is normalized, the number of function quantifiers is n, and the first quantifier is existential (universal).

Every analytical predicate is either arithmetical in Φ or Σ^1_n or Π^1_n in Φ for some $n \geq 1$. This classification of relativized analytical predicates is called the relativized analytical hierarchy.

A predicate is Δ^1_n in Φ iff it is both Σ^1_n in Φ and Π^1_n in Φ.

In the arithmetical hierarchy, Δ^0_1 is a distinguished layer which contains the recursive predicates. Therefore, they can be defined without number quantifiers. What about Δ_1^1 predicates in the analytical hierarchy? Can they also be defined without function quantifiers? It can be proven that they are not just the arithmetical relations which are only a proper subclass of layer Δ_1^1 in the analytical hierarchy. A complete characterization of Δ_1^1 relations needs an extension of the arithmetical hierarchy which is called the hyperarithmetic hierarchy.

The hyperarithmetic hierarchy started in a first layer with the recursive predicates. In the following layers, they are relativized to sets of indices which are generated step by step in a sequence of Turing jumps and finalized by computable unions of these sets (Church and Kleene, 1936):

Definition of hyperarithmetic hierarchy.

(1) The system σ of indices is partially ordered with bifurcating branches by relation $<_\sigma$:

$$1 <_\sigma 2$$

$$x <_\sigma y \rightarrow x <_\sigma 2^y, y <_\sigma 2^y$$

$$\forall n([y](n) <_\sigma [y](n+1)), \ \exists n(x <_\sigma [y](n)),$$

$$\lambda n[y](n) \text{ total} \rightarrow x <_\sigma 3 \cdot 5^y$$

(2) The Turing jump M' of a *set* M is the effective disjoint union of all sets recursively enumerable in M, i.e., the largest Turing degree which is recursively enumerable in the degree of M.

H-sets are defined by recursion on $<_\sigma$:
$$H_1 = \emptyset$$
$$H_{2^y} = H_y'$$
$$H_{3 \cdot 5^y} = \{\langle a, b \rangle | b <_\sigma 3 \cdot 5^y, a \epsilon H_b\} \text{ with } \langle a, b \rangle = 2^a \cdot 3^b$$

(3) A set or predicate M is hyperarithmetic iff M is recursive in some H-set. The Turing degree of H-set H_y is determined by index number y. Their equivalence classes generate the hyperarithmetic hierarchy.

The class HYP of all hyperarithmetic sets is proven to be equal to the class of Δ_1^1-sets in the analytical hierarchy, i.e., HYP $= \Delta_1^1$. At a first glance, Δ_1^1-sets seem to be far away from recursive and decidable sets and predicates. But, nevertheless, hyperarithmetic turns out to be an important part of constructive mathematics which we will consider in proof theory and foundations of mathematics in the next chapter (Kreisel, 1960; Mainzer, 1973). In general, the (relativized) analytical hierarchy is the bridge from the theory of computability to set theory as basis of mathematics in order to classify mathematical sets and structures according to their degrees of complexity.

But can the concept of a machine be generalized along the ladder of increasing complexity degrees in the analytical hierarchy? Is there computability beyond Turing computability of Δ_1^0?

Turing on machines beyond Turing machines. Historically, it was Turing himself who seems to have introduced a first machine beyond the Turing machine. In his Ph.D. thesis (Princeton, 1938), he introduced the concept of an "oracle machine", (Turing, 1939):

> "Let us suppose that we are supplied with some unspecified means of solving number-theoretic problems; a kind of oracle as it were. We shall not go any further into the nature of this oracle apart from saying that it cannot be a [Turing] machine. With the help of the oracle we could form a new kind of machine (call them o-machines), having as one of its fundamental processes that of solving a given number-theoretic problem."

The halting set of a Turing machine means the set of inputs which let the machine stop with a final result after finite steps. It is obvious that an o-machine with its halting set as oracle can compute the halting function. Actually, an o-machine with the halting set as oracle can compute all recursively enumerable functions.

Hierarchy of oracle machines. The halting sets of machines allow to distinguish different levels of computability which can be referred to the relativized theory of recursive functions. A function or predicate is computable by an oracle machine with oracle M iff it is recursive in M. Therefore, a recursive function is computable by an oracle machine with the empty set $\emptyset$ as oracle. No Turing machine can solve its own halting function according to the unsolvability of Turing's halting problem. In general, no oracle machine can solve its own halting function. But the unsolvability of the halting problem is no absolute limitation of computability. It must be relativized with respect to the allowed computational tools:

The halting set of an oracle machine with oracle M can be generated by the Turing jump M' of M. With Turing jumps, we get a sequence of increasingly powerful machines which, at least theoretically, could solve the halting problem of the previous machines:

- The oracle of a Turing machine which computes a recursive function is the empty set $\emptyset$.
- In the next step, the halting function for Turing machines with oracle $\emptyset$ is recursive in $\emptyset'$.
- In the next step, the halting function for machines with oracle $\emptyset'$ is recursive in $\emptyset''$.

$$\vdots$$

- In the next step, the halting function for machines with oracle $\emptyset^{(n)}$ is recursive in $\emptyset^{(n+1)}$.

$$\vdots$$

Oracle machines and arithmetical hierarchy

We already know by replacement lemma:

- A predicate P is Σ^0_{n+1} iff it is recursively enumerable in the set of Π^0_n predicates.
- A predicate is Δ^0_{n+1} iff it is recursive in the set of Π^0_n predicates.

It follows:

- A predicate P is Σ^0_{n+1} iff it is recursively enumerable in $\emptyset^{(n)}$.
- A predicate is Δ^0_{n+1} iff it is recursive in $\emptyset^{(n)}$.

From a theoretical point of view, the layers of arithmetical hierarchy can be characterized by o-machines with increasing computer power beyond Turing machines (Börger, 1989). For Δ-predicates, we get a hierarchy of oracles by Turing jumps:

- Δ_1^0 corresponds to *o*-machines with $\varnothing$ oracle.
- Δ_2^0 corresponds to *o*-machines with $\varnothing'$ oracle.

$$\vdots$$

- Δ_{n+1}^0 corresponds to *o*-machines with $\varnothing^{(n)}$ oracle.

$$\vdots$$

The recursively enumerable functions and predicates of Σ_1^0 can be characterized by a so-called accelerated Turing machine (Ord, 2006). Its concept dates back to old ideas of Greek mathematicians and philosophers which were later discussed by Russell (1936) and Weyl (1927) in modern versions. The Greek philosopher Zeno of Elea (ca. 490–430 BC), one of the Presocratic philosophers, irritated his contemporaries with the following paradox:

Achilles, the best runner of antiquity, and a tortoise are in a footrace. Achilles allows the tortoise a head start of, e.g., 100 meters. After some finite time, Achilles will have run 100 meters and arrive at the tortoise's starting point. During this time, the tortoise has run a much shorter distance, e.g., 10 meters. It will then take Achilles some further time to run that distance, by which time the tortoise will have advanced further. Therefore, whenever Achilles reaches somewhere the tortoise has been, he still has farther to go. In the end, because there are an infinite number of points Achilles must reach where the tortoise has already been, he can never overtake the tortoise.

Behind Zeno's paradox is the concept of processes that perform the first step in one unit of time and each subsequent step in a shorter time of the step before, for example, $1 + \frac{1}{2} + \frac{1}{4} + \frac{1}{8} + \cdots$. As the series $1 + \frac{1}{2} + \frac{1}{4} + \frac{1}{8} + \cdots$ is smaller than 2, a process of this kind could finish an infinite number of steps in two time units.

Definition of an accelerated Turing machine. An accelerated Turing machine TM^a is programmed to simulate an arbitrary Turing machine TM with arbitrary input:

- If the Turing machine TM halts on its input, TM^a changes the value of a square on its tape (for example, the first one) from 0 to 1.
- If TM does not halt, then TM^a leaves the square as 0.
- Anyway, after two time units, the first square on the tape of TM^a holds the value of the halting function of TM and its input.

At a first glance, the only difference between an accelerated Turing machine and a standard Turing machine seems to be the speed at which they operate. But, with respect to the arithmetical hierarchy, accelerated Turing machines correspond to more then Δ_1^0-predicates (Ord, 2002):

Proposition. *The class of predicates decided by accelerated Turing machines* $[TM^a] = \Sigma_1^0$.

Proof.

$$[\mathrm{TM}^a] \supseteq \Sigma_1^0 :$$

A Σ_1^0-predicate can be represented as $\exists y R(x, y)$ with R recursive.

For a given input x, an accelerated TM^a can compute $R(x, 1)$ in a finite number of steps:

The particular square of the Turing tape is marked if $R(x, 1)$ is true.

If $R(x, 1)$ is not true, TM^a can proceed to compute $R(x, 2)$, $R(x, 3), \ldots$, mark the particular square, and halt if it finds an example which is true.

After two external time units, the particular square will contain 1 iff $\exists y\, R(x, y)$ holds.

By that, it is proved that every Σ_1^0-predicate has a corresponding accelerated Turing machine TM^a.

$$[\mathrm{TM}^a] \subseteq \Sigma_1^0 :$$

Let TM^a be an accelerated Turing machine. We define a predicate $R_{\mathrm{TM}^a}(x, y)$ to be true iff TM^a, computing with input x, marks the particular square within y steps:

$\exists y\, R_{\mathrm{TM}^a}(x,y)$ is true iff TM^a prints 1 when computing with input x.
$\exists y\, R_{\mathrm{TM}^a}(x,y)$ is false iff TM^a prints 0 when computing with input x.

As $R_{\mathrm{TM}^a}(x,y)$ is recursive for all TM^a, $\exists y\, R_{\mathrm{TM}^a}(x,y)$ is a Σ_1^0-predicate for all TM^a.

Therefore, every accelerated Turing machine has a corresponding predicate in Σ_1^0. $\qquad\square$

From a technical point of view, the machine concepts beyond Turing machines only seem to be theoretical models without practical use. Anyway, they help to clarify mathematical statements which cannot be reduced to the level of physical computers. We will come back to the physical and epistemic aspects of computability beyond Turing computability later on.

Chapter 4

Constructive Proof Theory

In Euclid's time, mathematical proofs were realized by logical conclusions and constructive procedures. In Greek geometry, figures were constructed by compass and ruler which were rooted in everyday technologies of practical engineering. With the development of new curves in modern times, constructive tools were extended beyond compass and ruler. Descartes, for example, suggested new kinematic instruments to construct algebraic curves which he defined by algebraic equations. Later on, constructions were supported by machines and nowadays we use computers for applications of algorithms. Thus, since the very beginning, proofs have been referred to as tools of construction (Mainzer, 1980). In the previous chapter, hierarchies of computational degrees were suggested to distinguish different degrees of constructivity.

Proofs are logical procedures to verify the truth of statements. Thus, it depends on the accepted tools of construction whether a proof of a statement is accepted. A proof can be defined as a constructive procedure referring to the logical operations of a statement (Heyting, 1934; Kolmogorov, 1932):

Definition of Brouwer–Heyting–Kolmogorov–interpretation.

(1) A proof of $A \wedge B$ is given by a proof of A and a proof of B.
(2) A proof of $A \vee B$ is given either a proof of A or a proof of B.

(3) A proof of $A \to B$ is a construction to transform any proof of A into a proof of B.

(4) A contradiction $\perp$ has no proof. A proof of $\neg A :\leftrightarrow A \to \perp$ is (according to (3)) a construction which transforms any assumed proof of A into a proof of a contradiction.

(5) A proof of $\forall x A(x)$ is a construction which transforms a proof of the statement that a is an element of range M of the variable x into a proof of $A(a)$.

(6) A proof of $\exists x A(x)$ is given by a construction of an element a from M, and a proof of $A(a)$.

The Brouwer–Heyting–Kolmogorov (BHK) interpretation uses "construction" as an intuitive notion which could be interpreted by different accepted procedures. Anyway, it can be imagined as a kind of effective function mapping an element of a certain domain onto a function value of another domain. We represent functions by the λ-notation (Church, 1941): $\lambda a.b$ means a function mapping an element a onto the function value b with $\lambda a.b[a] = b$. In the following, proofs are represented by terms $a, b, c, \ldots$; statements are represented by $A, B, C, \ldots$. Some examples may help to illustrate the intended understanding of the BHK-interpretation (Troelstra and van Dalen, 1988, Vol. 1):

Example. $A \to (B \to A)$ is logically provable.

Proof. Suppose proof a proves A.

$\lambda b.a$ (constant function assigning a to each argument b) transforms an assumed proof b of B into a proof of A which means according to (3):

$$\lambda b.a \text{ proves } B \to A.$$

$\lambda a.(\lambda b.a)$ (mapping assigning constant function $\lambda b.a$ to each argument a) transforms a proof of A into a proof of $B \to A$ which, again, means according to (3):

$$\lambda a.(\lambda b.a) \text{ proves } A \to (B \to A). \qquad \square$$

The principle of excluded middle (PEM) only assumes two values of truth: A statement is either true or false — in Latin: *tertium non datur!*

Example. $A \lor \neg A$ is constructively not provable.

Proof. $A \lor \neg A$ is provable means either a proof of A or a proof of $\neg A$, i.e., according to BHK (2) and (4), a method for obtaining a contradiction from a hypothetical proof of A. If such a universal method were available, we could also decide a statement the truth of which has not been decided yet (e.g., Goldbach's conjecture). $\square$

In Euclid's time, proofs were represented by sequences of propositions which were deduced by logical rules. The propositions can graphically be represented by nodes which are ordered in branches of trees. In this sense, construction of proofs means construction of proof trees. Gentzen's natural deduction rules support the constructive understanding of proving (Gentzen, 1934). Consider, e.g., the construction of a proof of statement B from an assumed proof of A. According to BHK-interpretation (3), this means the introduction of implication $A \to B$. In this case, the assumption A is said to be cancelled. There is also an elimination of the implication: If proofs of $A \to B$ and A are given, we also can prove B. According to (3), a proof of $A \to B$ means a construction to transform an assumed proof of A into a proof of B. Consequently, as a proof of A is given, we get a proof of B.

Gentzen collected introduction rules (I-rules) and elimination rules (E-rules) for the logical operators. Deductions as sequences of propositions are labeled by letter d. The conclusion B with deduction d is indicated by

$$d \; \vdots \\ B$$

An assumption of A is indicated by brackets $[A]$. Therefore, a deduction d of B under assumption of A is denoted by

$$[A]$$
$$d \;\; \vdots$$
$$B$$

Variables of terms are denoted by t, s.

Definition of natural deduction rules.

$$
(\wedge\,\mathrm{I}) \quad \frac{\begin{array}{cc} d_1 \vdots & d_2 \vdots \\ A & B \end{array}}{A \wedge B}
\qquad\qquad
(\wedge\,\mathrm{E}_1) \quad \frac{d \vdots \\ A \wedge B}{A}
\qquad
(\wedge\,\mathrm{E}_2) \quad \frac{d \vdots \\ A \wedge B}{B}
$$

$$
(\rightarrow\,\mathrm{I}) \quad \frac{\begin{array}{c} [A] \\ d \vdots \\ B \end{array}}{A \rightarrow B}
\qquad\qquad
(\rightarrow\,\mathrm{E}) \quad \frac{\begin{array}{cc} d_1 \vdots & d_2 \vdots \\ A \rightarrow B & A \end{array}}{B}
$$

$$
(\vee\,\mathrm{I}_1) \quad \frac{d_1 \vdots \\ A}{A \vee B}
\qquad
(\vee\,\mathrm{I}_1) \quad \frac{d_2 \vdots \\ B}{A \vee B}
\qquad
(\vee\mathrm{E}) \quad \frac{\begin{array}{ccc} & [A] & [B] \\ d_1 \vdots & d_2 \vdots & d_3 \vdots \\ A \vee B & C & C \end{array}}{C}
$$

$$
(\forall\mathrm{I}) \quad \frac{d \vdots \\ A}{\forall y\, A[x/y]}
\qquad\qquad
(\forall\mathrm{E}) \quad \frac{d \vdots \\ \forall x\, A}{A[x/t]}
$$

$$
(\exists\mathrm{I}) \quad \frac{d \vdots \\ A[x/t]}{\exists x\, A}
\qquad\qquad
(\exists\mathrm{E}) \quad \frac{\begin{array}{cc} & [A] \\ d_1 \vdots & d_1 \vdots \\ \exists y A[x/y] & C \end{array}}{C}
$$

Example. The occurrence and cancellation of assumptions are indicated by numbers.

$$
\cfrac{
(3)\ \neg\neg\forall x\, A(x) \qquad
\cfrac{
\cfrac{
(2)\ \neg A(y) \qquad
\cfrac{(1)\ \forall x\, A(x)}{A(y)}\ (\forall E)
}{\bot}\ (\to E)
}{(1)\ \neg\forall x\, A(x)}\ (\to I)
}{\bot}\ (\to E)
$$

$$
\cfrac{
\cfrac{
\cfrac{
\cfrac{\bot}{(2)\ \neg\neg A(y)}\ (\to I)
}{\forall x\, \neg\neg A(x)}\ (\forall I)
}{(3)\ \neg\neg\forall x\, A(x) \to \forall x\, \neg\neg A(x)}\ (\to I)
}{}
$$

The deductions may be considered as labeled trees. The label attached to a node is a statement and the name of the applied rule. The example is illustrated by the following tree:

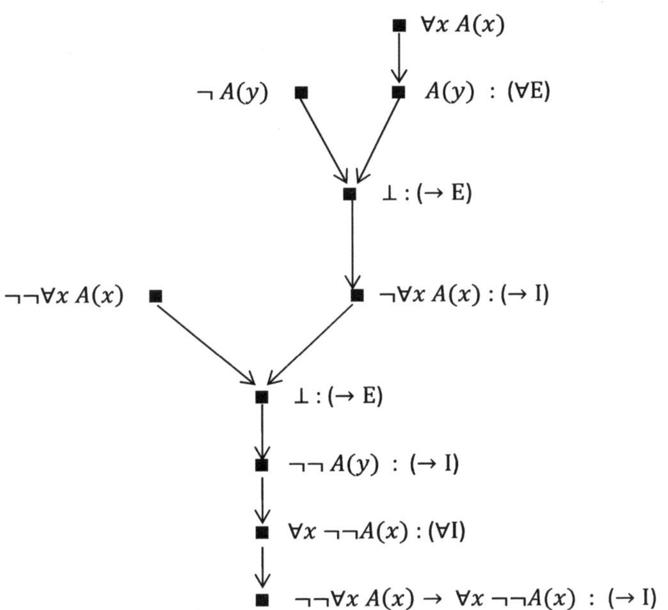

Natural numbers are generated by the basic process of counting. Counting is a construction of a successor function S step by step with $0, S0, SS0, \ldots$. The successor function S satisfies the properties

$$x = x \leftrightarrow Sx = Sy,$$

$$0 \neq Sy.$$

Statements $A(x)$ about all natural numbers x are proved by the following principle of induction:

$$A(0) \wedge \forall x(A(x) \rightarrow A(Sx)) \rightarrow \forall x A(x).$$

Intuitively, the principle of induction means that, given $A(0)$ and $\forall x(A(x) \rightarrow A(Sx))$, an inductive proof is built as a tree of deductions $A(0), A(S0), A(SS0), \ldots$ along the successive construction of natural numbers $0, S0, SS0, \ldots$.

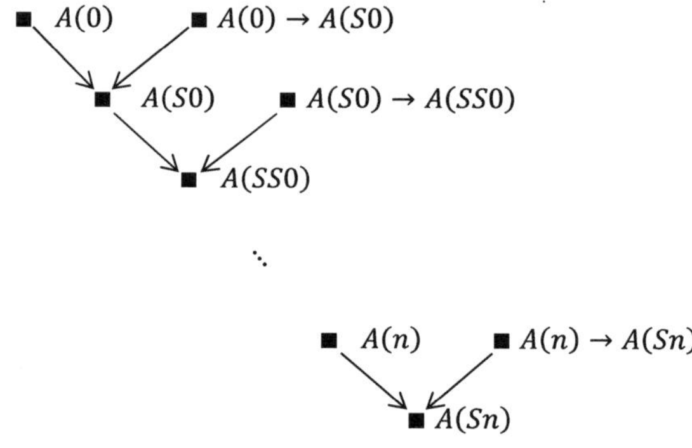

Besides successor, arithmetic includes addition and multiplication as constructive procedures which can be defined in an inductive manner: We define these operations for addition or multiplication with 0, and in an inductive step, given the results of addition and multiplication with n, we define the operations for the successor Sn. Along this line, we can accept the primitive recursive functions (compare the previous chapter) as constructive procedures of arithmetic.

David Hilbert believed that the consistency of mathematics can be reduced to this kind of constructive (i.e., primitive recursive) arithmetic (Detlefsen, 1986). Hilbert's program of proof theory aimed to justify the use of infinitistic mathematical principles in the proofs of finitistic mathematical statements by giving a finitistic proof that the infinitistic principles are consistent (Sieg, 2013). In his article "Über das Unendliche" (Hilbert, 1926), he illustrated his program by the famous claim: "No one will drive us from the paradise which Cantor created for us." Gödel's second incompleteness theorem implies that the consistency of Peano arithmetic cannot be proved by any finitistic method that can be formalized in a theory the consistency of which is provable in Peano arithmetic. An example of accepted finitistic procedures is the theory of primitive recursive functions. It can be proved to be consistent in Peano arithmetic. But the theory of primitive recursive functions cannot prove the consistency of Peano arithmetic.

According to Gödel's incompleteness theorem, a consistent formal system of arithmetic contains an unprovable, but true proposition (Gödel, 1931). Even if the theory is extended by the undecidable proposition as axiom, then there is another true, but unprovable proposition etc. ad infinitum. Therefore, no computer program can prove all and only the true propositions. Nevertheless, richer formal programs can be generated ad infinitum. The extension of much more richer formal systems will never exhaust all mathematical truths.

In the previous chapter, we approached the incompleteness of arithmetic from the side of computers. Turing proved that incompleteness directly follows from the undecidability of the halting problem: If there is a complete formal system with formal proofs for all mathematical truths, then there is a procedure of deciding whether a computer program will stop or not. A popular philosophical interpretation argues that Gödel's incompleteness theorem and Turing's unsolvability of the halting problem demonstrate absolute limitations of formal proofs and computation. That is definitely not the case. In the previous chapter, we introduced hierarchies of computational degrees beyond Turing complexity. In this chapter, we will study hierarchies of proof-theoretic measures beyond the Gödel results. Actually, Gödel's result was a first step of proof theory as a new mathematical discipline beyond Hilbert's program.

The beginning of modern proof theory is deeply connected with the contributions of Alan Turing and Gerhard Gentzen (1909–1945).

Gentzen discovered that the Gödel's obstacle to prove the consistency of number theory depends on the principle of induction which is restricted to natural numbers (Gentzen, 1936). It can be overcome by using transfinite induction up to a sufficiently great ordinal, Cantor's ε_0, the first ordinal greater than all ω, ω^ω, ω^{ω^ω}, ... or the least of the ε-numbers ξ with $\omega^\xi = \xi$. All possible proofs of number theory are well-ordered by ordinal induction numbers in an order in which they are being proved, just as the natural numbers used as induction numbers do in an ordinary induction.

Before going into the details, we remind the reader of some basics of ordinal numbers in the terminology of elementary set theory. Intuitively, the concept of ordinal numbers generalizes the concept of natural numbers. The idea is to arrange a collection of objects in order, one after another. Ordinals are distinct from cardinal numbers which determine how many objects are in a collection (Bachmann, 1967):

Reminder of ordinal numbers. *Let $x, y, z, \ldots$ be sets. $x \in y$ means that x is element of y. $x \subseteq y$ means that x is subset of y. Ordinal numbers are denoted by $\alpha, \beta, \gamma \ldots$.*

- A set z is called transitive iff $x \in y$ and $y \in z$ implies $x \in z$ (or equivalently: $y \in z$ implies $y \subseteq z$).
- A set x is called an ordinal number iff x and all its elements are transitive.
- Order on ordinals: $\alpha < \beta :\leftrightarrow \alpha \in \beta$, $\alpha \in \beta \leftrightarrow \alpha \subset \beta$.
- Successor of ordinal α: $s(\alpha) := \alpha \cup \{\alpha\}$.
- Limit ordinal: γ is not successor of any ordinal.

John von Neumann's model of ordinal numbers.

$$0 := \emptyset$$

$$1 := \{0\} = \{\emptyset\}$$

$$2 := \{0, 1\} = \{\emptyset, \{\emptyset\}\}$$

$$3 := \{0, 1, 2\} = \{\emptyset, \{\emptyset\}, \{\emptyset, \{\emptyset\}\}\}$$

$$\vdots$$

$$\omega := \{0, 1, 2, 3, \ldots\}$$

$$\omega + 1 := \{0, 1, 2, 3, \ldots, \omega\}$$

$$\omega + 2 := \{0, 1, 2, 3, \ldots, \omega, \omega + 1\}$$

$$\vdots$$

$$\omega + (n + 1) := \{0, 1, 2, 3, \ldots, \omega, \omega + 1, \omega + 2, \ldots, \omega + n\}$$

$$\vdots$$

$$\omega + \omega := \bigcup_n (\omega + n)$$

$$\vdots$$

The generation of ordinals needs formal definitions of addition, multiplication, and power. Contrary to inductive definitions on natural numbers, we also must consider the case of limit ordinals.

Definition of ordinal operations. Addition of ordinals:

$$\alpha + 0 := \alpha$$

$$\alpha + s(\beta) := s(\alpha + \beta)$$

$$\alpha + \gamma := \bigcup_{\beta < \gamma} \alpha + \beta$$

Multiplication of ordinals:

$$\alpha \cdot 0 := \alpha$$

$$\alpha \cdot s(\beta) := \alpha \cdot \beta + \alpha$$

$$\alpha \cdot \gamma := \bigcup_{\beta < \gamma} \alpha \cdot \beta$$

Power of ordinals:

$$\alpha^0 := 1$$

$$\alpha^{s(\beta)} := \alpha^\beta \cdot \alpha$$

$$\alpha^\gamma := \bigcup_{\beta < \gamma} \alpha^\beta$$

Limit ordinal ε_0: $\varepsilon_0 := \bigcup_n \omega^{\omega^{\cdot^{\cdot^{\cdot^\omega}}}}$

For abbreviation of series of power ordinals, we introduce

$$\omega_1 := \omega,$$

$$\omega_{n+1} := \omega^{\omega_n}.$$

There are several remarkable properties of ordinal operations: Ordinal addition and multiplication are not commutative. For example, it follows by definition

$$\omega + 1 \neq \omega,$$

$$1 + \omega = \bigcup_{\beta < \omega} 1 + \beta = \omega,$$

$$\omega \cdot 2 \neq \omega,$$

$$2 \cdot \omega = \bigcup_{\beta < \omega} 2 \cdot \beta = \omega.$$

All ordinals α with $0 < \alpha < \varepsilon_0$ can be generated by a finite number of additions and powers of ω. Therefore, each ordinal $\alpha \neq 0$ smaller than ε_0 can be represented as follows.

Cantor's normal form. $\alpha = \omega^{\alpha_m} \cdot c_m + \omega^{\alpha_{m-1}} \cdot c_{m-1} + \cdots + \omega^{\alpha_0} \cdot c_0$ with $\alpha_m > \alpha_{m-1} > \cdots > \alpha_0$ and $c_i \neq 0$ with $m \geq -1$. ($m = -1$ corresponds to the empty sum.)

The limit ordinal ε_0 is unbelievably huge, but, nevertheless, countable. It is the first ordinal which cannot be represented in Cantor's normal form. But, it is the first ordinal satisfying the fixed point property:

$$\omega^{\varepsilon_0} = \varepsilon_0.$$

The concept of levels of ordinals is useful for proofs with transfinite induction.

Definition of levels of ordinals.

$$\operatorname{level}(\alpha) := \begin{cases} 0, & \text{if } \alpha = 0 \text{ (empty sum)}, \\ \operatorname{level}(\alpha_m) + 1, & \text{if } \alpha = \omega^{\alpha_m} + \omega^{\alpha_{m-1}} + \cdots + \omega^{\alpha_0} \\ & \quad \text{with } \alpha_m \geq \alpha_{m-1} \geq \ldots \geq \alpha_0 \end{cases}$$

For ordinals of level k, $\omega_k \leq \alpha < \omega_{k+1}$:

$$\operatorname{level}(\alpha) = 0 : \omega_0 = 0 \leq \alpha < \omega = \omega_1$$

$$\operatorname{level}(\alpha) = 1 : \omega_1 = \omega \leq \alpha < \omega^\omega = \omega_2$$

$$\operatorname{level}(\alpha) = 2 : \omega_2 = \omega^\omega \leq \alpha < \omega^{\omega^\omega} = \omega_3$$

$$\vdots$$

$$\operatorname{level}(\alpha) = k : \omega_k = \omega^{\omega^{\cdot^{\cdot^{\cdot^\omega}}}} \leq \alpha < \omega^{\omega^{\cdot^{\cdot^{\cdot^{\omega^\omega}}}}} = \omega_{k+1}$$

Now, let us consider Gentzen's approach to overcome Gödel's incompleteness (Gentzen, 1938). Gentzen showed that the consistency of the first-order Peano axioms is provable in a theory of primitive recursive arithmetic extended by the transfinite induction up to the ordinal ε_0. To represent ordinals in the formal theory of arithmetic, an ordinal notation is needed, in order to assign natural numbers to ordinals less than ε_0. This can be done in various ways, one example is provided by Cantor's normal form theorem.

Definition of Gödel numbers of ordinals. The Gödel number $\lceil \alpha \rceil$ of ordinal α is inductively defined by

$$\lceil \alpha \rceil = \lceil \omega^{\alpha_m} \cdot c_m + \omega^{\alpha_{m-1}} \cdot c_{m-1} + \cdots \omega^{\alpha_0} \cdot c_0 \rceil := \left(\prod_{i \leq m} p_{\lceil \alpha_i \rceil}^{c_i} \right) - 1$$

with the nth prime number p_n (beginning with $p_0 = 0$). The corresponding ordinal notation $o(x)$ is inductively defined by

$$o\left(\left(\prod_{i \leq l} p_i^{q_i}\right) - 1\right) := \sum_{i \leq l} \omega^{o(i)} q_i$$

for non-negative integers. It can be proved by induction:

$$o(\lceil \alpha \rceil) = \alpha$$
$$\lceil o(x) \rceil = x$$

The following abbreviations are convenient:

$$x \prec y := o(x) < o(y)$$
$$\omega^x := \lceil \omega^{o(x)} \rceil$$
$$x \oplus y := \lceil o(x) + o(y) \rceil$$
$$xk := \lceil o(x)k \rceil$$
$$\omega_k := \lceil \omega_k \rceil$$

The formal theory of arithmetic $\mathbf{Z}$ is based on minimal logic which not only rejects PEM (*tertium non datur*) like intuitionistic logic, but also the principle of explosion (ex falso quodlibet): $\bot \to A$. Thus, $\mathbf{Z}$ should widely be accepted by different logical positions. The language of $\mathbf{Z}$ contains constants denoting computational functions, relations, and predicates. Examples are function constant S for successor function, relation constants $=$ for equality, and $\prec$ for the ordering of type ε_0 of natural numbers.

Terms are inductively generated by object variables $x, y, z, \ldots$ and function constants $f(t_1, \ldots, t_m)$. Terms $0, S0, SS0, \ldots, S^n 0, \ldots$, are called numerals representing the natural numbers. Formulas are also inductively introduced, beginning with $\bot$ and atomic formulas $R(t_1, \ldots, t_m)$ with relation constant R, which can be connected by the logical operators $A \to B$ and $\forall x A$ with formulas A and B. (Note: Negation is defined as $\neg A := A \to \bot$.)

Axioms of formal ordinal arithmetic Z. The axioms of **Z** include the Peano axioms of successor S and induction on numerals:

$$\text{PA1:} \quad Sx = Sy \to x = y$$

$$\text{PA2:} \quad Sx = 0 \to A$$

$$\text{PA3:} \quad A(x/0) \to \forall x(A \to A(x/Sx)) \to \forall x A(x).$$

The formal counterparts to the properties of ordinals and their operations up to ε_0 are given by axioms of the (irreflexive and transitive) relation constant $\prec$ and operation constants $\oplus$ and $\lambda xy\,.\,\omega^x y$:

$$\text{ord1:} \quad x \prec 0 \to A$$

$$\text{ord2:} \quad x \prec y \oplus \omega^0 \to (z \prec y \to A) \to (z = y \to A) \to A$$

$$\text{ord3:} \quad x \oplus 0 = x$$

$$\text{ord4:} \quad x \oplus (y \oplus z) = (x \oplus y) \oplus z$$

$$\text{ord5:} \quad 0 \oplus x = x$$

$$\text{ord6:} \quad \omega^x 0 = 0$$

$$\text{ord7:} \quad \omega^x(Sy) = \omega^x y \oplus \omega^x$$

$$\text{ord8:} \quad z \prec y \oplus \omega^{Sx} \to z \prec y \oplus \omega^{\mathrm{e}(x,y,z)}\mathrm{m}(x, y, z)$$

$$\text{ord9:} \quad z \prec y \oplus \omega^{Sx} \to \mathrm{e}(x, y, z) \prec Sx$$

with primitive recursive constants e and m.

At first, we can prove an initial case of transfinite induction in **Z** at least up to ω_n (Troelstra and Schwichtenberg, 2000, chapter 10.2). The proof illustrates how derivations in Z work.

Theorem. $\forall x(\forall y \prec x A(y) \to A(x)) \to \forall x \prec \omega_n A(x)$ *is derivable in* **Z** *for arbitrary formulas A.*

Proof. In the following, a formula $B(x)$ is called progressive iff

$$(*) \quad \forall x(\forall y \prec x B(y) \to B(x)).$$

At first, we assign a formula $A^+(x)$ to any formula $A(x)$:

$$(**) \quad A^+(x) :\equiv \forall y (\forall z \prec y A(z) \to \forall z \prec y \oplus \omega^x A(z))$$

and prove the auxiliary proposition.

Auxiliary proposition. *If $A(x)$ is progressive, then $A^+(x)$ is progressive.*

Auxiliary proof. We assume that $A(x)$ is progressive and want to prove that $A^+(x)$ is progressive, i.e.,

$$\forall x (\forall y \prec x A^+(y) \to A^+(x)).$$

We assume the premise

(1) $\forall y \prec x A^+(y)$

and have to prove $A^+(x)$. According to the definition $(**)$ of $A^+(x)$, we assume

(2) $\forall z \prec y A(z)$ and $z \prec y \oplus \omega^x$

and have to prove $A(z)$ by induction on x.

Case $x = 0$. If $z \prec y \oplus \omega^0$, then, according to ord2, then we have to derive $A(z)$ from $z \prec y$ as well as from $z = y$:

If $z \prec y$, then $A(z)$ follows from assumption (2).
If $z = y$, then $A(z)$ follows from the assumption that $A(x)$ is progressive $(*)$.

Case Sx. If $z \prec y \oplus \omega^{Sx}$, then $z \prec y \oplus \omega^{e(x,y,z)} m(x,y,z)$ (ord8) and $e(x,y,z) \prec Sx$ (ord9).

From assumption (1), we derive $A^+(e(x,y,z))$. The definition $(**)$ of $A^+(x)$ delivers

$$\forall u \prec y \oplus \omega^{e(x,y,z)} v A(u) \to \forall u \prec (y \oplus \omega^{e(x,y,z)} v) \omega^{e(x,y,z)} A(u).$$

It follows by ord4 and ord7 that

$$\forall u \prec y \oplus \omega^{e(x,y,z)} v A(u) \to \forall u \prec y \oplus \omega^{e(x,y,z)} Sv A(u).$$

From assumption (2) with ord6 and ord3, we get

$$\forall u \prec y \oplus \omega^{e(x,y,z)} 0 A(u).$$

With an appropriate instance of the induction scheme, we get

$$\forall u \prec y \oplus \omega^{\mathrm{e}(x,y,z)} \mathrm{m}(x,y,z) A(u)$$

and hence $A(z)$, which finishes the auxiliary proof.

We now prove the transfinite induction scheme up to ω_n, i.e.,

$$\forall x (\forall y \prec x A(y) \to A(x)) \to \forall x \prec \omega_n A(x)$$

The premise assumes that $A(x)$ is progressive (*). We have to prove $\forall x \prec \omega_n A(x)$ by induction on n:

Case $n = 0$. Let $x \prec \omega^0$. It follows $x \prec 0 \oplus \omega^0$ by ord5. By ord2, we have to derive $A(x)$ from $x \prec 0$ as well as from $x = 0$:

If $x \prec 0$, then $A(x)$ follows by ord1.
If $x = 0$, then $A(x)$ follows from the assumption that $A(x)$ is progressive (*).

Case $n + 1$. We assumed that $A(x)$ is progressive. By the auxiliary proposition, it follows that $A^+(x)$ is also progressive.

The induction hypothesis applied to $A^+(x)$ delivers

$$\forall x \prec \omega_n A^+(x).$$

It follows $A^+(\omega_n)$ because $A^+(x)$ is progressive (*).
With definition (**) of $A^+(x)$, ord1, and ord5, it follows that

$$\forall z \prec \omega^{\omega_n} A(z). \qquad \square$$

The preceding theorem can be restricted to subsystems $\mathbf{Z}_k$ of $\mathbf{Z}$ (Troelstra and Schwichtenberg, 2000, chapter 10.2). They are defined like $\mathbf{Z}$, but with the induction scheme PA3 restricted to formulas A with level $A \leq k$.

Definition of level of formulas.

$$\mathrm{level}\ (R(t_1, \ldots, t_m)) := \mathrm{level}(\bot) := 0$$

$$\mathrm{level}\ (A \to B) := \max(\mathrm{level}(A) + 1,\ \mathrm{level}(B))$$

$$\mathrm{level}\ (\forall x A) := \ \max(1, \mathrm{level}(A))$$

Theorem. *In $\mathbf{Z}_k$, transfinite induction can derived for any formula of level $\leq l$ up to $\omega_{k-l+2}[m]$ for any m with $\omega_1[m] := m$ and $\omega_{i+1}[m] := \omega^{\omega_i[m]}$, i.e.,*

$$\forall x(\forall y \prec x A(y) \to A(x)) \to \forall x \prec \omega_{k-l+2}[m]A(x).$$

But, the application of transfinite induction in $\mathbf{Z}$ and $\mathbf{Z}_k$ is bounded: In $\mathbf{Z}$, transfinite induction cannot be derived up to ε_0, and in $\mathbf{Z}_k$, transfinite deduction cannot be derived up to ω_{k+1}.

Theorem. *Transfinite induction up to ε_0 is underivable in $\mathbf{Z}$: $\mathbf{Z} \vdash \forall x(\forall y \prec x P(y) \to P(x)) \to \forall x P(x)$ is false. Transfinite induction up to ω_{k+1} is underivable in $\mathbf{Z}_k$:*

$$\mathbf{Z}_k \vdash \forall x(\forall y \prec x P(y) \to P(x)) \to \forall x \prec \omega_{k+1}P(x) \text{ is false.}$$

In his later version, Gentzen (1943) proved the initial cases of transfinite induction along $\prec$ and demonstrated the underivability of the transfinite induction up to ε_0. In an early version (Gentzen, 1936), he used the so-called reduction steps for deductions which preserve correctness. Ordinals less than ε_0 are assigned to deductions. Suitable reduction steps lower the ordinals assigned to a deduction. If a proof of a contradiction could be derived, then the reduction procedure would result in an infinite descending sequence of ordinals $< \varepsilon_0$.

Gentzen's proof was improved by replacing the induction axioms by a rule with infinitely many premises (Schütte, 1977; Lorenzen, 1980; Troelstra and Schwichtenberg, 2000). The so-called ω-rule concludes $\forall x A(x)$ from premises $A(0), A(1), A(2), \ldots$:

$$
\begin{array}{ccccc}
d_0 \vdots & d_0 \vdots & & d_0 \vdots & \\
A(0) & A(1) & \ldots & A(n) & \ldots \\
\hline
& & \forall x\, A(x) & & \omega\text{-rule}
\end{array}
$$

Deductions with the ω-rule can be illustrated as infinitely (countably) branching trees. Deductions are defined inductively. By that, the corresponding trees are well-founded without an infinitely descending branch:

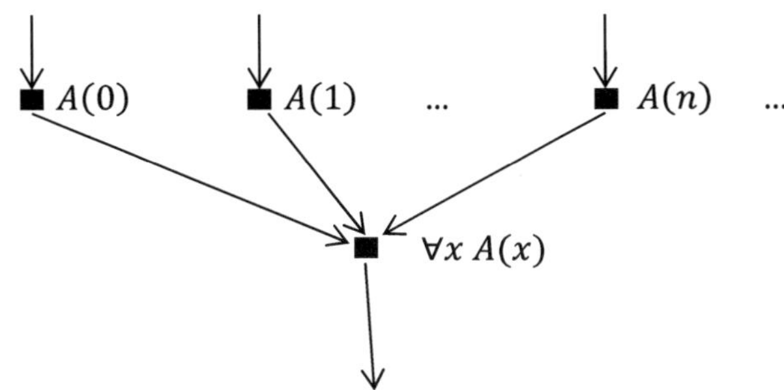

Any deduction in **Z** can be translated into an infinitary deduction of the ω-arithmetic $\mathbf{Z}^\infty$ with the ω-rule. $\mathbf{Z}^\infty$ has the language of **Z** with the same axioms apart of the induction axioms which are replaced by the ω-rule.

After Gentzen, we consider Turing's approach to overcome Gödel's incompleteness (Turing, 1939). According to Gödel's incompleteness, each formal axiomatic theory T (containing arithmetic) is associated with a true but unprovable proposition A_T. Turing started with statements of the Π_1^0 form $\forall x R(x)$ expressing that a certain (primitive) recursive property R holds for all integers x. He tried to overcome Gödel's incompleteness by constructive degrees of incompleteness. Therefore, he extended the incomplete theories by the corresponding unprovable propositions (as axioms) in a progression along an effective construction of ordinals, i.e., beginning with the initial theory T_1 and its unprovable A_1,

$$T_2 = T_1 \cup \{A_1\}, \ldots, T_n = T_1 \cup \{A_1, \ldots, A_{n-1}\}, \ldots,$$

$$T_\omega = T_1 \cup \{A_1, \ldots, A_{n-1}, \ldots\}, \ldots.$$

In order to maintain an effective generation, one must pass to a transfinite limit ordinal α and theory T_α only when α is the limit of an effectively presented sequence $\alpha_1, \ldots, \alpha_n, \ldots$ and the theories T_{α_n} are already obtained. Effectiveness is made precise by an effective representation of ordinals in the integers which can be realized in different ways. In 1936, Church and Kleene defined the index system $\mathcal{O}$ which we already considered in the previous chapter. It can be used to introduce constructive ordinals. Constructivity comes in by machine numbers representing effective procedures (Goodstein, 1964).

Church–Kleene definition of constructive ordinals. The index set $\mathcal{O}$ is partially ordered by $<_{\mathcal{O}}$.

An ordinal α is constructive iff it has an effective notation $a \in \mathcal{O}$ with $\alpha = |a|$ with

$$|1| = 0,$$

$$|2^a| = |a| + 1,$$

$$|3 \cdot 5^e| = \sup_{n < \omega}(|[e](n)|).$$

It follows for ordering $<_{\mathcal{O}}$ that

$$b <_{\mathcal{O}} a \rightarrow |b| < |a|,$$

$\forall n([e](n) <_{\mathcal{O}} [e](n+1))$ for limit ordinals.

Addition and multiplication can be defined as partial recursive operations:

$$|a +_{\mathcal{O}} b| = |a| + |b|,$$

$$|a \cdot_{\mathcal{O}} b| = |a| \cdot |b|.$$

The Church–Kleene ordinal ω_1^{CK} denotes the least non-constructive ordinal.

In his Ph.D. thesis, Turing (1939) wanted to distinguish degrees of completeness with sequences $\Lambda = \{T_a | a \in \mathcal{O}\}$ of logical theories which are associated with indices a of constructive ordinals.

Turing's definition of ordinal progression of axiomatic theories. The ordinal progression $\Lambda_T = \{T_a | a \in \mathcal{O}\}$ of an initially incomplete theory T (e.g., Peano's system) is defined on the indices of constructive ordinals:

$$T_a = T,$$

$$T_{2^a} = T_a \cup \{A_a\},$$

$$T_{3 \cdot 5^e} = \bigcup_n T_{[e](n)}.$$

A theorem A is provable in the ordinal progression Λ_T iff a $a \in \mathcal{O}$ with A is provable in T_a.

A theorem A is deeper than a theorem B iff the least constructive ordinal $\alpha = |a|$ of theory T_a for proving A is greater than the least constructive ordinal $\beta = |b|$ of T_b for proving B.

Indices a increase with increasing ordinals of a, thus overcoming incompleteness at least gradually at each state a. One of Turing's main results was that Λ is complete for true Π_1^0 statements. He intended to strengthen this completeness result for Π_2^0 statements. These sentences have the form $\forall x \exists y \, R(x, y)$ with a primitive recursive predicate R. The motivation of his interest in Π_2^0 predicates is the fact that this layer of arithmetical hierarchy includes very interesting mathematical statements (e.g., the Riemann hypothesis Turing was strongly interested in).

According to Gödel, terms and statements can be coded by numbers (e.g., statement A by Gödel number $\lceil A \rceil$). Turing defined a logical theory very generally as any recursively enumerable set T of code numbers of Π_2^0 sentences A, such that $\lceil A \rceil \in T$ implies that A is true.

A logical derivation of A in T (abbreviation $A \vdash T$) means $\lceil A \rceil \in T$. A theory T' of this kind is said to be at least as complete as T if $T \subseteq T'$. T' is said to be more complete than T if T is a proper subset with $T \subset T'$. In general, Turing expected to get a classification of Π_2^0 theorems according to their logical depth. A theorem which required a (constructive) ordinal α to prove it would be deeper than the one which could be proved by the use of a (constructive) ordinal β less than α (Enderton and Luckham, 1964).

But, in 1962, Solomon Feferman proved that Λ is incomplete for Π_2^0 sentences. Anyway, besides Gentzen, Turing's Ph.D. thesis was the second initiative of ordinal proof theory which has analyzed measures of proof complexity with ordinals. An important extension of ordinal proof theory aims at the analysis of predicative mathematics which is closely connected with the Δ_1^1-layer of analytical hierarchy and hyperarithmetic beyond Turing computability (Feferman, 1964, 1968a,b).

Predicativity dates back to historical debates on paradoxes with circular arguments. Famous mathematicians and philosophers were involved in these debates. In 1905, the French mathematician Jules Richard described the following example of a semantic paradox.

Richard's paradox of impredicativity. Real numbers which are defined by English phrases can be arranged in an infinite list of phrases with dictionary order. This ordering yields an infinite list of corresponding real numbers r_1, r_2, $r_3, \ldots, r_n, \ldots$. Then, we can define a real number r^* in an English phrase which is different to each number of this list by a diagonal construction. (For example, the integer part of r^* is 0, the nth decimal place of r is 1 if the nth decimal place of r_n is not 1, and the nth decimal place of r is 2 if the nth decimal place of r_n is 1.) This is a contradiction.

Henri Poincaré argued that there is a circular argument, because the definition of number r^* is referred to the supposed total class of decimal numbers which can be defined by English phrases. In short, r^* is an element of the class of objects which is used to define r^*. In 1908, Russell suggested a syntactical version of the circular argument: Whatever contains an apparent variable must not be a possible value of that variable. A famous example is Russell's paradox of the set of all sets which are not elements of themselves.

Russell's paradox of impredicativity. Cantor's (unrestricted) comprehension principle states

$$(^*) \quad \forall P\, \exists y \forall x (x \in y \leftrightarrow P(x)).$$

As example, we define $P(x) :\leftrightarrow \neg x \in x$. Then, by $(^*)$, there is a set M with

$$\forall x (x \in M \leftrightarrow \neg x \in x).$$

For instance $x := M$, we get the contradiction

$$M \in M \leftrightarrow \neg M \in M.$$

Obviously, the definition of M refers to M itself.

Russell avoided impredicative concepts with circular arguments by his type theory. In 1918, Weyl (and later Lorenzen, 1955, 1965) criticized set-theoretical foundations of mathematics and suggested the introduction of mathematical analysis through a theory of real numbers which is only based on the construction of natural numbers, integers, and rational numbers step by step in an inductive manner without circular arguments. We will come back to this approach later on in this book. In this chapter, we used the hierarchies of recursion theory and degrees of computability to analyze predicative definitions and proofs. Actually, ordinals can be determined, up to which degree of complexity predicative definitions and proofs are possible.

Kleene started to refer definitions of sets of natural numbers to arithmetical and analytical hierarchies. The sets which are obtained by finite iteration of Turing jumps are, up to relative recursiveness, all the arithmetically definable sets. In the next step, Kleene extended this hierarchy by the index system $\mathcal{O}$ of Church–Kleene notations for constructive ordinals. The least ordinal α which cannot be indicated by an index $a \in \mathcal{O}$ as $\alpha = |a|$ is denoted ω_1^{CK}. It can also be proven that recursive well-orderings on the natural numbers (Spector, 1955) have order types as ordinals which are less than ω_1^{CK}. We already introduced the hyperarithmetic hierarchy along increasing ordinals $< \omega_1^{CK}$: A set is called hyperarithmetic iff it is recursive in a set H_a for some $a \in \mathcal{O}$. The class HYP of all hyperarithmetic sets is exactly the Δ_1^1-layer of the analytical hierarchy, i.e., $\text{HYP} = \Delta_1^1$.

HYP is referred to predicative definitions by the ramified type theory which is applied to the analytical hierarchy restricted to constructive ordinals (Mainzer, 1973). At first, we introduce the hierarchy $(\mathcal{A}_k)_{k<\omega}$ generating predictively definable sets step by step along the natural numbers starting with the class $\mathcal{A}$ of arithmetic predicates:

$$\mathcal{A}_0 := \mathcal{A},$$

$A_{k+1} := A_k^*$ with predicates the second order quantifiers of which are relativized to $\mathcal{A}_k$,

$$A_\omega := \bigcup_{k<\omega} \mathcal{A}_k.$$

The ramified hierarchy is extended for constructive ordinals.

Definition of the ramified analytical hierarchy.

$$\forall y \in \mathcal{O} \; A_y := \begin{cases} \mathcal{A} & \text{if } y = 1, \\ \mathcal{A}^*_{(y)o} & \text{if } y = 2^{(y)o}, \\ \displaystyle\bigcup_n \mathcal{A}_{[(y)2](n)} & \text{if } y = 3 \cdot 5^{(y)2}. \end{cases}$$

In 1955, Kleene proved that $\mathrm{HYP} = \mathcal{A}_{\omega_1^{CK}}$.

In 1960, Kreisel suggested to identify the predicatively definable sets with hyperarithmetic sets from HYP. An ordinal is called predictively definable if it is the order type of a predicatively definable well-ordering $\prec$ of the natural numbers. Obviously, all constructive ordinals are predicative. Consequently, a set is considered predictively definable iff it belongs to $\mathcal{A}_y$ for some $y \in \mathcal{O}$. Because of $\mathrm{HYP} = \mathcal{A}_{\omega_1^{CK}}$, all hyperarithmetic sets are accepted to be predicative. For the converse, the predicatively definable sets do not go beyond HYP, because they do not go beyond the recursive well-ordering with order type (ordinal) less than ω_1^{CK}.

In the case of predicatively definable sets, we considered the hierarchy of hyperarithmetic sets. Provability is referred to formal systems with certain comprehension axioms which are crucial for predicativity. Therefore, in order to determine predicative provability, we start with a formal theory $\mathcal{H}$ of second-order arithmetic, the comprehension axiom of which is refined to hyperarithmetic Δ_1^1-sets:

Δ_1^1-comprehension axiom. $\forall P \in \Delta_1^1 \exists X \forall x (x \in X \leftrightarrow P(x))$, where $\forall x (P(x) \leftrightarrow Q(x))$ with any Π_1^1-formula $P(x)$ and any Σ_1^1-formula $Q(x)$.

In analogy to index system $\mathcal{O}$ for predicatively definable sets, an index system N of ordinal numbers was introduced to represent all predicatively provable well-orderings. A recursive progression of theories $\mathcal{H}_a$ (with an element of N) starts with $\mathcal{H}$ and is extended step by step: If formula $F(x)$ is provable in system $\mathcal{H}_a$ for all x, then $\forall x F(x)$ is provable in $\mathcal{H}_{a \oplus 1}$ (with N-addition $\oplus$). In formal theories, numbers (including ordinals) are represented by their formal counterparts. The enumerable predicate $pr(a\lceil F(x)\rceil)$ means that formula F (with Gödel number $\lceil F(x)\rceil$) is provable in $\mathcal{H}_a$ (Feferman, 1964, 1968a).

Hyperarithmetic theory progression.

$\mathcal{H}_0 := \mathcal{H}.$

$\mathcal{H}_{a \oplus 1} := \mathcal{H}_a \cup \{\forall x \, pr(a, \lceil F(x)\rceil) \to \forall x F(x) | F$ formula with free variable of numerals$\}.$

$\mathcal{H}_a := \bigcup_{b \prec a} \mathcal{H}_b$ with a numeral of limit ordinal.

In order to guarantee predicative probability, the ordinal indices are restricted to those which are autonomous to $(\mathcal{H}_a)_a$, i.e.,

(i) 0 is an autonomous ordinal index.

(ii) If b is an autonomous ordinal index and a codes an ordinal provable in $\mathcal{H}_b$, then a is an autonomous ordinal index.

(iii) $\mathrm{Aut}(\mathcal{H}) :=$ the less ordinal α with $|c| < \alpha$ for all c autonomous to $(\mathcal{H}_a)_a$.

$(\mathcal{H}_a)_a$ starts with a predicatively justified theory $\mathcal{H}$ with predicatively definable predicates and a hyperarithmetic comprehension principle of sets. The justification of predicative provability is continued by the condition of autonomous ordinal index along the theory progression $(\mathcal{H}_a)_a$. In this sense, proofs in $(\mathcal{H}_a)_a$ are predicative. A well-ordering can be predicatively proven if its ordinal type is smaller than $\mathrm{Aut}(\mathcal{H})$.

But, actually, we can argue for more: All predicative proofs can be characterized by $(\mathcal{H}_a)_a$. This argument reminds us of Church's thesis: All Turing-computable functions are computable in an intuitive sense. But, why are intuitively computable functions Turing computable? Church argued with many different approaches of computability which are all intuitively computable. The decisive point is that these approaches are mathematically equivalent with Turing computability in a rigorous sense. Analogously, several approaches to predicative probability were suggested, which are equivalent to the hyperarithmetic approach in a rigorous sense. Therefore, predicative proofs are assumed to be characterized by the hyperarithmetic theory progression $(\mathcal{H}_a)_a$ (Feferman, 1968a,b; Mainzer, 1973).

An example of another approach is a progression of ramified theories the ordinal indices of which are also restricted by a condition of autonomy. It can be proven that these approaches have the same ordinal bound, i.e., the least ordinal α with $|c| < \alpha$ for all c autonomous to the considered theory progression. Further on, under certain conditions, the proofs in one of these theory progressions can be translated into appropriate proofs of another equivalent theory progression. All these arguments support the assumption that predicative provability is exactly realized in the approaches equivalent to hyperarithmetic theory progression $(\mathcal{H}_a)_a$.

The limit ordinal of the predicatively provable ordinals can be independently characterized by a hierarchy $(\chi_\alpha)_\alpha$ of critical functions of ordinals α. For $\alpha \neq 0$, critical function χ_α enumerates the set of fixed points of all χ_β for each $\beta < \alpha$ in order of size (Veblen, 1908; Feferman, 1968a; Schütte, 1965, 1977).

Definition of Veblen hierarchy of critical functions.

$\chi_\alpha(\xi) := \omega^\xi$, if $\alpha = 0$.

χ_α enumerates the common fixed points of χ_β with $\beta < \alpha$, if $\alpha \neq 0$.

$\Gamma_\alpha := \alpha$ th fixed point ξ of $\chi_\xi(0) = \xi$.

Γ_0 is the smallest ξ such that $\chi_\xi(0) = \xi$. It is exactly the ordinal bound of predicatively provable well-orderings which is sometimes called the Feferman–Schütte ordinal (Feferman, 2002; Pohlers, 1989). Transfinite inductions up to ordinals smaller than Γ_0 can be proven in one of the equivalent theory progressions, e.g., in $(\mathcal{H}_a)_a$. But transfinite induction up to Γ_0 is excluded. Therefore, Γ_0 can also be interpreted as the smallest only impredicatively provable well-ordering.

Chapter 5

Computational Mathematics and Digital Information Systems

Mathematics cannot be completely reduced to sets and functions of numbers. Applied mathematics in natural and engineering sciences use functionals with number-theoretic functions as arguments (e.g., observables in quantum physics) or even spaces of functionals (e.g., differential geometry in the theory of relativity). In order to evaluate these abstract mathematical objects by computers, the concept of computability must be generalized for functionals in higher types. The question arises how to compute these abstract entities with a huge amount of data which are obviously needed in ordinary mathematics and scientific applications. Therefore, in the following chapter, a general concept of digital information systems is introduced to solve these problems (Scott, 1982).

In physical computers, computations must be finite. Therefore, in any evaluation of functionals $\Phi(\varphi)$, the argument φ can be called only finitely many times, i.e., the value of $\Phi(\varphi)$ (if defined) must be determined by some finite subfunction of φ. Let $\mathcal{G}$ be a partial functional of type 3, mapping type-2 functionals Φ to natural numbers. If Φ is given and $\mathcal{G}(\Phi)$ evaluates to a defined value, evaluation must be finite. Hence the argument Φ can only be called on finitely many functions φ. Each φ must be presented to Φ in a finite form (i.e., as a set of ordered pairs) (Schwichtenberg and Wainer 2012, Chapter 6).

Finite approximation of functionals. A finite approximation of a functional Φ is a finite set X of pairs (φ_0, n) such that

- (i) φ_0 is a finite function.
- (ii) $\Phi(\varphi_0)$ is defined with value n.
- (iii) If (φ_0, n) and (φ_0', n') belong to X (where φ_0 and φ_0' are consistent), then $n = n'$.

A functional Φ is viewed as the union of all its finite approximations. It is a type-2 argument of type-3 functional $\mathcal{G}$ which should satisfy the following principle.

Principle of finite support. If $\mathcal{G}(\Phi)$ is defined with value n, then there is a finite approximation Φ_0 of Φ such that $\mathcal{G}(\Phi_0)$ is defined with value n.

If $\mathcal{G}(\Phi)$ is evaluated, then we obtain the same value independent of any extension of argument Φ. Functional Φ' extends Φ if for any piece of data (φ_0, n) in Φ there exists another (φ_0', n) in Φ' such that φ_0 extends φ_0'. With the notion of functional extension, we can formalize the following principle.

Principle of monotonicity. If $\mathcal{G}(\Phi)$ is defined with value n and Φ' extends Φ, then also $\mathcal{G}(\Phi')$ is defined with value n.

From these principles, we can conclude the following proposition.

Proposition. *Any functional is determined by its set of finite approximations.*

Proof. Let functionals Φ, Φ' have the same finite approximations and $\mathcal{G}(\Phi) = n$ for some value n,

$\implies \mathcal{G}(\Phi_0) = n$ for some finite approximation Φ_0 (principle of finite support),

$\implies \mathcal{G}(\Phi') = n$ for some extension Φ' (principle of monotonicity),

$\implies \mathcal{G}(\Phi) = \mathcal{G}(\Phi')$ for all $\mathcal{G}$. $\qquad\qquad\square$

The concept of finite approximation motivates a first definition of generalized computability.

Definition of generalized effectivity. A generalized mathematical object (e.g., functional) is computable iff its set of finite approximations is (primitive) recursively enumerable (Σ_1^0-definable).

The computability of number-theoretic functions was defined by Turing machines. Turing machines (and their technical realizations as computers) are examples of information systems. We introduce a general concept of information systems to compute finite approximations of generalized mathematical objects like functionals and spaces (Scott, 1982). From a practical point of view, information systems seem to be universal in a digital world. They are needed everywhere to process the increasing mass of data in bits and bytes. Especially, general concepts of mathematical objects like functionals contain a huge amount of data. These considerations motivate the following mathematical definition of an information system.

In order to describe approximations of abstract objects like functionals by finite ones, we use an information system with a countable set A of bits of data ("tokens"). Approximations need finite sets U of data which are consistent with each other. An "entailment relation" expresses the fact that the information of a consistent set U of data is sufficient to compute a bit of information ("token").

Definition of an information system. An information system is a structure $\mathcal{A} := (A, \mathrm{Con}, \vdash)$ with a countable set A("tokens"), non-empty set Con of finite ("consistent") subsets of A and subset $\vdash$ of $\mathrm{Con} \times A$ ("entailment relation") with

 (i) $U \subseteq V \in \mathrm{Con} \implies U \in \mathrm{Con}$.
 (ii) $\{a\} \in \mathrm{Con}$.
 (iii) $U \vdash a \implies U \cup \{a\} \in \mathrm{Con}$.
 (iv) $a \in U \in \mathrm{Con} \implies U \vdash a$.
 (v) $U, V \in \mathrm{Con} \implies \forall a \in V(U \vdash a) \implies (V \vdash b \implies U \vdash b)$.

The ideals ("objects") of an information system $\mathcal{A} := (A, \mathrm{Con}, \vdash)$ are defined as subjects x of A with

 (i) $U \subseteq x \implies U \in \mathrm{Con}$ (x is consistent).
 (ii) $x \supseteq U \vdash a \implies a \in x$ (x is deductively closed).

Example. The deductive closure $\bar{U} := \{a \in A | U \vdash a\}$ of $U \in \mathrm{Con}$ is an ideal.

The set of all ideals of information system $\mathcal{A}$ is denoted by $|\mathcal{A}|$.

Let us illustrate this abstract concept with some examples of information systems.

Example: Flat information system. Every countable set A can be turned into a flat information system $\mathcal{A} := (A, \mathrm{Con}, \vdash)$ with A set of tokens,

$$\mathrm{Con} := \{\emptyset\} \cup \{\{a\} | a \in A\},$$

$$U \vdash a := \equiv a \in U.$$

It follows that the ideals are all elements of Con.

For $A = \mathbb{N}$, the Con-sets of this example can be represented in the following tree:

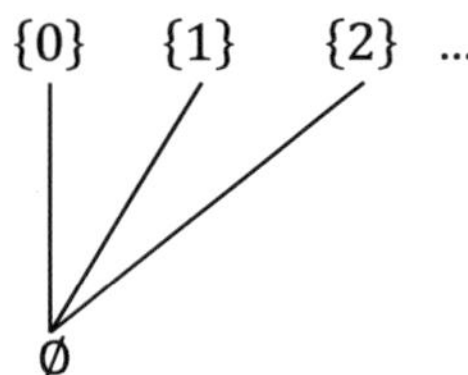

Example: Approximations of functions from a countable set A into a countable set B. Let $A \times B$ be a countable set of tokens (a,b) with $a \in A$ and $b \in B$, $\mathrm{Con} := \{\{\{a_i, b_i\} | i < k\} | \forall i, j < k(a_i = a_j \rightarrow b_i = b_j)\}$ and $U \vdash (a,b) := \equiv (a,b) \in U$. It follows that the ideals are the graphs of all partial functions from A to B.

Example: Approximations of a fixed partial functional Φ. Any (fixed) partial functional Φ can be considered as an information system. Let the pair (φ_0, n) be a token with finite function φ_0 and $\Phi(\varphi_0) := n$. We take Con to be the set of all finite sets of tokens. $U \vdash (\varphi_0, n)$ means that φ_0 extends some φ_i with $U := \{(\varphi_i, n_i) | i = 1, \ldots, k\}$.

It follows that the ideals are all sets x of tokens with the property $(\varphi_0, n) \in x \implies (\varphi_0', n) \in x$ with φ_0' extending φ_0.

Finally, we want to define the computability of functionals with information systems. In a first step, we introduce the class of approximable maps between two information systems which can intuitively be considered as a kind of "function space". A "function space" between information systems is also an information system which can be used to characterize (continuous) functionals. In the sense of information systems, these functionals are sets of tokens. Therefore, they can be called computable iff they (i.e., sets of tokens) are recursively enumerable.

Definition of function spaces between information systems. The function space $\mathcal{A} \to \mathcal{B} := (C, \mathrm{Con}, \vdash)$ of information systems $\mathcal{A} := (A, \mathrm{Con}_A, \vdash_A)$ and $\mathcal{B} := (B, \mathrm{Con}_B, \vdash_B)$ is defined by

$$C := \mathrm{Con}_A \times B,$$

$$\{(U_i, b_i) | i \in I\} \in \mathrm{Con}$$

$$:\equiv \forall \mathcal{J} \subseteq I \left(\bigcup_{j \in \mathcal{J}} U_j \in \mathrm{Con}_A \to \{b_j | i \in \mathcal{J}\} \in \mathrm{Con}_B \right),$$

$$W \vdash (U, b) :\equiv WU \vdash_B b.$$

with application $\{(U_i, b_i) | i \in I\}\, U := \{b_i | U \vdash_A U_i\}$ of $W := \{(U_i, b_i) | i \in I\} \in \mathrm{Con}$ to $U \in \mathrm{Con}_A$.

$\mathcal{A} \to \mathcal{B}$ can be proven to be an information system for $\mathcal{A}, \mathcal{B}$ information systems.

The approximable maps between two information systems can exactly be identified as the ideals of the function space (Schwichtenberg and Wainer, 2012, p. 255).

Definition of approximable maps in functional spaces. Let $\mathcal{A} := (A, \mathrm{Con}_A, \vdash_A)$ and $\mathcal{B} := (B, \mathrm{Con}_B, \vdash_B)$ be information systems. A relation $r \subseteq \mathrm{Con}_A \times B$ is an approximable map iff

(i) $r(U, b_1), \ldots, r(U, b_n) \implies \{b_1, \ldots, b_n\} \in \mathrm{Con}_B,$
(ii) $r(U, b_1), \ldots, r(U, b_n), \{b_1, \ldots, b_n\} \vdash_B b \implies r(U, b),$
(iii) $r(U, b), U \vdash_A U' \implies r(U, b).$

Theorem. *Let $\mathcal{A}$ and $\mathcal{B}$ be information systems. Then the ideals of functional space $\mathcal{A} \to \mathcal{B}$ are exactly the approximable maps from $\mathcal{A}$ to $\mathcal{B}$.*

Alternatively, a function space of information systems can be characterized by continuous maps with an appropriate topology on the data set of an information system (Scott, 1982).

Definition of Scott typology. Let $\mathcal{A} := (A, \mathrm{Con}_A, \vdash_A)$ be an information system with $U \in \mathrm{Con}$. We define subsets $\mathcal{O}_U$ of the set $|\mathcal{A}|$ of all ideals of $\mathcal{A}$ with $\mathcal{O}_U := \{x \in |\mathcal{A}| \,\|\, U \subseteq x\}$. Since the ideals $x \in |\mathcal{A}|$ are deductively closed, $x \in \mathcal{O}_U$ implies that the deductive closure is a subset of x, i.e., $\bar{U} \subseteq x$.

Lemma. *The system of all (open sets) $\mathcal{O}_U$ with $U \in \mathrm{Con}$ forms the basis of a topology on $|\mathcal{A}|$.*

Proof. Let be $U, V \in \mathrm{Con}$ and $x \in \mathcal{O}_U \cap \mathcal{O}_V$. According to the axioms of a topology, we must find $W \in \mathrm{Con}$ with $x \in \mathcal{O}_W \subseteq \mathcal{O}_U \cap \mathcal{O}_V$. We choose $W := U \cup V$. $\qquad\square$

Definition of continuity. Let $\mathcal{A}$ and $\mathcal{B}$ be informations systems with function $f : |\mathcal{A}| \to |\mathcal{B}|$. f is continuous iff for any basic open set $\mathcal{O}_V \subseteq |\mathcal{B}|$ (with $V \in \mathrm{Con}_B$), the set $f^{-1}(\mathcal{O}_V) = \{x | V \subseteq f(x)\}$ is open in $|\mathcal{A}|$.

Lemma of continuous functions. *Let $\mathcal{A}$ and $\mathcal{B}$ be information systems with map $f\colon |\mathcal{A}| \to |\mathcal{B}|$. The following statements are equivalent:*

(a) *f is continuous with respect to the Scott topology.*
(b) *f is monotone, i.e., $x \subseteq y$ implies $f(x) \subseteq f(y)$ for all $x, y \in |\mathcal{A}|$, and satisfies the principle of finite support, i.e., $b \in f(x) \Longrightarrow b \in f(\bar{U})$ for some $U \subseteq x$.*
(c) *f is monotone and commutes with directed unions, i.e., for every directed union $D \subseteq |\mathcal{A}|$ is $f(\bigcup_{x \in D} x) = \bigcup_{x \in D} f(x)$.*

The lemma guarantees that continuous maps $f: |\mathcal{A}| \to |\mathcal{B}|$ are those that can be characterized by finite approximations of the abstract objects (ideals) $x \in |\mathcal{A}|$ and $f(x) \in |\mathcal{B}|$. If there is a finite approximation V to the value $f(x)$, then there is a finite approximation U to the argument x such that $f(\bar{U})$ contains the information in V. It follows from the principle of monotonicity that $f(\bar{U}) \subseteq f(x)$.

Theorem: Approximable mappings as continuous functions.
Let $\mathcal{A} := (A, \mathrm{Con}_A, \vdash_A)$ and $\mathcal{B} := (B, \mathrm{Con}_B, \vdash_B)$ be information systems. Then the ideals of $\mathcal{A} \to \mathcal{B}$ are in a natural bijective correspondence with the continuous functions from $|\mathcal{A}|$ to $|\mathcal{B}|$ as follows:

(a) *With any approximable map $r : \mathcal{A} \to \mathcal{B}$, a continuous function $|r| : |\mathcal{A}| \to |\mathcal{B}|$ is associated by $|r|(z) := \{b \in \mathcal{B} | r(U, b)$ for some $U \subseteq z\}$ ("application of r to z").*
(b) *With any continuous function $f : |\mathcal{A}| \to |\mathcal{B}|$, an approximable map $\hat{f} : \mathcal{A} \to \mathcal{B}$ is associated by $\hat{f}(U, b) := (b \in f(\bar{U}))$.*

These assignments are inverse to each other, i.e., $f = |\hat{f}|$ and $r = |\hat{r}|$.

There are functions and functionals of different degrees of complexity which can be distinguished by the introduction of types. Especially in computer languages, types are essential to recognize the different domains of variables. With respect to our characterization of functionals with information systems, we also can introduce types of information systems.

Types are built from base types by the formation of function types $\rho \to \sigma$. For every type ρ the information system $\mathcal{C}_\rho = (C_\rho, \mathrm{Con}_\rho, \vdash_\rho)$ can be defined. The ideals $x \in |\mathcal{C}_\rho|$ are the partial continuous functionals of type ρ . Since $\mathcal{C}_{\rho \to \sigma} = \mathcal{C}_\rho \to \mathcal{C}_\sigma$, the partial continuous functionals of type $\rho \to \sigma$ correspond to the continuous functions from $|\mathcal{C}_\rho|$ to $|\mathcal{C}_\sigma|$ with respect to the Scott topology.

Definition of functional computability. A partial continuous functional $x \in |\mathcal{C}_\rho|$ of type ρ is computable iff it is recursively enumerable as a set of tokens.

For example, the following figure illustrates tokens and entailments for the algebra N of natural numbers. For tokens, b, an entailment $\{a\} \vdash b$ means that there is a path from a (up) to b (down):

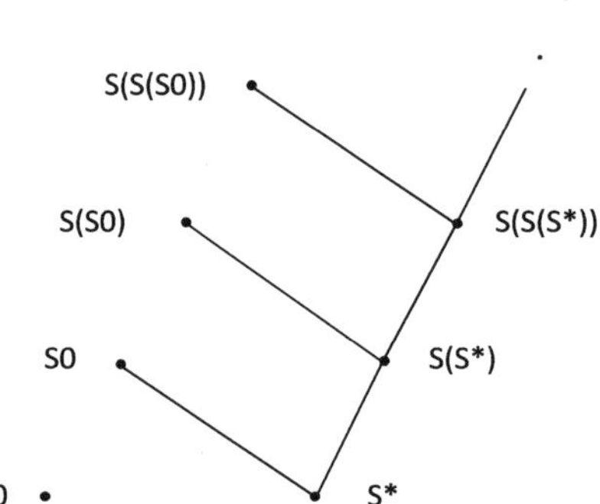

Partial continuous functionals of type ρ can be used as semantics of a formal functional programming language (in the style of Plotkin, 1977): Every closed term of type ρ in the programming language denotes a computable partial continuous functional of type ρ, i.e., a recursive enumerable consistent and deductively closed set of tokens. Another approach uses recursive equations to define computable functionals (Berger *et al.*, 2003). We will discuss the characterization of functionals by types later in the book.

<h1 style="text-align:center">Chapter 6</h1>

<h1 style="text-align:center">Intuitionistic Mathematics and
Human Creativity</h1>

In computational mathematics, proofing and computing is reduced to digital machines evaluating mathematical data and information. Contrary to machine-oriented foundations, intuitionistic mathematics is deeply rooted in a philosophy of human creativity. The Dutch mathematician L. E. J. Brouwer (1881–1966) started with human intuitions which enable mathematical activities and constructions. But his philosophy of the human subject does not only exclude non-constructive abstractions in classical mathematics. His radical principles of construction even lead to theorems which are not accepted in classical mathematics (Weyl, 1921). A crucial point of Brouwer's analysis is his intuitionistic understanding of the continuum and infinity as basis of ordinary mathematics. This discussion culminates in his fan theorem which plays a dominant role in intuitionistic mathematics. How far can it be accepted in computational and constructive mathematics? We begin with the intuitionistic philosophy of a creative subject.

Kant's philosophy of intuition as root of Brouwer's intuitionism

According to Brouwer, mathematical truth is founded by constructions of a "creative subject" (Brouwer, 1907, 1975). He followed Kant's explanation of human understanding by human subjects.

87

Kant assumed that human understanding is made possible by two forms of intuition which are given to the subject before (*a priori*) any empirical experience (Kant, 1787):

(1) Human subjects have an intuition of spatial forms which are constructed step by step by certain rules (e.g., circle and ruler in geometry).

(2) Human subjects have an intuition of sequential temporal points which are constructed step by step by adding a unit according to the rule of counting in arithmetic.

On the one side, the spatial and temporal forms of intuition enable the human subject to understand empirical objects and events in space and time. For Kant, human subjects are born with an intuitive understanding of space and time which makes possible their orientation in the world. On the other side, the spatial and temporal forms of intuition provide the schemes of geometric and arithmetic constructions and, by that, the foundations of mathematics (Mainzer, 1970, 1973). Kant illustrates the temporal form of intuition by an unlimited sequence of points $\cdot, \cdot\cdot, \cdot\cdot\cdot, \ldots$, which are extended step by step by a unique point. In our temporal intuition, these points represent a sequence of present moments ("now") which are passing in a linear order. Formally, this process corresponds to the construction of the natural numbers $1, 1+1, 1+1+1, \ldots$ or in the usual abbreviation of decimal numbers $1, 2, 3, \ldots$ according to the rule of counting.

After Kant, the mathematician Leopold Kronecker claimed that the natural numbers are "made by God", but "all the rest" is made by the human being (according to a quotation of Weber, 1893). Independent of Kronecker's reference to God, Kant argues on the same line: The natural numbers are given to the subject by the temporal form of intuition and its arithmetical scheme of counting. "All the rest" must be reduced to the fundamental form of arithmetical construction. Thus, for Kant and Kronecker, infinity is not given, but only a *façon de parler* (Poincaré) or "regular idea" (Kant) for the unlimited process of counting. Brouwer derived radical consequences for the truth of theorems and their proofs: In temporal intuition, only finite sequences of natural numbers can be constructed. Thus, for Brouwer, mathematical truth depends on finite stages of realization in time by a creative subject. Georg Kreisel suggested a formal definition (Kreisel, 1967; Sundholm, 2014).

Intuitionistic principle of creative subject. A creative subject has a proof of proposition A at stage m (abbreviation: $\sum \vdash_m A$) iff

(CS1) For any proposition A, $\sum \vdash_m A$ is a decidable function of A, i.e., $\forall x \in \mathbb{N}(\sum \vdash_x A \vee \neg \sum \vdash_x A)$,

(CS2) $\forall x, y \in \mathbb{N}(\sum \vdash_x A \to (\sum \vdash_{x+y} A))$,

(CS3) $\exists x \in \mathbb{N}(\sum \vdash_x A) \leftrightarrow A$.

A weaker version of CS3 is Kreisel's "Axiom of Christian Charity" (1967):

(CC) $\neg \exists x \in \mathbb{N}(\sum \vdash_x A) \to \neg A$.

The idea that only finite initial segments of infinite sequences are given leads to Brouwer's concept of choice sequences. A choice sequence means a process which is not necessarily predetermined by some law or algorithm (Troelstra, 1968; Dummett, 1977).

Definition of choice sequences.

(i) α lawless sequence $:\equiv$ At any stage of $\alpha 0, \alpha 1, \alpha 2, \ldots$ only finitely many values of α are known.

(ii) α lawlike sequence $:\equiv$ All values of α are known by a law (i.e., algorithm).

Lawless sequences can be illustrated by sequences of casts with a die after some already realized casts in the beginning. The die can be thrown in arbitrarily many times. Nevertheless, at any stage, only finitely many casts are known and the following cast is unknown. In lawlike sequences, all stages are predetermined by a law. An example is the sequence of even natural numbers with law $\alpha n = 2n$ or the sequence of decimal places of real number π.

The radical difference to classical mathematics came in, when choice sequences were allowed in real analysis. If real numbers are defined by fundamental sequences to be given by choice sequences, then the statement "all total functions from $\mathbb{R}$ to $\mathbb{R}$ are continuous" can be proven intuitionistically. This statement is false in classical mathematics.

In intuitionist mathematics, infinite objects are considered as ever growing and never finished. Therefore, sets need a new foundation. The intuitionistic analogue of a set is a so-called spread which is defined as a countably branching tree labeled with natural numbers or other finite objects and containing only infinite paths. A fan is a finitely branching spread. A branch is an intuitionistic choice sequence, i.e., an infinite sequence of numbers (or finite objects) created step by step by a law (algorithm) or without law (e.g., coin). A lawless sequence is ever unfinished. The only available information about a lawless sequence at any stage is the initial segment of the sequence created thus far. In order to formalize these concepts, we introduce basic concepts of graph theory (Troelstra and van Dalen, 1988, 186 f.).

Definition of equality and ordering of sequences. Let $\alpha, \beta,$ $\gamma, \ldots$ be variables ranging over sequences of natural numbers. They are mappings $\alpha, \beta, \gamma, \ldots : \mathbb{N} \to \mathbb{N}$ with values, e.g., $\alpha 1, \alpha 2, \alpha 3, \ldots$:

$$\alpha = \beta :\equiv \forall x(\alpha x = \beta x),$$

$$\alpha \leq \beta :\equiv \forall x(\alpha x \leq \beta x).$$

Definition of concatenation and initial segments of sequences. An empty sequence is denoted by $\langle\,\rangle$. A sequence with value αx is denoted by $\langle \alpha x \rangle$. The concatenation of a finite sequence with $\langle \alpha x \rangle$ is denoted by $*$. An initial segment of a sequence is defined inductively:

$$\bar{\alpha}0 \ := \ \langle\,\rangle = 0,$$

$$\bar{\alpha}(x+1) \ := \ \bar{\alpha}x * \langle \alpha x \rangle = 0,$$

$$\alpha \in n :\equiv \exists x(\bar{a}x = n)(\alpha \text{ has initial segment } n).$$

Concatenation can be extended to concatenation of a finite sequence with an infinite one. In this case, $n * \alpha$ means "n followed by α":

$$(n * \alpha)(x) := \begin{cases} (n)_x & \text{if } x < lth(n) \\ \alpha(x - lth(n)) & \text{if } x \geq lth(n). \end{cases}$$

Definition of a tree. A tree T is defined as an inhabited, decidable set of finite sequences of natural numbers which are closed under the operation of predecessor:

(i) $\langle \, \rangle \in T$ (inhabited),
(ii) $\forall n(n \in T \vee n \notin T)$ (decidable),
(iii) $\forall n, m(n \in T \wedge m \prec n \to m \in T)$ (closed under predecessor).

$\langle \alpha x \rangle$ denotes the node of tree T with number αx. The empty node is denoted by $\langle \, \rangle$.

Definition of spreads, fans, and branches.

T spread:

A spread T is a tree in which each node has at least one successor:

$$\forall n \in T \exists x(n * \langle x \rangle \in T).$$

T fan:

A fan is a finitely branching spread:

$$\forall n \in T \exists z \, \forall x(n * \langle x \rangle \in T \to x \leq z).$$

α branch:

A sequence α is a branch of tree T iff all segments of α belong to T:

$$\alpha \in T := \forall x(\bar{a}x \in T).$$

T_u universal spread/tree := Spread of all finite sequences of natural numbers;

T_{01} binary tree := Tree of all binary sequences.

In order to introduce the concept of continuity in intuitionistic mathematics, we define a topology on spreads.

Reminder of the axioms of a topology. $\mathcal{T}$ topology of set X satisfies the following axioms:

(i) $\varnothing \in \mathcal{T}$,
(ii) $X \in \mathcal{T}$,
(iii) $X_1, \ldots, X_n \subset X \to \cap_{1 \leq i \leq n} X_i \in \mathcal{T}$,
(iv) $X_1, \ldots, X_n, \ldots \subset X \to \cup_i X_i \in \mathcal{T}$.

A topological space is a pair $(X, \mathcal{T})$ of a set X with a topology $\mathcal{T}$. The elements of X are the points of the space. The elements of $\mathcal{T}$ are the open sets of the space.

A basis $\mathcal{B}$ of typology $\mathcal{T}$ is a subset $\mathcal{B} \subset \mathcal{T}$ such that

$$\forall Y \in \mathcal{T} : Y = \bigcup \{X | X \epsilon \mathcal{B} \wedge X \subset Y\}.$$

It is well known that for a basis the following properties are satisfied:

(i) $U, U' \in \mathcal{B}, x \in U \cap U' \Rightarrow \exists U'' \in \mathcal{B}(x \in U'' \subset U \cap U')$,
(ii) $x \in X \Rightarrow \exists U \in \mathcal{B}(x \in U)$.

Definition of a topology of spread T. Infinite branches of spread T as elements ("open sets") of topology $\mathcal{T}$:

$$\mathcal{T} = \{\alpha | \alpha \text{ infinite branch of } T\}.$$

Basis $\mathcal{B}$ of topology $\mathcal{T}$ as all sets

$$V_n = \{\alpha | \alpha \in n \wedge \alpha \in T\} \text{ for each } n \in T.$$

Reminder of the topological definition of continuity.

Mapping $f \colon (X, \mathcal{T}) \to (X', T')$ is continuous

$$:\equiv \forall U' \in \mathcal{T}' : f^{-1}(U') \in \mathcal{T} \text{ or } \forall U' \in \mathcal{B}' : f^{-1}(U') = \cup_{U \in \mathcal{B}} U$$

Examples of continuous functionals.

$\Phi : \mathbb{N}^{\mathbb{N}} \to \mathbb{N}$ is continuous $\equiv \forall\alpha\exists x\,\forall\beta \in \bar{\alpha}x(\Phi\alpha = \Phi\beta)$

$\psi : \mathbb{N}^{\mathbb{N}} \to \mathbb{N}^{\mathbb{N}}$ is continuous $\equiv \forall x\,\forall\alpha\exists y\,\forall\beta \in \bar{\alpha}y((\psi\alpha)(x) = (\psi\beta)(x))$

Let us suppose that functional $\Phi : \mathbb{N}^{\mathbb{N}} \to \mathbb{N}$ is defined for all choice sequences. In this case, we call Φ a total functional. For a choice sequence, only an initial segment is known at any given moment. Therefore, we suppose that Φ is continuous on $\mathbb{N}^{\mathbb{N}}$, i.e.,

$$\forall\alpha\exists x\,\forall\beta \in \bar{\alpha}x(\Phi\alpha = \Phi\beta).$$

Functionals $\Phi : \mathbb{N}^{\mathbb{N}} \to \mathbb{N}$ can be used for the following principle of choice $\mathrm{AC}^{1,0}$-N:

$$\forall\alpha\exists x A(\alpha, x) \to \exists\Phi : \mathbb{N}^{\mathbb{N}} \to \mathbb{N}\,\forall\alpha A(\alpha, \Phi\alpha).$$

If we combine $\mathrm{AC}^{1,0}$-N with the supposition that Φ is continuous on $\mathbb{N}^{\mathbb{N}}$, we get the principle of weak ("local") continuity for numbers:

Principle of weak ("local") continuity WC-N.

$$\forall\alpha\exists x A(\alpha, x) \to \forall\alpha\exists x\exists y\,\forall\beta \in \bar{\alpha}x A(\beta, y).$$

WC-N refers to "weak" or "local" continuity, because the existence of some y is only demanded for a neighborhood of α for each α.

WC-N can be applied to the constructive BHK-interpretation of logical disjunction

$$A \vee B \equiv \exists x((x = 0 \wedge A) \vee (x \neq 0 \wedge B)).$$

Combined with WC-N, we get the following version WC-N$^{\vee}$:

$$\forall\alpha(A\alpha \vee B\alpha) \to \forall\alpha\exists x(\forall\beta \in \bar{\alpha}x A(\beta) \vee \forall\beta \in \bar{\alpha}x B(\beta)).$$

It is remarkable that the intuitionistic principle of weak continuity conflicts with classical logic. We consider the version $\forall$-PEM of principle of excluded middle (PEM):

$$\forall\alpha(\forall x\,\alpha x = 0 \vee \neg\forall x\,\alpha x = 0).$$

Proposition. WC-N$^\vee$ *refutes* $\forall$-PEM.

Proof.

$$
\begin{aligned}
&\quad\ \forall\text{-PEM} &&\text{assumption}\\
\Rightarrow\ &\forall\alpha(\forall x\,\alpha x = 0 \vee \neg\forall x\,\alpha x = 0) &&\text{definition}\\
\Rightarrow\ &\forall\alpha\exists y(\forall\beta \in \bar\alpha x\,\forall x(\beta x = 0) \vee \forall\beta \in \bar\alpha x\,\neg\forall x(\beta x = 0) &&\text{WC-N}^\vee
\end{aligned}
$$

Define $\alpha := \lambda x.0$:

$\Rightarrow\ \forall\beta \in n\,\forall x(\beta x = 0) \vee \forall\beta \in n\,\neg\forall x(\beta x = 0)$ for some y with $n = \overline{(\lambda x.0)}\,(y)$

$\Rightarrow\ \bot,$

because the first disjunct is false for any $\beta \in n * \langle 1 \rangle$ and the second disjunct is false by choosing $\beta = \alpha = \lambda x.0$. $\qquad\qquad\square$

Local continuity does not exhaust our full intuition of continuity: If $\Phi : \mathbb{N}^\mathbb{N} \to \mathbb{N}$ is any continuous functional, we assume to know whether an initial segment of α suffices to compute $\Phi\alpha$. In this case, we expect to associate a neighborhood function $\varphi : \mathbb{N} \to \mathbb{N}$ such that

$$
\varphi(\overline{\alpha}n) = \begin{cases} 0 & \text{iff } \overline{\alpha}n \text{ is not sufficiently long to compute } \Phi\alpha \\ m + 1 & \text{iff } \overline{\alpha}n \text{ is sufficiently long to compute} \\ & \qquad \Phi\alpha \text{ and } \Phi\alpha = m \end{cases}
$$

Principle of global continuity for numbers C-N.

$$
\forall\alpha\exists x A(\alpha, x) \to \exists\gamma \in K\,\forall\alpha A(\alpha, \gamma(\alpha))
$$

with K as the class of neighborhood functions.

Brouwer's intuitionistic mathematics is motivated by two fundamental intuitions: The first intuition is the scheme of counting as principle of natural numbers in order to form proofs and theorems of arithmetic. The second intuition is the concept of choice sequences in order to capture the idea of the continuum and to form the proofs and theorems of analysis. Especially, he wanted to guarantee the theorem that a continuous function from a closed interval to $\mathbb{R}$ is uniformly continuous (Berger, 2005). Historically, Brouwer derived this theorem from his bar theorem (Brouwer, 1975, 1981). Following Troelstra and van Dalen (1988) and Troelstra (1974), we start with the so-called fan theorem (which was considered by Brouwer as a corollary of his bar theorem) as axiom. Actually, a fan as finitary spread is more convenient and all known important applications of the bar theorem can also be derived from the fan theorem.

For decidable properties, the fan principle means intuitively: If T is a fan, and for each infinite branch α in T there is an initial segment $\overline{\alpha}x$ satisfying the decidable property A, then there is a uniform upper bound to the involved x:

Fan principle for decidable properties $\mathbf{FAN}_D$.

$$\forall n_T(An \vee \neg An) \wedge \forall \alpha_T \exists x A(\overline{\alpha}x) \rightarrow \exists z \, \forall \alpha_T \exists y \leq z A(\overline{\alpha}y) \text{ for fan } T.$$

FAN_D leads to a restricted form of compactness of fan T. Again, we choose a topology generated by the basis of all sets $V_n := \{\alpha \in T | \alpha \in n\}$ for all $n \in T$. Obviously, any decidable set of basis elements covering T has a finite subcover: The decidable set of basis elements is the set $\{V_n | A(n) \wedge n \in T\}$, the finite subcover is $\{V_n | A(n) \wedge n \in T \wedge lth(n) \leq z\}$ with z given by FAN_D.

In classical logic, fan principle FAN_D is the contraposition of König's lemma and therefore equivalent:

Proposition. *In classical logic,* $\mathrm{FAN}_D \Leftrightarrow$ *König's lemma.*

Proof.

$(`` \Leftarrow ")$ $\quad \forall \alpha_T \exists x A(\overline{\alpha}x)$ with any fan T $\qquad$ assumption

$\quad \Rightarrow T_A := \{m | \forall m' \preccurlyeq m \neg Am'\}$

$\qquad\qquad$ is a tree with all branches finite.

$\qquad\quad \neg \exists z \, \forall \alpha_T \exists y \leq z A(\overline{\alpha}y)$ $\qquad$ assumption

$\quad \Rightarrow \forall z \exists \alpha_T \forall y \leq z \neg A(\overline{\alpha}y)$ $\qquad$ classical logic

$\quad \Rightarrow T_A$ has arbitrarily long finite branches

$\quad \Rightarrow$ There is an infinite branch in T, *i.e.,* $\qquad$ König's lemma

$\qquad \exists \alpha_T \forall z \neg A(\overline{\alpha}z)$

$\quad \Rightarrow \perp$ $\qquad\qquad\qquad\qquad\qquad\qquad$ contradiction

$\qquad\qquad\qquad\qquad\qquad\qquad\qquad\qquad$ to assumption

$\qquad\qquad\qquad\qquad\qquad\qquad\qquad\qquad \forall \alpha_T \exists x A(\overline{\alpha}x)$

$(`` \Rightarrow ")$ FAN_D assumption

$\Rightarrow \forall \alpha_T \exists x A(\overline{\alpha}x) \rightarrow \exists z \, \forall \alpha_T \exists y \leq z A(\overline{\alpha}y)$

for decidable A

$\Rightarrow \neg \exists z \, \forall \alpha_T \exists y \leq z A(\overline{\alpha}y) \rightarrow \neg \forall \alpha_T \exists x A(\overline{\alpha}x)$ contraposition

$\Rightarrow \forall z \exists \alpha_T \forall y \leq z \neg A(\overline{\alpha}y) \rightarrow \exists \alpha_T \forall x \neg A(\overline{\alpha}x)$ classical logic

$\Rightarrow$ König's lemma

$\square$

FAN_D can intuitively be illustrated for the binary tree T_{01} of 01-sequences: We denote variables of finite 01-sequences by n, m and infinite 01-sequences by α, β. For any decidable predicate A with $\forall \alpha \exists x A(\overline{\alpha}x)$, set $A^* := \{n | \forall m \preccurlyeq n \neg A(m)\}$ is well-founded. The situation is illustrated in Fig. 7: Following any branch of 01-sequence through the binary tree, we will always meet the bar $P := \{n | A(n) \wedge \forall m \prec n \neg A(m)\}$. In Fig. 7, the bar is illustrated by black nodes for $A(n) := lth\,(n) > 2^{(n)_0 + (n)_1}$.

A stronger version is the fan principle FAN.

Fan principle FAN. $\forall \alpha_T \exists x A(\overline{\alpha}x) \rightarrow \exists z \, \forall \alpha_T \exists y \leq z A(\overline{\alpha}y)$

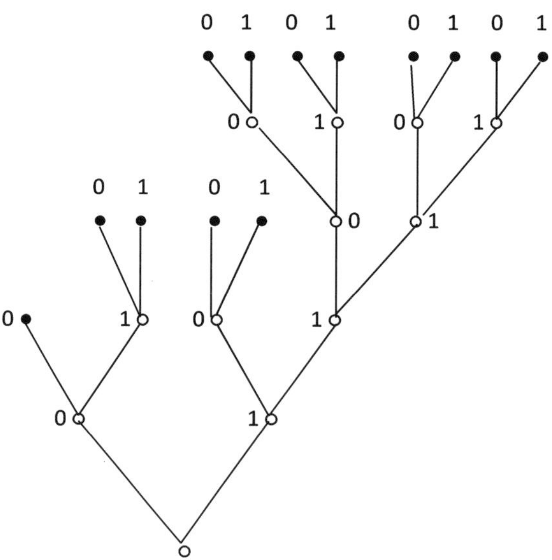

Fig. 7. Bar (black nodes) in binary tree T_{01}.

Fan principle FAN is stronger than FAN_D because it is an expression of full compactness of the fan T. FAN is still classically valid. But, that is no longer true for the following generalization.

Generalized fan principle FAN_G.

$$\text{fan}(T) \wedge \forall \alpha_T \exists x A(\alpha, x) \to \exists z \, \forall \alpha_T \exists y \le z A(\alpha, y)$$

Obviously, FAN_G is classically false. For example, consider the classically valid sentence

$$\forall \alpha_T \exists x (\exists y (\alpha y = 0) \to \alpha x = 0)$$
$$\Rightarrow \exists z \, \forall \alpha_T (\exists y (\alpha y = 0) \to \exists y \le z (\alpha y = 0)) \qquad \text{FAN}_G$$
$$\Rightarrow \bot.$$

Brouwer distinguished "lawlike" and "lawless" choice sequences. According to Church's thesis, it is plausible to identify lawlike functions with recursive functions. This assumption can be expressed by an axiom.

Church's thesis CT. $\forall \alpha \exists x \, \forall y \exists z (T(x, y, z) \wedge U(z) = \alpha(y))$ with Kleene's T-predicate.

An arithmetical form of Church's thesis only uses numerical quantifiers.

Arithmetical Church's thesis CT_0.

$$\forall n \exists m A(n, m) \to \exists k \, \forall n \exists m (A(n, U(m)) \wedge T(k, n, m))$$

The assumption of this axiom means that the functions in an intuitionistic proof are always recursive. An extended Church's thesis is the following assumption with an almost negative condition A:

Extended Church's thesis ECT_0. $\forall x (A(x) \to \exists y B(x, y)) \to \exists z \, \forall x (A(x) \to \exists u (T(z, x, u) \wedge B(x, U(u))))$

Extended Church's thesis ECT_0 and Markov's principle (MP) imply the continuity of functionals from $\mathbb{N}^{\mathbb{N}}$ to $\mathbb{N}$. The fan principle FAN_D is compatible with the continuity principle C-N, but not with Church's thesis CT.

Theorem. FAN_D *is not compatible with CT.*

Proof. In the sense of CT, we define a recursive tree T well-founded with respect to all recursive 01-sequences:

$T := \{n \,|\, \forall m \preccurlyeq n\, R(m)\}$ with recursive R such that $\forall nm(R(n * m) \to R(n))$ and

(1) $\forall\alpha\exists x\neg R(\overline{\alpha}x)$,
(2) $\forall x\exists n(lth(n) = x \wedge R(n))$.

But (2) implies that there is no upper bound to the length of branches in the tree which contradicts FAN_D.

A recursive predicate R satisfying (1) and (2) can be defined with Kleene's T-predicate by

$$R(\overline{\alpha}x) :\equiv \forall y < x(T(y, y, u) \wedge U(u) = 0) \to \alpha(y) = 1) \wedge$$

$$(\exists u < x(T(y, y, u) \wedge U(u) = 1) \to \alpha(y) = 0)).$$

Condition (2) is satisfied by R:

For a given x, we take the recursive 01-sequence α with

$$\alpha(y) = 1 \leftrightarrow (y < x \wedge \exists u < x(T(y, y, u) \wedge U(u) = 0)).$$

Condition (1) is satisfied by R:

Let $\alpha = [y]$ and suppose $T(y, y, u)$. We will prove that the assumption $R(\overline{\alpha}(u + y + 1))$ leads to a contradiction. Therefore, it follows $\neg R(\overline{\alpha}(u + y + 1))$.

The contradiction of $R(\overline{\alpha}(u + y + 1))$ is obvious. By definition of R, there are two possibilities:

First case: $U(u) = 0$ assumption
$\Rightarrow \exists v < u + y + 1(T(y, y, v) \wedge U(v) = 0)$
$\Rightarrow \alpha(y) = 1$
$\Rightarrow [y](y) = 1$
$\Rightarrow \bot.$

Second case: $U(u) = 1$ assumption
$\Rightarrow \bot.$ similarly

$\square$

Brouwer's famous bar induction (for decidable predicates P) can be understood as an induction principle over trees (Brouwer, 1975; Mainzer, 1977). Decidability of P with $\forall n(P(n) \vee \neg P(n))$ and $\forall \alpha \exists x P(\overline{\alpha}x)$ imply that $\{n | \forall m \prec n \neg P(m)\}$ is a regular well-founded tree. The bar of this tree is the set $\{n | P(n) \wedge \forall m \prec n \neg P(m)\}$. The implications $\forall n(P(n) \rightarrow Q(n))$ and $\forall n(\forall y Q(n * \langle y \rangle) \rightarrow Q(n)$ express that Q is a property which holds on a bar and is transferred upwards. If Q holds for all immediate successors of a node n, then it also holds for n. At that point, Brouwer's bar induction concludes that Q holds for the empty node $\langle \ \rangle$.

Brouwer's bar induction (for decidable properties) $\mathbf{BI}_D$

$$\forall \alpha \exists x P(\overline{\alpha}x) \wedge \forall n(P(n) \vee \neg P(n)) \wedge \forall n(P(n) \rightarrow Q(n)) \wedge$$

$$\forall n(\forall y Q(n * \langle y \rangle) \ \rangle \ Q(n)) \ \rangle \ Q(n)) \ \rangle \ Q(\langle \ \rangle)$$

Intuitionism draws rigorous consequences for proofs on infinitely growing trees which are not only important entities for mathematicians, but also for computer scientists. Obviously, bounds ("bars") in search trees spare us time and costs in the sense of Ockham's principle of parsimony. Intuitionistic mathematics yields controllable and reliable tools to approximate the infinity by finitude (cf. lawless choice sequences). But, the question arises whether Brouwer's belief in human intuitions is necessary. The digital information system which we introduced in the previous chapter also approximated the infinity by finite means. In short: Humans seem only to be biological examples of an information system to process data and information. We will discuss this question in Chapter 11 where brains and neural networks are considered complex systems which are equivalent to appropriate automata and machines.

Chapter 7

Proof Mining Bridging Logic, Mathematics, and Computer Science

Since antiquity, proofs are not only verifications of truth, but also construction. Applied mathematicians, natural scientists, and engineers have been interested in effective solutions of problems with geometric constructions in the past and computations with computers in modern times. Mathematical proofs sometimes seem to be abstract, ineffective, and beyond computational solutions. But, often, they include "hidden" effective information such as computable bounds which must be extracted only from the given proof. In this chapter, we consider the foundational research program of "proof mining" which aims at general logical tools to find ("unwind") effective procedures and algorithms in mathematical proofs of ordinary and practical mathematics. Are there even logical metatheorems to solve these tasks in general for whole classes of theorems and theories? In this case, logic would no longer be an isolated enterprise of only theoretical and philosophical interest separated from mathematics and computer science. Proof mining would actually bridge logic, mathematics, and computer science.

A historically nice example of "proof mining" is yielded by a proof of Euclid which he published in Book IX as proposition 20 of his work *The Elements of Euclid* (Euclid, 1782, p. 63; Aigner and Ziegler, 2001, p. 3).

Euclid's Theorem. *There are infinitely many prime numbers.*

Euclid's Proof (*reductio ad absurdum*). Assume there are finitely many prime numbers $p \leq x$.

We define $a := 1 + \prod_{\substack{p \leq x \\ p \text{ prime}}} p$

$\Rightarrow$ a cannot be prime (because $a > p$ for all $p \leq x$),

$\Rightarrow$ a contains a prime factor (by the decomposition of every number into prime factors) $q \leq a$ with $q > x$ (otherwise q is a prime factor $q \leq x$),

$\Rightarrow$ contradiction to assumption! $\qquad\qquad\qquad\qquad\qquad\square$

A careful analysis of Euclid's proof reveals a computable bounding function. At first, we note that the predicate $P(x) \equiv$ "x is a prime number" can be expressed in a quantifier-free (QF) way as primitive recursive predicate. In the next step (Kohlenbach, 2008, p. 15), we introduce the computable bound

$$g(x) := 1 + x! \geq 1 + \prod_{\substack{p \leq x \\ p \text{ prime}}} p,$$

which practically can be approximated by the Stirling formula

$$g(x) \approx 1 + (2\pi x)^{\frac{1}{2}} \left(\frac{x}{e}\right)^x = 1 + \sqrt{2\pi} \cdot e^{x \ln x - x + \frac{1}{2} \ln x} \ldots$$

But, actually, we aim at an upper bound on the $(r+1)$th prime number p_{r+1} which only depends on r instead of $x \geq p_r$. Euclid's proof yields

$$p_{r+1} \leq a := p_1 \cdot \ldots \cdot p_r + 1.$$

By induction on r, we can prove $p_r < 2^{2^r}$ for all $r \geq 1$ which is a computable upper bound.

In the history of mathematics, even much more better bounds were delivered (e.g., Euler's proof) (Pinasco, 2009). All these examples underline that proofs can include more information than the verification of truth. They also show that effective procedures can be extracted with different degrees of computability. Obviously, the extraction of effective procedures from given proofs is deeply rooted in the practice of mathematics.

The logician and mathematician Georg Kreisel brought it to the point with his claim for "unwinding of proofs" (Feferman, 1996): "What more do we know if we have proved a theorem by restricted means than if we merely know that it is true?" Later on, Kreisel's unwinding of proofs was called "proof mining" (Kohlenbach, 2008, p. 13) which reminds us of "data mining" as an important tool in the age of Big Data.

Anyway, in the past, the extraction of effective procedures from given proofs only depends on ordinary mathematics (e.g., number-theoretic knowledge in the case of Euclid's proof). What is new and exciting in Kreisel's suggestion? He favored a research strategy which uses logical and proof-theoretical tools to extract effective procedures from given proofs. General metatheorems can be proved in order to extract effective procedures from a whole class of theorems and proofs.

Finally, the research strategy of proof mining aims at an automated extraction of effective procedures which can be realized by computers. These ideas have far-reaching consequences if computer programs are considered as formulized proofs. In this case, proof mining leads to program verification which is a crucial issue in a world with increasing automation: How can security and reliance of algorithms and computer programs be guaranteed?

Anyway, logic and proof theory are no longer an isolated research of some specialists with no influence to the rest of the scientific world. With the research program of proof mining, logic and proof theory become a crucial link between mathematics and computer science with interdisciplinary consequences for practical applications. Above all, it opens new avenues in the philosophical debate on truth, effectivity, and constructivity.

In mathematics, the claims for truth and constructivity clash in existential theorems. Let us consider an existential theorem with the logical form $A \equiv \exists x B(x)$: A weaker requirement is to construct a list of terms $t_1, \ldots, t_n$ which are candidates for A, such that $B(t_1) \vee \ldots \vee B(t_n)$ holds. More general, we are often confronted with a Π_2^0-statement in order to extract an effective procedure.

Logical form of extracting effective procedures. If $A \equiv \forall x \exists y B(x, y)$, then one can ask for an algorithm p such that $\forall x B(x, p(x))$ holds or (in a weaker requirement) for a bounding function b such that $\forall x \exists y \leq b(x) B(x, y)$.

In general, the logical form of an existential theorem gives now hints how to extract effective procedures for computing solutions or, at least, bounds. But, there is a weakened version of an existential statement to determine possible candidates (Herbrand, 1967):

Definition of Herbrand Normal Form. Let A be a sentence in the prenex normal form

$$A \equiv \forall y_0 \exists x_1 \forall y_1 \ldots \exists x_n \forall y_n A_0(y_0, x_1, y_1, \ldots, x_n, y_n).$$

The Herbrand normal form of A is defined by

$$A^H :\equiv \forall y_0 \exists x_1, \ldots, x_n A_0(y_0, x_1, f_1(x_1), \ldots, x_n, f_n(x_1, \ldots, x_n))$$

with Herbrand index functions f_1, ..., f_n.

Obviously, Herbrand's normal form eliminates universal quantifiers. The following example illustrates the procedure step by step:

$$\forall y_0 \exists x_1 \forall y_1 \exists x_2 \forall y_2 A_0(y_0, x_1, y_1, x_2, y_2)$$
$$\Leftrightarrow \forall y_0 \exists x_1 \exists x_2 \forall y_2 A_0(y_0, x_1, f_1(x_1), x_2, y_2) \text{ with index function } f_1,$$
$$\Leftrightarrow \forall y_0 \exists x_1 \exists x_2 A_0(y_0, x_1, f_1(x_1), x_2, f_2(x_1, x_2)) \text{ with index functions } f_1, f_2.$$

In general, prenex sentences A and their Herbrand normal form A^H are (semantically) equivalent with respect to logical validity. We denote the statement "A is true" with $\models A$ which must be distinguished from $\vdash A$("A is derivable in a formal system"):

$$\models A \Leftrightarrow \models A^H,$$

but not (syntactically) with respect to formal predicate logic PL because, in general,

$$\mathrm{PL} \nvdash A^H \to A.$$

But, for the formal extension of $\mathrm{PL}^2_{-=}$ by n-ary function variables and function quantifiers, and with the axiom of choice

$$\mathrm{AC} : \forall \underline{x} \exists y A(\underline{x}, y) \to \exists f \forall \underline{x} A(\underline{x}, f(\underline{x}))(\underline{x} = x_1, \ldots, x_n),$$

we get $\mathrm{PL}^2_{-=} + \mathrm{AC} \vdash A \leftrightarrow A^H$.

The Herbrand normal form must not be confused with the Skolem normal form (Skolem, 1967).

Definition of Skolem normal form. Let A be a sentence in the prenex normal form

$$A \equiv \forall x_1 \exists y_1 \ldots \forall x_n \exists y_n A_0(x_1, y_1, \ldots, x_n, y_n).$$

Skolem's normal form of A is defined by

$$A^S :\equiv \forall x_1, \ldots, x_n A_0(x_1, f_1(x_1), \ldots, x_n, f_n(x_1, \ldots, x_n))$$

with Skolem functions $f_1, \ldots, f_n$.

In this case, the existential quantifiers are replaced by functions depending on the universally quantified variables from the universal quantifiers to the left. An example illustrates Skolem's procedure:

$\forall x_1 \exists y_1 \forall x_2 \exists y_2 A_0(x_1, y_1, x_2, y_2)$
$\Leftrightarrow \forall x_1 \forall x_2 \exists y_2 A_0(x_1, f_1(x_1), x_2, y_2)$ with Skolem function f_1,
$\Leftrightarrow \forall x_1 \forall x_2 A_0(x_1, f_1(x_1), x_2, f_2(x_1, x_2))$ with Skolem functions f_1, f_2.

But how can we find possible candidates to satisfy an existential quantifier with the help of Herbrand's normal form? Let us consider the following example (Kohlenbach, 2008, p. 24).

Consider the sentence

$A \equiv \forall x \exists y \forall z A_0(x, y, z) \equiv \forall x \exists y \forall z (P(x, y) \vee \neg P(x, z))$ with
$A_0(x, y, z) :\equiv P(x, y) \vee \neg P(x, z)$.

In its Herbrand normal form

$$A^H :\equiv \forall x \exists y A_0(x, y, g(y)) \equiv \forall x \exists y (P(x, y) \vee \neg P(x, g(y))),$$

the universal quantifier "$\forall z$" is eliminated by a Herbrand index function g. The existential quantifier "$\exists y$" can now be interpreted by a list of possible candidates $y = c$ and $y = g(c)$ for any constant c, i.e., logically a disjunction D:

$$A^{H,D} :\equiv \forall x A_0(x, c, g(c)) \equiv \forall x (P(x, c) \vee \neg P(x, g(c)))$$
$$\vee (P(x, g(c)) \vee \neg P(x, g(g(c)))).$$

Obviously, $A^{H,D}$ is a logical tautology even if $g(c)$ is replaced by y and $g(g(c))$ by z:

$$A^D :\equiv \forall x A_0(x, c, g(c))$$
$$\equiv \forall x (P(x, c) \vee \neg P(x, y)) \vee (P(x, y) \vee \neg P(x, z)))$$

is still a tautology. By simple applications of logical rules with $\forall$-introduction and $\exists$-introduction, we can derive A from A^D:

$$
\begin{array}{ll}
P(x, c) \vee \neg P(x, y) \vee (P(x, y) \vee \neg P(x, z)) & \\
\Rightarrow P(x, c) \vee \neg P(x, y) \vee \forall z (P(x, y) \vee \neg P(x, z)) & \forall\text{-introduction} \\
\Rightarrow P(x, c) \vee \neg P(x, y) \vee \exists y \forall z (P(x, y) \vee \neg P(x, z)) & \exists\text{-introduction} \\
\Rightarrow \forall y (P(x, c) \vee \neg P(x, y)) \vee \exists y \forall z (P(x, y) \vee \neg P(x, z)) & \forall\text{-introduction} \\
\Rightarrow \exists u \forall y (P(x, u) \vee \neg P(x, y)) \vee \exists y \forall z (P(x, y) \vee \neg P(x, z)) & \exists\text{-introduction} \\
\Rightarrow \exists y \forall z (P(x, y) \vee \neg P(x, z)) & \text{contraction} \\
\Rightarrow \forall x \exists y \forall z (P(x, y) \vee \neg P(x, z)) & \forall\text{-introduction.}
\end{array}
$$

Our example leads us immediately to the following theorem which seems to yield a logical scheme to construct a list of terms satisfying the existential quantifiers in a given sentence (Kohlenbach, 2008, p. 25):

Herbrand's Theorem. *Let* $A \equiv \exists x_1 \forall y_1 \ldots \exists x_n \forall y_n A_0(x_1, y_1, \ldots, x_n, y_n)$. *Then it holds:* $\text{PL}_{-=} \vdash A$ *iff there are terms* $t_{1,1}, \ldots, t_{1,k_1}, \ldots, t_{n,1}, \ldots, t_{n,k_n}$ *(built out of the constants, free variables and function symbols of* A *and the index functions used for the formation of* A^H) *such that*

$$A^{H,D} \quad :\equiv \quad \bigvee_{j_1=1}^{k_1} \cdots \bigvee_{j_n=1}^{k_n} A_0(t_{1,j_1}, f_1(t_{1,j_1}), \ldots, t_{n,j_n}, f_n(t_{1,j_1}, \ldots,$$
$$t_{n,j_n})) \text{ is a tautology.}$$

The terms $t_{i,j}$ can be extracted constructively from a given $\mathrm{PL}_{-=}$ proof of A and conversely one can construct a $\mathrm{PL}_{-=}$ proof for A out of a given tautology of $A^{H,D}$.

The proof of A under the assumption of given terms of the tautology $A^{H,D}$ follows the steps in the previous example (Shoenfield, 1967, p. 244). It is more difficult to construct these terms under the assumption of A. But, in general, Herbrand's theorem does not hold for extensions of PL such as PA (Peano arithmetic), when non-logical axioms come in. Anyway, Herbrand's theorem inspires interpretations of existential quantifiers by computable functionals instead of the construction of terms.

Example: Realization of existential quantifiers by functionals. Again, we consider the sentence

$A \equiv \forall x \exists y \forall z A_0(x, y, z) \equiv \forall x \exists y \forall z (P(x, y) \vee \neg P(x, z))$ with $A_0(x, y, z)$:
$\equiv P(x, y) \vee \neg P(x, z)$ and decidable prime formula $P(x, y) :\equiv T(x, x, y)$ (e.g., Kleene's T-predicate).

In the Herbrand normal form

$$A^H \colon \equiv \forall x \exists y A_0(x, y, g(y)) \equiv \forall x \exists y (T(x, x, y) \vee \neg T(x, x, g(y))),$$

the universal quantifier "$\forall z$" is eliminated by a Herbrand index function g. Instead of a disjunction with a list of possible candidates, the existential quantifier "$\exists y$" is now realized by a computable functional Φ of type 2 with

$$\Phi(x, g) := \begin{cases} x, & \text{if } \neg T(x, x, g(x)) \\ g(x), & \text{otherwise.} \end{cases}$$

It follows that

$$\forall x \forall g A_0(x, \Phi(x, g), g(\Phi(x, g)))$$
$$\equiv \forall x \forall g (T(x, x, \Phi(x, g)) \vee \neg T(x, x, g(\Phi(x, g)))).$$

This example leads us to the following general procedure.

Definition of functional interpretation by no-counterexample interpretation. Let $A \equiv \exists x_1 \forall y_1 \ldots \exists x_n \forall y_n A_0(x_1, y_1, \ldots, x_n, y_n)$.

If functionals $\Phi_1, \ldots, \Phi_n$ realizes the Herbrand normal form A^H of A, i.e., $\forall \underline{f} A_0(\Phi_1(\underline{f}), f_1(\Phi_1(\underline{f})), \ldots, \Phi_n(\underline{f}), f_n(\Phi_1(\underline{f}), \ldots, \Phi_n(\underline{f})))$ is true (with $\underline{f} = f_1, \ldots, f_n$), then $\underline{\Phi} = \Phi_1, \ldots, \Phi_n$ is said to satisfy the no-counterexample interpretation of A.

Why is that functional interpretation called "no-counterexample"? Consider the negation $\neg A$ of our above example with

$$\neg A \equiv \forall x_1 \exists y_1 \ldots \forall x_n \exists y_n \neg A_0(x_1, y_1, \ldots, x_n, y_n).$$

A counterexample to A is given by functions $f_1, \ldots, f_n$ such that

$$(*) \quad \forall \underline{x} \neg A_0(x_1, f_1(x_1), \ldots, x_n, f_n(x_1, \ldots, x_n))$$

holds. Therefore, functionals $\underline{\Phi}$ satisfying the no-counterexample interpretation of sentence A produce a counterexample to the negation $\neg A$, i.e., to the existence of functionals $f_1, \ldots, f_n$ in formula $(*)$.

No-counterexample interpretation can be realized for many classical theories. For example, Σ_1^0-formulas in Peano arithmetic have a no-counterexample interpretation with primitive recursive functionals. Full Peano arithmetic requires primitive recursive functionals of higher types (compare the definition of primitive recursive functions in Chapter 2).

Definition of primitive recursive functionals. A functional F is primitive recursive (of type ≤ 2) iff it can be defined by the following procedures for $\underline{x} = x_1, \ldots, x_{p-1}$ and $\underline{f} = f_1, \ldots, f_{q-1}$ with $p, q \geq 1$:

(i) <u>Projection:</u>
$$F(\underline{x}, \underline{f}) = \begin{cases} x_i, & \text{if } i < p, \\ 0, & \text{otherwise.} \end{cases}$$

(ii) <u>Function application:</u>
$F(\underline{x}, \underline{f}) = f_i(x_{j_0}, \ldots, x_{j_{l-1}})$ for $i < q$ and $j_0, \ldots, j_{l-1} < p$ and f_i of arity l.

(iii) <u>Successsor:</u>
$F(\underline{x}, \underline{f}) = x_i + 1$ for $i < p$.

(iv) <u>Substitution:</u>
$F(\underline{x}, \underline{f}) = G(H_0(\underline{x}, \underline{f}), \ldots, H_{l-1}(\underline{x}, \underline{f}), \lambda y.K_0(y, \underline{x}, \underline{f}), \ldots, K_{j-1}$
$(y, \underline{x}, \underline{f}))$ for $G, H_0, \ldots, H_{l-1}, \lambda y.K_0(y, \underline{x}, \underline{f}), \ldots, K_{j-1}(y, \underline{x}, \underline{f})$
primitive recursive.

(v) <u>Primitive recursion:</u>
$F(0, \underline{x}, \underline{f}) = G(\underline{x}, \underline{f})$,
$F(y + 1, \underline{x}, \underline{f}) = H(F(y, \underline{x}, \underline{f}), y, \underline{x}, \underline{f})$ for G, H primitive
recursive.

In order to demonstrate the practical use of functional interpretations, we consider an application of no-counterexample interpretation
to real analysis. It is well known that real numbers can be represented
by Cauchy sequences. We will come back to this topic in a later
chapter in more details. For our application, it is sufficient to have a
non-increasing sequence $(r_n)_{n\in\mathbb{N}}$ of rational numbers in the interval
[0,1]. This sequence can also be understood as a number-theoretic
function. Operations and ordering between rational numbers are
primitive recursive. A Cauchy sequence is defined by the following
proposition.

Definition of Cauchy criterion. $(r_n)_{n\in\mathbb{N}}$ is a Cauchy sequence iff
$\forall k \in \mathbb{N}\exists n \in \mathbb{N}\forall m \in \mathbb{N}(|r_{n+m} - r_n| <_\mathbb{Q} 2^{-k})$.

Even for certain primitive recursive sequences $(r_n)_{n\in\mathbb{N}}$ there is in
general no computable bound $f(k)$ of n (Specker, 1949). But, at least,
we can find a primitive recursive functional bound $\Phi(g, k)$ depending
on k and number-theoretic functions g (Kohlenbach, 2008, p. 31).

**Proposition: No-counterexample interpretation of the
Cauchy criterion.** *There is a primitive recursive functional Φ satisfying the no-counterexample interpretation of the Cauchy criterion:*

$$\forall k \in \mathbb{N}\forall g \in \mathbb{N}^\mathbb{N}\exists n \leq \Phi(g, k)(|r_{n+g(n)} - r_n| <_\mathbb{Q} 2^{-k}).$$

Proof. For number-theoretic functions $g : \mathbb{N} \to \mathbb{N}$, we define $\tilde{g} :$
$\mathbb{N} \to \mathbb{N}$ with $\tilde{g}(n) := n + g(n)$ and start with the proof of an

auxiliary statement

$$\forall k \in \mathbb{N} \forall g \in \mathbb{N}^{\mathbb{N}} \exists i \leq 2^k \big(r_{\tilde{g}^{(i)}(0)} - r_{\tilde{g}^{(i+1)}(0)} <_{\mathbb{Q}} 2^{-k} \big).$$

On the contrary, let us assume

$$\forall i \leq 2^k \big(r_{\tilde{g}^{(i)}(0)} - r_{\tilde{g}^{(i+1)}(0)} \geq_{\mathbb{Q}} 2^{-k} \big)$$

for some $k \in \mathbb{N}$ and $g \in \mathbb{N}^{\mathbb{N}}$. Note that $\tilde{g}^{(0)}(0) = 0$. Then, it follows that $r_0 - r_{\tilde{g}^{(2^k+1)}(0)} \geq_{\mathbb{Q}} (2^k + 1) \cdot 2^{-k} > 1$ which is a contradiction and proves the auxiliary statement.

Remember that sequence $(r_n)_{n \in \mathbb{N}}$ is non-increasing. Therefore, the auxiliary statement implies

$$\forall k \in \mathbb{N} \forall g \in \mathbb{N}^{\mathbb{N}} \exists i \leq 2^k \big(|r_{\tilde{g}^{(i)}(0)} - r_{\tilde{g}^{(i)}(0)+g(\tilde{g}^{(i)}(0))}| <_{\mathbb{Q}} 2^{-k} \big).$$

As functional bound, we now define the maximum of all indices $\tilde{g}^{(i)}(0)$ with $i \leq 2^k$, i.e.,

$$\Phi(g, k) := \tilde{g}^{(2^k)}(0). \qquad \qquad \Box$$

In the next step, the bound $\Phi(g, k)$ allows us to determine a primitive recursive realizer $\Psi((r_n)_{n \in \mathbb{N}}, g, k)$ for the existential quantifier "$\exists n$". But, the no-counterexample interpretation cannot be universally applied because it is restricted to functionals of type 2. We will consider the *modus ponens* conclusion in logic in order to explain that functionals of type 2 are, in general, not sufficient whereas functionals of higher types solve the problem (Kohlenbach, 2008, p. 24):

Example: A no-counterexample interpretation of *modus ponens* fails! In language $\mathcal{L}(\mathrm{PA})$ of Peano arithmetic, the *modus ponens* rule is given by the scheme

$$\frac{A \quad A \to B}{B}$$

for sentences A, B in prenex normal form and QF prime formulas A_0, B_0 with

$$A :\equiv \forall x \exists y \forall z A_0(x, y, z),$$

$$B :\equiv \forall u \exists v B_0(u, v).$$

The prenex normal form $A \to B$ is

$$(A \to B)^{pr} :\equiv \forall u \exists x \forall y \exists z, v(A_0(x, y, z) \to B_0(u, v)).$$

The no-counterexample interpretation of A and $(A \to B)^{pr}$ asks for functionals realizing the Herbrand normal forms

$$A^H :\equiv \forall x, g \exists y A_0(x, y, g(y)),$$

$$((A \to B)^{pr})^H :\equiv \forall u, f \exists x, z, v(A_0(x, f(x), z) \to B_0(u, v)).$$

Assume functionals $\varphi_0, \varphi_1, \varphi_2, \varphi_3$ realizing the Herband normal forms of the premises of *modus ponens* with

$$A^H : \forall x, g A_0(x, \varphi_0(x, g), g(\varphi_0(x, g))),$$
$$((A \to B)^{pr})^H : \forall u, f(A_0(\varphi_1(u, f), f(\varphi_1(u, f)), \varphi_2(u, f)) \to$$
$$B_0(u, \varphi_3(u, f))).$$

In order to realize the conclusion of *modus ponens* by functional interpretation, one has to solve the following equations for x, f, g:

$$x = \varphi_1(u, f),$$

$$\varphi_0(x, g)) = f(\varphi_1(u, f)),$$

$$g(\varphi_0(x, g)) = \varphi_2(u, f).$$

There is no primitive recursive functional in $u, \varphi_0, \varphi_1, \varphi_2, \varphi_3$ to solve these equations. The solution requires so-called bar recursion (of type 0) (Spector, 1962) which goes beyond primitive recursive functionals.

Functional interpretation beyond no-counterexample interpretation

We interpret $A \to B$ as $A^H \to B$, i.e.,

$$\forall x, g \exists y A_0(x, y, g(y)) \to \forall u \exists v B_0(u, v).$$

The axiom of choice (AC) yields

$$\exists Y \forall x, g A_0(x, Y(x, g), g(Y(x, g))) \to \forall u \exists v B_0(u, v)$$

with prenex normal form

$$(*) \quad \forall u, Y \exists x, g, v(A_0(x, Y(x, g), g(Y(x, g))) \rightarrow B_0(u, v)).$$

The equivalence of $A \rightarrow B$ and $(*)$ depends on the axiom of choice. We assume functionals $\Phi_0, \Phi_1, \Phi_2, \Phi_3$ realizing the premises of *modus ponens*:

$$A : \forall x, g A_0(x, \Phi_0(x, g), g(\Phi_0(x, g))),$$
$$(+) \; A \rightarrow B : \forall u, Y \exists x, g, v(A_0(\Phi_1(u, Y), Y(\Phi_1(u, Y)), \Phi_2(u, Y)) \rightarrow$$
$$B_0(u, \Phi_3(u, Y))).$$

In this case, a solution of the *modus ponens* problem is given by

$$Y := \Phi_0, \quad x := \Phi_1(u, \Phi_0), \quad g := \Phi_2(u, \Phi_0)$$

for the conclusion

$$B : \forall u B_0(u, \Phi_3(u, \Phi_0)).$$

But, functionals $\Phi_0 : \mathbb{N} \times \mathbb{N}^{\mathbb{N}} \rightarrow \mathbb{N}$ of type 2 and $\Phi_3 : N \times \mathbb{N}^{(\mathbb{N} \times \mathbb{N}^{\mathbb{N}})} \rightarrow \mathbb{N}$ of type 3 go beyond no-counterexample interpretation (Kohlenbach, 2008, p. 35).

In Chapter 4, we already introduced the BHK-interpretation of logical constants. The BHK-interpretation explains the meaning of the logical constants $\bot, \wedge, \vee, \rightarrow, \forall, \exists$ in terms proof constructions. The disadvantage of the BHK-interpretation is the unexplained notion of construction resp. constructive proof. Gödel wanted that constructive proofs of existential theorems provide explicit realizers. Therefore, he replaced the notion of constructive proof by the more definite and less abstract concept of computable functionals of finite type. In Chapter 4, the notion of a computable functional of finite type was mathematically defined as an ideal in an information system. Gödel's proof interpretation is largely independent of a precise definition of computable functionals: One only needs certain basic functionals as computable (e.g., primitive recursion in finite types) and their closure under composition.

The introduction of functionals of higher types requires to enrich the formal language with appropriate variables and quantifiers. They are especially needed to formalize proofs in real analysis in following chapters (Kohlenbach, 2008, Chapter 3.3).

Definition of finite types of functionals. The set T of all finite types (over $\mathbb{N}$) is generated inductively by

(i) $0 \in$ T,
(ii) $\rho, \tau \in$ T $\Rightarrow \tau(\rho) \in$ T.

The type 0 is the type of natural numbers. Objects of type $\tau(\rho)$ are functions which map objects of type ρ to objects of type τ. Sometimes, it is written $\tau(\rho) \equiv \rho \to \tau$.

The set P $\subset$ T of pure types is defined by

(i) $0 \in$ P,
(ii) $\rho \in$ P $\Rightarrow 0(\rho) \in$ P.

Pure types are often denoted by natural numbers:

$$0(n) := n + 1 \ (\text{e.g.,} \ 00 = 1, \ 0(00) = 2).$$

The type level or degree $\deg(\rho)$ of type ρ is defined as

$$\deg(0) := 0,$$

$$\deg(\tau(\rho)) := \max \ (\deg(\tau), \ \deg(\rho) + 1).$$

Objects of type ρ with $\deg(\rho) > 1$ are called functionals.
   In the next step, we introduce:

Formal language $\mathcal{L}(\text{E-HA}^\omega)$ of (extensional) Heyting arithmetic (E-HA$^\omega$) in all finite types.

$\mathcal{L}(\text{E-HA}^\omega)$ is based on a many-sorted version of intuitionistic logic IL$^\omega$

Variables:	x^ρ, y^ρ, z^ρ
Quantifiers:	$\forall x^\rho, \exists x^\rho$ for every type ρ
Constants:	0^0 (zero),
	S^{00} (successor),
	$\Pi_{\rho,\tau}^{\rho\tau\rho}$ (projector),
	$\Sigma_{\delta,\rho,\tau}$ (combinator of type $\tau\delta(\rho\delta)(\tau\rho\delta)$),

$$\underline{R}_\rho = (R_1)_\rho, \ldots, (R_k)_\rho \ \text{(simultaneous recursor}$$
constants, where R_i has type $\rho_i(\rho_k 0\underline{\rho}^t)\ldots(\rho_1 0\underline{\rho}^t)\underline{\rho}^t 0$
for all $\delta, \rho, \tau, \underline{\rho}(= (\rho_1)\ldots(\rho_k)), \underline{\rho}^t(:= (\rho_k)\ldots(\rho_1))$

Predicate symbol: $=_0$ (equality between objects of type 0)

Terms and formulas of E-HA$^\omega$.

<u>Terms</u> are defined inductively:

(i) constants c^ρ and variables x^ρ of type ρ are terms of type ρ.
(ii) $t^{\tau\rho}$ term of type $\tau\rho$, s^ρ term of type $\rho \Rightarrow t(s)$ term of type τ.

Abbreviation: $ts_1\ldots s_k$ or $t(s_1, \ldots, s_k)$ instead of $t(s_1)\ldots(s_k)$.

<u>Formulas</u> are defined inductively:

(i) atomic (prime) formulas $s =_0 t$ are formulas (with terms s^0, t^0 of type 0).
(ii) A, B formulas $\Rightarrow (A \wedge B), (A \vee B), (A \to B)$ formulas.
(iii) A formula, x^ρ variable of type $\rho \Rightarrow (\forall x^\rho A), (\exists x^\rho A)$ formulas.

Abbreviations:

(a) Higher type equations $s =_\rho t$ between s, t of type $\rho = 0\rho_k\ldots$ $\rho_1(k \geq 1)$ are abbreviations for $\forall y_1^{\rho_1}, \ldots, y_k^{\rho_k}(sy_1\ldots y_k =_0 ty_1 \ldots y_k)$ with variables $y_1\ldots y_k$ which do not occur in s, t.
(b) $\neg A :\equiv A \to \bot, \bot :\equiv (0 =_0 1), A \leftrightarrow B :\equiv (A \to B) \wedge (B \to A)$.

Axioms and rules of E-HA$^\omega$.

(i) all axioms and rules of $\mathrm{IL}^\omega_{\underline{=}=}$ (intuitionistic logic without $=$),
(ii) equality axioms for $=_0$:

    (a) $x =_0 x$,
    (b) $x =_0 y \to y =_0 x$,
    (c) $x =_0 y \wedge y =_0 z \Rightarrow x =_0 z$.

(iii) higher type extensionality:

$$E_\rho : \forall z^\rho, x_1^{\rho_1}, y_1^{\rho_1}, \ldots, x_k^{\rho_k}, y_k^{\rho_k}(\wedge_{i=i}^k x_i =_{\rho_i} y_i) \to z\underline{x} =_0 z\underline{y}$$

with $\rho = 0\rho_k \ldots \rho_1$,

(iv) successor axioms:

(a) $S^{00}(x) \neq 0$,

(b) $S^{00}(x) = S^{00}(y) \rightarrow x = y$,

(v) inductive scheme:

$$\text{IA} : A(0) \wedge \forall x^0(A(x) \rightarrow A(Sx)) \rightarrow \forall x^0 A(x)$$

with $A(x^0)$ formula of E-HA$^\omega$.

Axioms for $\Pi_{\rho,\tau}$, $\Sigma_{\delta,\rho,\tau} R_\rho$:

(Π): $\Pi_{\rho,\tau} x^\rho y^\tau =_\rho x^\rho$,

(Σ): $\Sigma_{\delta,\rho,\tau} xyz =_\tau xz(yz)(x^{\tau\rho\delta}, y^{\rho\delta}, z^\delta)$,

(R): $(R_i)_{\underline{\rho}} yz =_{\rho_i} y_i$,

$\quad (R_i)_{\underline{\rho}}(Sx^0)\underline{y}z =_{\rho_i} z_i(\underline{R}_\rho x\underline{y}\underline{z})x \quad (i = 1, \ldots, k)$,

where $\underline{\rho} = \rho_1, \ldots \rho_k$, $\underline{y} = y_1, \ldots, y_k$, $\underline{z} = z_1, \ldots, z_k$ with y_i of type ρ_i and z_i of type $\rho_i 0\underline{\rho}^t$.

In 1945, Kleene proposed a variant to the BHK-interpretation. He understood a construction as a machine procedure which can be coded by a machine number. As in the BHK-interpretation, an existential statement $\exists x A(x)$ is considered to be an incomplete information. The statement $\exists x A(x)$ can only be made if x can be constructed or "realized" by a Turing machine or, in general, by an information system (or, more precise, by a functional which can be understood as an ideal of an information system). Kleene's historical concept of realizability was restricted to (partial) recursive functions and their machine numbers. Before generalizing his concept, let us introduce his historical concept which is also comparable to a classical truth definition: It interprets arithmetic sentences in the mathematical structure $\mathcal{N} = (\mathbb{N}, 0, S, +, \cdot)$ of natural numbers. The elements (numbers) of $\mathbb{N}$ are denoted by $n, m, \ldots$, the corresponding numerals in formal language by $\bar{n}, \bar{m}, \ldots$. The statement "n realizes A" is defined by induction on the formal complexity of formula A (Kleene, 1945). $(n)_i$ denotes (computable) components i of an encoded tupel of instructions of a machine with machine number n, e.g., for $n = \langle a, b \rangle$ is $(n)_1 = a$ and $(n)_2 = b$:

Kleene's definition of realizability.

(i) n realizes $(t = s)$ iff $t = s$ holds in $\mathcal{N}$.

(ii) n realizes $A \wedge B$ iff $(n)_1$ realizes A and $(n)_2$ realizes B.

(iii) n realizes $A \vee B$ iff $(n)_1 = 0$ and $(n)_2$ realizes A, or $(n)_1 \neq 0$ and $(n)_2$ realizes B.

(iv) n realizes $A \to B$ iff for each m realizing A, $[n](m)$ is defined and realizes B.

(v) n realizes $\exists x A(x)$ iff $(n)_2$ realizes $A(x/\overline{(n)_1})$.

(vi) n realizes $\forall x A(x)$ iff for all m, $[n](m)$ is defined and realizes $A(\bar{m})$.

Instead of a semantical definition of realizability, we could reformulate realizability as a syntactical mapping assigning a formula xrA of Heyting arithmetic HA to each formula of HA. A generalized version for functionals was introduced by Kreisel (with notation $\underline{y}\underline{x} := y_1\underline{x}$, $\ldots, y_n\underline{x}$, where $\underline{y} = y_1, \ldots, y_n$ and $\underline{x} = x_1, \ldots, x_n$ are tuples of functionals of certain types and $y_i\underline{x} := y_i x_1 \ldots x_n$) (Kreisel, 1959; Kohlenbach, 2008, Chapter 5):

Definition of modified realizability in $\mathcal{L}(\text{E-HA}^\omega)$. For each formula A of $\mathcal{L}(\text{E-HA}^\omega)$, formula $\underline{x}\, mr A$("$\underline{x}$ modified realizes A") is defined in $\mathcal{L}(\text{E-HA}^\omega)$:

(i) $\underline{x}\, mr A :\equiv A$ with empty tuple $\underline{x}$ and A prime formula $\underline{x}$,

(ii) $\underline{x}, \underline{y}\, mr(A \wedge B) :\equiv \underline{x}\, mr A \wedge \underline{y}\, mr B$,

(iii) $z^0, \underline{x}, \underline{y}\, mr(A \vee B) :\equiv (z =_0 0 \to \underline{x}\, mr A) \wedge (z \neq_0 0 \to \underline{y}\, mr B)$,

(iv) $\underline{y}\, mr(A \to B) :\equiv \forall \underline{x}(\underline{x}\, mr A \to \underline{y}\underline{x}\, mr B)$,

(v) $\underline{x}\, mr(\forall y^\rho A(y)) :\equiv \forall y^\rho(\underline{x}y\, mr A(y))$,

(vi) $z^\rho, \underline{x}\, mr(\exists y^\rho A(y)) :\equiv \underline{x}\, mr A(z)$.

In order to prove the soundness of functional interpretation with modified realizability with respect to extensional Heyting arithmetic E-HA^ω, we need two further principles.

Axiom of choice.

$\text{AC}^{\rho,\tau} : \forall x^\rho \exists y^\tau A(x, y) \to \exists Y^{\tau\rho} \forall x^\rho A(x, Yx)$
$\text{AC} := \bigcup_{\rho,\tau \in \text{T}} \{\text{AC}^{\rho,\tau}\}.$ with formula A of E-HA^ω.

The independence-of-premise principle.

$\mathrm{IP}^{\rho}_{ef} : (A \to \exists x^{\rho} B(x)) \to \exists x^{\rho}(A \to B(x))$
$\mathrm{IP}^{\omega}_{ef} := \bigcup_{\rho \in T}\{\mathrm{IP}^{\rho}_{ef}\}$ with A $\exists$-free and without x free.

In general, an interpretation of a formal system is sound, if any derivable formula of the formal system is true with respect to the interpretation. Actually, an interpretation is a mapping from a formal language into another one. Thus, the interpreted formula is "true" if it is derivable in the formal system the formula was mapped into. In order to prove soundness of modified realization in $\mathcal{L}(\mathrm{E\text{-}HA}^{\omega})$, the formal system $\mathrm{E\text{-}HA}^{\omega}$ is extended by the principles $\mathrm{AC}^{\rho,\tau}$ and IP^{ρ}_{ef} (Troelstra, 1973).

Soundness theorem of modified realization in $\mathcal{L}(\mathrm{E\text{-}HA}^{\omega})$.
Let A be a formula in $\mathcal{L}(\mathrm{E\text{-}HA}^{\omega})$, Δ_{ef} set of $\exists$-free sentences. Then the following rule holds:

$$\mathrm{E\text{-}HA}^{\omega} + \mathrm{AC} + \mathrm{IP}^{\omega}_{ef} + \Delta_{ef} \vdash A \Rightarrow \mathrm{E\text{-}HA}^{\omega} + \Delta_{ef} \vdash \underline{t}\,mr\,A,$$

where t is a tuple of terms of $\mathrm{E\text{-}HA}^{\omega}$ with $FV(\underline{t}) \subseteq FV(A)$ which can be extracted from proof of A.

Proof (induction on the length of the derivation of A).
Induction on the length of the derivation of A means that the logical axioms, the logical rules, and the non-logical axioms must be realized in $\mathrm{E\text{-}HA}^{\omega}$.

(1) Logical axioms:

    For example, we prove the soundness of axiom $A \vee A \to A$:

    We have to determine an appropriate term $\underline{t}$ with $\mathrm{E\text{-}HA}^{\omega}+\Delta_{ef} \vdash \underline{t}\,mr\,A \vee A \to A$ under the condition of $\mathrm{E\text{-}HA}^{\omega} + \mathrm{AC} + \mathrm{IP}^{\omega}_{ef} + \Delta_{ef} \vdash A \vee A \to A$.

       Assume $z^0, \underline{x}, \underline{y}\,mr\,(A \vee A) \Rightarrow (z =_0 0 \to \underline{x}\,mrA) \wedge (z \neq_0 0 \to \underline{y}\,mrA)$ Def. (iii)

Define

$$\underline{t}z^0\underline{x}\underline{y} := \begin{cases} \underline{x}, & \text{if } z = 0, \\ \underline{y}, & \text{if } z \neq 0. \end{cases}$$

$$\Rightarrow \underline{t}z^0\underline{x}\underline{y}\,mr\,A$$

$$\Rightarrow \underline{t}\,mr\,(A \vee A \to A) \hspace{4cm} \text{Def. (iv)}$$

(2) <u>Logical rules:</u>

For example, we prove the soundness of the *modus ponens* rule $\frac{A\ \ A\to B}{B}$:

Assume $\underline{t}\,mr\,A$ and $\underline{s}\,mr\,(A \to B) \Rightarrow \underline{s}\,\underline{t}\,mr\,B$ \hspace{2cm} Def. (iv)

We define term $\underline{r}$ by replacing all free variables in $\underline{st}$ (by constants $\underline{O}^\rho := \lambda x_1^{\rho_1}, \ldots, \lambda x_k^{\rho_k}.0^0$ for $\rho = 0(\rho_k)\ldots 0(\rho_1)$) which occur in A but not in B. It follows $\underline{r}\,mr\,B$.

(3) <u>Axioms for</u> $=_0, E_\rho, S^{00}, \Pi_{\rho,\tau}, \Sigma_{\delta,\rho,\tau}, R_\rho$:

These axioms are all $\exists$-free and, by that, realized by themselves.

(4) <u>Induction scheme:</u>

Let the two premises of the axiom be modified realized by $\underline{x}\,mr\,A(0)$ and $\underline{y}\,mr\,\forall z^0(A(z) \to A(z+1))$. We have to determine a term $\underline{t}$ with $\underline{t}\,mr\,\forall z^0 A(z)$. We define $\underline{t}$ primitive recursively by

$$\underline{t}\,\underline{x}\,\underline{y}\,0 = \underline{x}$$

$$\underline{t}\,\underline{x}\,\underline{y}(z + 1) = \underline{y}z(\underline{t}\,\underline{x}\,\underline{y}z).$$

The formula $\underline{t}\,\underline{x}\,\underline{y}z\,mr\,A(z)$ is proved by induction on z.

$\Rightarrow \underline{t}\,\underline{x}\,\underline{y}\,mr\,\forall z A(z)$ \hspace{6cm} Def. (v).

(5) <u>Axioms AC and</u> IP^ω_{ef} are not needed to verify modified realizability.

(6) <u>Axioms of</u> Δ_{ef}:

For $\exists$-free formulas A as elements of Δ_{ef}, it holds that $(\underline{x}\,mr\,A) \equiv A$ with $\underline{x}$ empty tuple. $\hspace{4cm} \square$

Truth of a statement is characterized by the existence of a (modified) realizer of the statement. That can be formally derived in the formal system of (extensional) Heyting arithmetic E-HA$^\omega$ extended by the axioms of choice AC and premise-independence IP$^\omega_{ef}$.

Characterization theorem for modified realizability. Let A be a formula of $\mathcal{L}$(E-HA$^\omega$). Then

$$\text{E-HA}^\omega + \text{AC} + \text{IP}^\omega_{ef} \vdash A \leftrightarrow \exists \underline{x}(\underline{x}\, mr\, A).$$

Proof (induction on the logical structure of A). We have to prove the statement for

(1) $A \equiv s =_0 t$ prime formula
(2) $A, B \Rightarrow A * B$ with $* = \wedge, \vee, \rightarrow$:
   For example, assume the induction hypotheses

$$A \leftrightarrow \exists \underline{x}(\underline{x}\, mr\, A)$$

$$B \leftrightarrow \exists \underline{x}(\underline{x}\, mr\, B)$$

$$\Rightarrow A \wedge B \leftrightarrow \exists \underline{x}(\underline{x}\, mr\, A) \wedge \exists \underline{x}(\underline{x}\, mr)B$$

$$\leftrightarrow \exists \underline{x}(\underline{x}\, mr\, (A \wedge B)) \qquad \text{Def. (ii)}$$

(3) $A, x^\rho \Rightarrow \forall x^\rho A, \exists x^\rho A$

which follows straightforward by the definition of modified realizability. $\qquad\qquad\square$

The soundness and characterization theorems allow us to extract an effective procedure (program) from a given proof which is the final goal of proof mining (Kohlenbach, 2008, p. 100):

Main theorem of program extraction by modified realization. Let $\forall x^\rho \exists y^\tau A(x, y)$ be a sentence of $\mathcal{L}$(E-HA$^\omega$) with types ρ, τ, and Δ_{ef} set of $\exists$-free sentences. Then the following rule holds:

$$\text{E-HA}^\omega + \text{AC} + \text{IP}^\omega_{ef} + \Delta_{ef} \vdash \forall x^\rho \exists y^\tau A(x, y)$$
$$\Rightarrow \text{E-HA}^\omega + \text{AC} + \text{IP}^\omega_{ef} + \Delta_{ef} \vdash \forall x^\rho A(x, tx),$$

where t is a closed term of $\mathcal{L}$(E-HA$^\omega$) which is extracted from a proof of the premise by modified realizability.

It follows that

$$\mathcal{L}^\omega \models \Delta_{ef} \Rightarrow \mathcal{L}^\omega \models \forall x^\rho A(x, tx).$$

Proof.

$$\vdash \forall x^\rho \exists y^\tau A(x, y)$$
$$\Rightarrow \vdash t_1, t_2, \ldots, t_n \ mr \ \forall x^\rho \exists y^\tau A(x, y) \qquad \text{soundness theorem}$$
$$\Rightarrow \vdash t_1, t_2, \ldots, t_n \ mr \ \exists t \forall x^\rho A(x, tx) \qquad \text{AC}$$
$$\Rightarrow \vdash t_2, \ldots, t_n \ mr \ \forall x^\rho A(x, t_1 x) \qquad \text{Def. (vi)}$$
$$\Rightarrow \vdash \forall x^\rho A(x, t_1 x) \qquad \text{characterization theorem} \ \square$$

Extraction of programs from given proofs can be automated by software. An example of automatic program extraction is MINLOG which was developed by Helmut Schwichtenberg and his research group in Munich. MINLOG is an interactive proof system which is equipped with tools to extract functional programs directly from proof terms (Schwichtenberg, 2006). The system is supported by automatic proof search and normalization by evaluation as an efficient term rewriting device. It uses minimal rather than classical or intuitionistic logic. Therefore, MINLOG can be extended to applications in intuitionistic as well as classical logic.

Example: Existence proof for a reversal list. We aim at a program to realize the existence of a list w with reverse order of a given list v (Schwichtenberg and Wainer, 2012, p.346). Obviously, this is a Π_2^0-statement of the kind "$\forall v \exists w$ (elements of list w have reverse order of list v)". For a convenient description of the program, we introduce some abbreviations:

Write vw for the result $v * w$ of appending the list w to the list v. Write vx for the result $v * x$ of appending the list w to the list v, vx for the result $v * x$: of appending the one element list x: to the list v. Write xv for the result $x :: v$ of writing one element x in front of a list v. The following formulas including a predicate variable R express how a computation of the reversed list should proceed:

Init Rev : $R(\text{nil}, \text{nil})$
GenRev : $\forall v, w, x(Rvw \to R(vx, xw))$

Proposition. $\forall v \in T \exists w \in T\, Rvw.$

Proof. The <u>informal existence proof</u> for the reversal list runs in the following way:

Induction on the length of v uses a lemma `ListInitLastNat` stating that every non-empty list can be written in the form vx. In the step case, the list is non-empty and can be written vx. Since v has smaller length, the induction hypothesis yields its reversal w. Then xw can be written.

For <u>automated program extraction</u>, we have to extract a normalized extracted term `neterm` from a formalization of the informal proof with variables f for unary functions on lists and p for pairs of lists and numbers, square brackets in [x] as notation for λ-abstraction λx:

```
[x0]
 (Rec nat => list nat => list nat) x0 ([v2](Nil nat))
 ([x2,f3,v4]
   [if v4
     (Nil nat)
     ([x5,v6][let p7 (cListInitLastNat v6 x5)
             (right p7::f3 left p7)])])
```

The constant `cListInitlastNat` denotes the content of the lemma.

In the step, the function defined recursively calls itself via f3. The underlying algorithm defines an auxiliary function g by

$$g(0, v) := \mathrm{nil}$$
$$g(n + 1, \mathrm{nil}) := \mathrm{nil}$$
$$g(n + 1, xv) := \mathrm{let}\ wy = xv\ \mathrm{in}\ y :: g(n, w)$$

and gives the result by applying g to $\mathrm{lh}(v)$ and v.

The term obtained by applying `neterm` to the length of a list and the list itself must be normalized via `nt` ("normalized term"). The result is yielded via pp ("pretty print"). Constants denote the computational content of lemmata that are "animated" (enabled) or "deanimated" (blocked). Id means the "identity lemma" Id: $P \to P$

with realizer of the form $\lambda fx.fx$:

```
(animate ''ListInitLastNat'')
(animate ''Id'')
(pp(nt(mk-term-in-app-form
      neterm (pt ''4'')(pt ''1::2::3::4:''))))
```

The returned list is the reversed list $4::3::2::1:$.   □

What is the practical use of soundness proofs and program extraction? In business, economy, and science, customers want software which solves a problem. Thus, they require a proof that it works. Suppliers answer with a proof of the existence of a solution to the specification of the problem. The proof has been automatically extracted from the formal specification of the problem by a proof mining software (e.g., MINLOG). But, the question arises whether the extraction mechanism of the proof is itself, in general, correct. The soundness theorem guarantees that every formal proof can be realized by a normalized extracted term.

For practical and theoretical reasons, it is useful to know the computational growth and bounds of functionals realizing existence proofs. Therefore, we will study the majorization of functionals and combine it with modified realization and program extraction (Kohlenbach, 2008, chapter 6).

Definition of majorization of functionals. $x^* \, maj_\rho \, x$ (x^* majorizes x) between functionals of type ρ is defined by induction on ρ:

$$x^* \, maj_0 \, x :\equiv x^* \geq_0 x$$

$$x^* \, maj_{\tau\rho} \, x :\equiv \forall y^*, y(y^* \, maj_\rho \, y \to x^* y^* \, maj_\tau \, xy).$$

In the formal system WE-HA$^\omega$ (weakly extensional) Heyting arithmetic with an extensionality axiom weakened to a QF rule of extensionality), we can prove a general majorization of functional terms:

Howard's majorization proposition. *For each closed term t^ρ of* WE-HA$^\omega$ *one can construct a closed term $t^{*\rho}$ of* WE-HA$^\omega$ *such that* WE-HA$^\omega \vdash t^* \, maj_\rho \, t$.

Proof (induction on the structure of t (Howard, 1973)). With this proposition, one can find a functional bound which is extractable from a given proof in the formal system $H^\omega := \text{E-HA}^\omega \pm \text{AC} \pm \text{IP}^\omega$.

Extraction theorem of functional bound. *Let s be a closed term, $A(x, y, z)$ a formula containing only x, y, z as free variables and $\deg(\tau) \leq 2$. Then the following rule holds:*

$$H^\omega \vdash \forall x^1 \forall y \leq_\rho sx \, \exists z^\tau A(x, y, z)$$
$$\Rightarrow H^\omega \vdash \forall x^1 \forall y \leq_\rho sx \, \exists z \leq_\tau txA(x, y, z),$$

where t is a suitable closed term which can be extracted from a given proof of the assumption.

Proof (by majorization theorem).

A corollary of the extraction theorem is the fan rule which is remarkable with respect to intuitionistic mathematics (Troelstra, 1977). We will come back to this topic later on.

Fan rule. Let A be a formula of language $\mathcal{L}(\text{WE-HA}^\omega)$ containing only free variables of levels ≤ 1. Then for H^ω the following rule holds:

$$H^\omega \vdash \forall y^1 \exists n^0 A(y, n)$$
$$\Rightarrow H^\omega \vdash \forall x^1 \exists m^0 \forall y \leq_1 x \exists n \leq_0 mA(y, n).$$

Now, we combine the majorization technique with functional interpretation by modified realizability which is called monotone modified realizability.

Definition of monotone modified realizability. Instead of modified realization with a closed term t of

$$\forall \underline{a}(\underline{t}\,\underline{a}\,mrA(\underline{a})),$$

monotone modified realization extracts a closed term t^* such that

$$\exists \underline{z}(\underline{t^*}\,maj\,\underline{z} \wedge \forall \underline{a}(\underline{z}\underline{a}\,mrA(\underline{a}))).$$

Similarly to modified realizability, one can prove a soundness theorem of monotone modified realizability in order to get a main theorem of program extraction, again. In the following, we use set Θ of sentences

$\exists \underline{v} \leq_\sigma \underline{r} B_{ef}(\underline{v})$ (where B_{ef} is $\exists$-free) for the formal system $\mathrm{H}^\omega :=$ E-HA$^\omega \pm$ AC $\pm$ IP$^\omega_{ef}$ $+ \Theta$ (Kohlenbach, 2008, 116 f.).

Soundness of monotone modified realizability. Let $A(\underline{a})$ be a formula in $\mathcal{L}(\text{E-HA}^\omega)$ containing only $\underline{a}$ free. The following rules hold

$$\mathrm{H}^\omega \vdash A(\underline{a}) \Rightarrow E\text{-HA}^\omega + \Theta \vdash \exists \underline{x}(t^* maj\ \underline{x} \wedge \forall \underline{a}(\underline{x}\underline{a}\, mr\, A(\underline{a}))),$$

where t^* is a suitable tuple of closed terms of E-HA$^\omega$ which can be extracted from a given proof of A.

For the following theorem of program extraction, we distinguish set Ω of sentences $\forall \underline{u}^\delta(C(u) \to \exists \underline{v} \leq_\sigma \underline{r}\,\underline{u}\neg B(\underline{u}, \underline{v}))$ with formulas B, C and set $\tilde{\Omega}$ containing the sentences $\exists \underline{V} \leq \underline{r}\forall \underline{u}(C(\underline{u}) \to \neg B(\underline{u}, \underline{V}\ \underline{u}))$. The corresponding formal system is $\mathrm{H}^\omega := $ E-HA$^\omega \pm$ $AC \pm$ IP$^\omega_\neg$.

Main theorem of uniform bound extraction by monotone modified realizability. *Let s be a closed term, $A(x^1, y^\rho, z^\tau)$ a formula containing only x, y, z as free variables and $\deg(\tau) \leq 2$. Then the following rule holds:*

$$\mathrm{H}^\omega + \Omega \vdash \forall x^1 \forall y \leq_\rho sx \exists z^\tau A(x, y, z)$$
$$\Rightarrow \mathrm{H}^\omega + \tilde{\Omega} \vdash \forall x^1 \forall y \leq_\rho sx \exists z \leq_\tau txA(x, y, z),$$

where t is a suitable closed term of E-HA$^\omega$ *which can be extracted from a given proof of the assumption.*

The previous functional interpretations have different disadvantages which are overcome by Gödel's functional ("Dialectia") interpretation. The no-counterexample interpretation needs formulas in prenex normal form. But logical transformation into a prenex normal form can, in general, only be applied in classical theories and not in intuitionistic ones. Further on, Gödel's functional interpretation is more convenient with respect to logical deduction rules (*modus ponens*) than no-counterexample interpretation. In contrast to the modified realizability interpretation, Gödel's functional interpretation also satisfies the Markov principle M^ω which is fundamental for mathematical applications in, e.g., real analysis.

Markov's principle M^ω : $\neg\neg\exists\underline{x}A_0(\underline{x}) \to \exists\underline{x}A_0(\underline{x})$ with A_0 QF.

There is also a form of M^ω with functional variables α:

$$\neg\neg\exists\underline{x}\alpha(\underline{x}) = 0 \to \exists\underline{x}\alpha(\underline{x}) = 0.$$

In constructive recursive mathematics, α is an algorithmic or recursive function and A_0 is a recursively decidable predicate. In this case, the Markov principle means: If there is an algorithm for testing A_0, and if we know by an indirect proof that the non-existence of an $\underline{x}$ must be refuted (i.e., we cannot avoid encountering $\underline{x}$ with $A_0(\underline{x})$), then, actually, we can find $\underline{x}$ with $A_0(\underline{x})$. The algorithm α is the procedure for finding $\underline{x}$. For an intuitionistic mathematician, the assumption $\neg\neg\exists\underline{x}\alpha(\underline{x}) = 0$ still provides no constructive procedure of finding a natural number n with $\alpha(n) = 0$. But in recursive mathematics, the acceptance of a proof $\neg\neg\exists\underline{x}\alpha(\underline{x}) = 0$ together with an algorithm α is sufficient for finding $\underline{x}$.

Gödel's historical motivation of his functional ("Dialectica") interpretation was a relative consistency proof for Heyting's intuitionistic arithmetic (Gödel, 1958). Peano's classical arithmetic can be embedded into the intuitionistic arithmetic system by negative translation which will be introduced later on. By that, consistency of Peano's arithmetic is also guaranteed.

Gödel's interpretation contains a formula translation and a proof translation. The formula translation describes how each statement A of Heyting arithmetic is mapped to a QF formula $A_D(\underline{x}, \underline{y})$ (with variables $\underline{x}, \underline{y}$ not free in A). Intuitively, A is interpreted by an existential statement $\exists\underline{x}\forall\underline{y}A_D(\underline{x}, \underline{y})$. The proof translation shows how a proof of A can be transformed into a closed term $\underline{t}$ and a proof of $A_D(\underline{t}, \underline{y})$.

Definition of Gödel's functional ("Dialectia") interpretation. To every formula A of $\mathcal{L}(\text{WE-HA}^\omega)$ a translation $A^D \equiv \exists\underline{x}\forall\underline{y}A_D(\underline{x}, \underline{y})$ is designed in the same language. The free variables of

A^D are that of A. The types and length of $\underline{x}, \underline{y}$ only depend on the logical structure of A. A_D is a QF formula:

(i) $A^D :\equiv A_D :\equiv A$ for prime formulas A.

Let $A^D \equiv \exists \underline{x} \forall \underline{y} A_D(\underline{x}, \underline{y})$ and $B^D \equiv \exists \underline{u} \forall \underline{v} B_D(\underline{u}, \underline{v})$. Then

(ii) $(A \wedge B)^D :\equiv \exists \underline{x}, \underline{u} \forall \underline{y}, \underline{v}[A \wedge B]_D$

$$:\equiv \exists \underline{x}, \underline{u} \forall \underline{y}, \underline{v}[A_D(\underline{x}, \underline{y}) \wedge B_D(\underline{u}, \underline{v})]$$

(iii) $(A \vee B)^D :\equiv \exists z^0, \underline{x}, \underline{u} \forall \underline{y}, \underline{v}[A \vee B]_D$

$$:\equiv \exists z^0, \underline{x}, \underline{u} \, \forall \underline{y}, \underline{v}[(z = 0 \to A_D(\underline{x}, \underline{y}))$$

$$\wedge(z \neq 0 \to B_D(\underline{u}, \underline{v}))]$$

(iv)　$(A \to B)^D :\equiv \exists \underline{U}, \underline{Y} \forall \underline{x}, \underline{v}(A \to B)_D$

$$:\equiv \exists \underline{U}, \underline{Y} \forall \underline{x}, \underline{v}(A_D(\underline{x}, \underline{Y}\underline{x}\underline{v}) \to B_D(\underline{U}\underline{x}, \underline{v}))$$

(v) $(\exists z^\rho A(z))^D :\equiv \exists z, \underline{x} \, \forall \underline{y}(\exists z A(z))_D :\equiv \exists z, \underline{x} \forall \underline{y} A_D(\underline{x}, \underline{y}, z)$

(vi) $(\forall z^\rho A(z))^D :\equiv \exists \underline{X} \forall z, \underline{y}(\forall z A(z))_D :\equiv \exists \underline{X} \forall z, \underline{y} A_D(\underline{X}z, \underline{y}, z)$

For proofs of soundness, characterization, and program extraction by Gödel's D (Dialectica)-interpretation, we need a scheme of premise independence of existential statements.

Definition of independence of premise-scheme $\mathrm{IP}_\forall^\omega$. $\mathrm{IP}_\forall^\omega$ for universal premises is the union (for all types) of

$$\mathrm{IP}_\forall^\rho : (\forall \underline{x} A_0(\underline{x}) \to \exists y^\rho B(y)) \to \exists y^\rho (\forall \underline{x} A_0(\underline{x}) \to B(y)).$$

Gödel's soundness theorem of functional (Dialectica) interpretation. *Let $\mathcal{P}$ be a set of purely universal sentences $\forall \underline{x}^\sigma B_0(\underline{x})$ (B_0 QF) of $\mathcal{L}(\mathrm{WE\text{-}HA}^\omega)$ and $A(\underline{a})$ be a formula of $\mathcal{L}(\mathrm{WE\text{-}HA}^\omega)$ containing only $\underline{a}$ free.*

Then the following rule holds:

$$\mathrm{WE\text{-}HA}^\omega + \mathrm{AC} + \mathrm{IP}_\forall^\omega + \mathrm{M}^\omega + \mathcal{P} \vdash A(\underline{a})$$

$$\Rightarrow \mathrm{WE\text{-}HA}^\omega + \mathcal{P} \vdash \forall \underline{y} A_D(\underline{t}\underline{a}, \underline{y}, \underline{a}),$$

where $\underline{t}$ is a suitable tuple of closed terms of $\mathrm{WE\text{-}HA}^\omega$ which can be extracted from a given proof of the assumption.

Proof (Induction on the length of derivation).

(1) <u>Logical axioms:</u>
For example, we prove the soundness of axiom $A \vee A \to A$:

We have to determine an appropriate term $\underline{t}$ with

$$\text{WE-HA}^\omega + \mathcal{P} \vdash \forall \underline{y}(A \vee A \to A)_D(\underline{ta}, \underline{y}, \underline{a})$$

under the condition of

$$\text{WE-HA}^\omega + \text{AC+IP}_\forall^\omega + \text{M}^\omega + \mathcal{P} \vdash (A \vee A \to A)(\underline{a}).$$

By Defs. (iii) and (iv), we get

$$(A \vee A \to A)^D$$
$$= \exists \underline{Y}, \underline{Y}', \underline{X}'' \forall z^0, \underline{x}, \underline{x}', \underline{y}''((z =_0 0 \to A_D(\underline{x}, \underline{Y}z\underline{x}\ \underline{x}'\underline{y}'', \underline{a}))$$
$$\wedge (z \neq_0 0 \to A_D(\underline{x}, \underline{Y}'z\underline{x}\ \underline{x}'\underline{y}'', \underline{a})) \to A_D(\underline{X}''z\underline{x}\ \underline{x}', \underline{y}'', \underline{a}))$$

Define

$$\underline{t}_{\underline{X}''} := \begin{cases} \lambda \underline{a}, z, \underline{x}, \underline{x}'.\underline{x}, & \text{if } z = 0, \\ \lambda \underline{a}, z, \underline{x}, \underline{x}'.\underline{x}', & \text{if } z \neq 0. \end{cases}$$

$$\underline{t}_{\underline{Y}} := \underline{t}_{\underline{Y}'} := \lambda \underline{a}, z, \underline{x}, \underline{x}', \underline{y}''.\underline{y}''.$$

The terms $\underline{t}_{\underline{Y}}$, $\underline{t}_{\underline{Y}'}$, and $\underline{t}_{\underline{X}''}$ realize the functional existence quantifiers $\exists \underline{Y}, \underline{Y}'\underline{X}''$ and satisfy Gödel's functional interpretation of the axiom.

(2) <u>Logical rules:</u>
For example, we prove the soundness of the *modus ponens* rule $\frac{A \quad A \to B}{B}$:

We assume

$$\forall \underline{y} A_D(\underline{t}_1\underline{a}, \underline{y}, \underline{a}) \quad \text{and} \quad \forall \underline{x}, \underline{v} A_D(\underline{x}, \underline{t}_2 \tilde{\underline{a}}\ \underline{x}\ \underline{v}, \underline{a}) \to B_D(\underline{t}_3 \tilde{\underline{a}}\ \underline{x}, \underline{v}, \underline{a}')$$

with

$$\tilde{\underline{a}} := \{\underline{a}\} \cup \{\underline{a}'\}.$$

We must find a term $\underline{t}_4$ satisfying $\forall \underline{v} B_D(\underline{t}_4 \underline{a}', \underline{v}, \underline{a}')$:

If we apply the second premise to $\underline{x} := \underline{t}_1 \underline{a}$, then we get

$$\forall \underline{v}(A_D(\underline{t}_1 \underline{a}, \underline{t}_2(\tilde{\underline{a}}, \underline{t}_1 \underline{a}, \underline{v}), \underline{a}) \to B_D(\underline{t}_3(\tilde{\underline{a}}, \underline{t}_1 \underline{a}), \underline{v}, \underline{a}')).$$

If we apply the first premise to $\underline{y} := \underline{t}_2(\tilde{\underline{a}}, \underline{t}_1 \underline{a}, \underline{v})$, then we get

$$\forall \underline{v} A_D(\underline{t}_1 \underline{a}, \underline{t}_2(\tilde{\underline{a}}, \underline{t}_1 \underline{a}, \underline{v}), \underline{a}).$$

It follows that $\forall \underline{v}\, B_D(\underline{t}_3(\tilde{\underline{a}}, \underline{t}_1 \underline{a}), \underline{v}, \underline{a}'))$.

Replace all variables a_i in $\underline{t}_3(\tilde{\underline{a}}, \underline{t}_1 \underline{a})$ which do not occur in $\underline{a}'$ by $\mathcal{O}$ of appropriate type. The result is denoted by $\underline{t}[\underline{a}']$. Obviously, $\underline{t}_4 := \lambda \underline{a}' . \underline{t}[\underline{a}']$ satisfies the functional interpretation of *modus ponens*.

(3) <u>Non-logical axioms:</u>

For example, we prove the soundness of the induction rule. We denote $B(y^0)^D \equiv \exists \underline{u} \forall \underline{v} B_D(\underline{u}, \underline{v}, y, \underline{a})$ and assume that the following condition is already proved:

$$\forall \underline{v} B_D(\underline{t}_1 \underline{a}, \underline{v}, 0, \underline{a}),$$
$$\forall \underline{u}, \underline{w}(B_D(\underline{u}, \underline{t}_2 y \underline{a}\ \underline{u}\ \underline{w}, y, \underline{a}) \to B_D(\underline{t}_3 y \underline{a}\ \underline{u}, \underline{w}, y+1, \underline{a})).$$

We define term $\underline{t}$ by primitive recursion in higher types with

$$\underline{t}(\underline{a}, 0) := \underline{t}_1 \underline{a},$$
$$\underline{t}(\underline{a}, y+1) := \underline{t}_3(y, \underline{a}, \underline{t}(\underline{a}, y)).$$

It follows that

$$\forall \underline{v} B_D(\underline{t}(\underline{a}, 0), \underline{v}, 0, \underline{a}),$$
$$\forall \underline{w}(B_D(\underline{t}(\underline{a}, y), \underline{t}_2(y, \underline{a}, \underline{t}(\underline{a}, y), \underline{w}), y, \underline{a})$$
$$\to B_D(\underline{t}(\underline{a}, y+1), \underline{w}, y+1, \underline{a})),$$

and hence

$$\forall \underline{v} B_D(\underline{t}(\underline{a}, 0), \underline{v}, 0, \underline{a}),$$
$$\forall \underline{v} B_D(\underline{t}(\underline{a}, y), \underline{v}, y, \underline{a}) \to \forall \underline{v} B_D(\underline{t}(\underline{a}, y+1), \underline{v}, y+1, \underline{a}).$$

The induction rule provides $\forall \underline{v} B_D(\underline{t}(\underline{a}, y), \underline{v}, y, \underline{a})$.

The functional interpretations of AC, M^ω and premise-independence principle $IP^\omega_\forall$ are examples of $(A \to A)^D$ which means that the functional interpretations of the premise and the conclusions are identical. M^ω needs the stability $(\neg\neg A \to A)^D$ of QF formulas. $\qquad\square$

The soundness theorem does only hold in a weakened version of extensional intuitionistic arithmetic WE-HA$^\omega$ with functional interpretation of the Markov principle. But the Markov principle is not accepted in intuitionistic arithmetic. Thus, the previous soundnesss theorem does not hold in E-HA$^\omega$.

If two formulas are equivalent in WE-HA$^\omega$ extended with the axiom of choice, Markov's principle and the premise-independence scheme, then their D-interpretation is also equivalent in WE-HA$^\omega$, i.e., formally for formulas A, B in language $\mathcal{L}(\text{WE-HA}^\omega)$ (Kohlenbach, 2008, p. 136):

$$\text{WE-HA}^\omega + \text{AC} + IP^\omega_\forall + M^\omega \vdash A \leftrightarrow B$$

$$\Rightarrow \text{WE-HA}^\omega \vdash A^D \leftrightarrow B^D.$$

Proof. We assume

$$
\begin{aligned}
&\text{WE-HA}^\omega + \text{AC} + IP^\omega_\forall + M^\omega \vdash A \leftrightarrow B \\
\Rightarrow\ &\text{WE-HA}^\omega \vdash (A \to B)^D \wedge (B \to A)^D && \text{soundness theorem} \\
&\text{WE-HA}^\omega \vdash (A \to B)^D \to (A^D \to B^D) \\
\Rightarrow\ &\text{WE-HA}^\omega \vdash (A^D \to B^D) \wedge (B^D \to A^D) \\
\Rightarrow\ &\text{WE-HA}^\omega \vdash A^D \leftrightarrow B^D && \square
\end{aligned}
$$

Truth of a statement is characterized by the existence of a functional realizer of the statement. That can be formally derived in the formal system of (weakened extensional) Heyting arithmetic WE-HA$^\omega$ extended by the axiom of choice, Markov's principle, and the premise-independence principle IP^ω_{ef}.

Characterization theorem of Gödel's functional interpretation. *For all formulas A of* WE-HA$^\omega$ *one has*

$$\text{WE-HA}^\omega + \text{AC} + IP^\omega_\forall + M^\omega \vdash A \leftrightarrow A^D.$$

Proof (induction on the logical structure of *A* (Yasugi, 1963; Troelstra, 1973)). The soundness and characterization theorems of D-interpretation allow us to extract an effective procedure (program) from a given proof which is the final goal of proof mining (Kohlenbach, 2008, p, 137).

Main theorem on program extraction of Gödel's functional interpretation. *Let $\mathcal{P}$ be a set of purely universal sentences $\forall \underline{a}^{\underline{\alpha}} P_0(\underline{a}) (P_0 \ QF)$ of $\mathcal{L}(\text{WE-HA}^\omega)$ and $A_0(x^\rho, u^\delta)$ be quantifier-free formula containing only x, u free and $B(x^\rho, y^\tau)$ an abitrary formula containing only x, y free and ρ, δ, τ are abitrary types. Then the following rule holds.*

$$\text{WE-HA}^\omega + \text{AC} + \text{IP}^\omega_\forall + \text{M}^\omega + \mathcal{P} \vdash \forall x^\rho (\forall u^\delta A_0(x, u) \to \exists y^\tau \, B(x, y))$$
$$\Rightarrow \text{WE-HA}^\omega + \text{AC} + \text{IP}^\omega_\forall + \text{M}^\omega + \mathcal{P} \vdash \forall x^\rho (\forall u^\delta A_0(x, u) \to B(x, tx)),$$

where t is a closed term of $\mathcal{L}(\text{WE-HA}^\omega)$ which is extracted from a proof of the premise by Gödel's functional interpretation.

Proof. We assume

$$\text{WE-HA}^\omega + \text{AC} + \text{IP}^\omega_\forall + \text{M}^\omega + \mathcal{P} \vdash \forall x^\rho (\forall u^\delta A_0(x, u) \to \exists y^\tau B(x, y))$$
$$\Rightarrow \text{WE-HA}^\omega + \text{AC} + \text{IP}^\omega_\forall + \text{M}^\omega + \mathcal{P} \vdash \forall x^\rho \exists y^\tau (\forall u^\delta A_0(x, u) \to B(x, y))$$
$$\text{IP}^\omega_\forall$$

We define

$$C(x, y) :\equiv \forall u^\delta A_0(x, u) \to B(x, y)$$

$\Rightarrow C^D(x, y) \equiv \exists \underline{a} \forall \underline{b} C_D(\underline{a}, \underline{b}, x, y).$	D-interpretation
$\Rightarrow (\forall x \exists y C(x, y))^D \equiv \exists Y, \underline{A} \forall x, \underline{b} \, C_D(\underline{A}x, \underline{b}, x, Yx)$	D-interpretation
$\Rightarrow \text{WE-HA}^\omega + \mathcal{P} \vdash \forall x, \underline{b} \, C_D(\underline{s}x, \underline{b}, x, tx)$	soundness
with closed terms $t, \underline{s}$	theorem
$\Rightarrow \text{WE-HA}^\omega + \mathcal{P} \vdash \forall x \exists \underline{a} \forall \underline{b} C_D(\underline{a}, \underline{b}, x, tx)$	
$\Rightarrow \text{WE-HA}^\omega + \mathcal{P} \vdash \forall x C^D(x, tx)$	definition of C^D
$\Rightarrow \text{WE-HA}^\omega + \text{AC} + \text{IP}^\omega_\forall + \text{M}^\omega + \mathcal{P}$	characterization
$\quad \vdash C^D(x, tx) \leftrightarrow C(x, tx)$	theorem

$$\Rightarrow \text{WE-HA}^\omega \text{AC} + \text{IP}_\forall^\omega + \text{M}^\omega + \mathcal{P} \vdash \forall x C(x, tx)$$

$$\Rightarrow \text{WE-HA}^\omega + \text{AC} + \text{IP}_\forall^\omega + \text{M}^\omega + \mathcal{P}$$

$$\vdash \forall x^\rho (\forall u^\delta A_0(x, u) \to B(x, tx)) \qquad \text{definition of } C \qquad \square$$

Example: Euclid's existence proof of the greatest common divisor.

Gödel's functional "Dialectica" interpretation can be used to extract functional programs from non-constructive proofs. For example, we consider Euclid's classical existence proof of the greatest common divisor with a quantifier-free kernel which does not contain an algorithm. The greatest common divisor (gcd) of two or more integers, when at least one of them is not zero, is the largest positive integer that divides the numbers without a remainder (Rem). In arithmetic, the remainder is the integer left over after dividing one integer by another to produce an integer quotient. In the case of gcd, the remainder is zero. (For example, the gcd of 4 and 16 is 4 with remainder 0). Thus, the gcd of the natural numbers a_1 and a_2 is a linear combination of the two numbers. Gödel's Dialectica (D) interpretation can be used to extract a formal term from the classical proof representing a computer program (Moschovakis, 1995). Extraction of programs from given proofs can be automated by software. An example of automatic program extraction is MINLOG which was already mentioned previously. For applying D-interpretation, we must analyze the informal existence proof in all details in order to apply formal program extraction (Schwichtenberg and Wainer, 2012, 389 f.).

Euclid's theorem. $\forall a_1 \forall a_2 > 0 \exists k_1, k_2,$

$$(0 < |k_1 a_1 - k_2 a_2| \land \text{Rem}(a_1, |k_1 a_1 - k_2 a_2|)$$

$$= \text{Rem}(a_2, |k_1 a_1 - k_2 a_2|) = 0).$$

Proof. Assume $0 < a_2$. Let $A(k_1, k_2) := (0 < |k_1 a_1 - k_2 a_2|)$.

There are k_1, k_2 with $A(k_1, k_2)$: take $k_1 := 0$, $k_2 := 1$.

The minimum principle for $A(k_1, k_2)$ with measure $|k_1 a_1 - k_2 a_2|$ provides k_1, k_2 with

$(*)$ $A(k_1, k_2)$
$(**)$ $\forall l_1, l_2(0 < |l_1 a_1 - l_2 a_2| < |k_1 a_1 - k_2 a_2| \to A(l_1, l_2) \to \bot)$.

We assume

$$\forall k_1, k_2(0 < |k_1 a_1 - k_2 a_2| \to \mathrm{Rem}(a_1, |k_1 a_1 - k_2 a_2|) = 0$$

$$\to \mathrm{Rem}(a_2, |k_1 a_1 - k_2 a_2|) = 0 \to \bot)$$

We must show $\bot$:

Since $(0 < |k_1 a_1 - k_2 a_2|)$ (because of $(*)$), it suffices to prove
$\mathrm{Rem}(a_i, |k_1 a_1 - k_2 a_2|) = 0 \ (i = 1, 2)$.
For $q := \mathrm{Quot}(a_1, |k_1 a_1 - k_2 a_2|)(\text{``quotient''})$
and $r := \mathrm{Rem}(a_1, |k_1 a_1 - k_2 a_2|)(\text{``remainder''})$,
we get $a_1 = q|k_1 a_1 - k_2 a_2| + r$ with $r < |k_1 a_1 - k_2 a_2|$.

With the step lemma (which is proved below), it follows that

$$r = |\mathrm{step}(a_1, a_2, k_1, k_2, q)a_1 - qk_2 a_2| < |k_1 a_1 - k_2 a_2|.$$

We abbreviate

$l_1 := \mathrm{step}\ (a_1, a_2, k_1, k_2, q),$
$l_2 := qk_2$
$\Rightarrow A(l_1, l_2) \to \bot$ $\qquad\quad$ $(**)$
$\Rightarrow 0 = |l_1 a_1 - l_2 a_2| = r.$

$\hfill \Box$

Step lemma of the informal existence proof of gcd.

$$a_1 = q|k_1 a_1 - k_2 a_2| + r \to r = |\mathrm{step}\ (a_1, a_2, k_1, k_2, q)a_1 - qk_2 a_2|$$

with step

$$(a_1, a_2, k_1, k_2, q) := \begin{cases} qk_1 - 1, & \text{if } k_2 a_2 < k_1 a_1 \text{ and } 0 < q, \\ qk_1 + 1, & \text{otherwise.} \end{cases}$$

Proof. We assume $0 < q$:

<u>First case</u>: $k_2 a_2 < k_1 a_1$

$$\Rightarrow a_1 = q(k_1 a_1 - k_2 a_2) + r$$

$$r = (1 - qk_1)a_1 + qk_2 a_2$$

$$= -(qk_1 - 1)a_1 + qk_2 a_2$$

$$= -\text{step}(a_1, a_2, k_1, k_2, q)a_1 + qk_2 a_2$$

$$= |\text{step}(a_1, a_2, k_1, k_2, q)a_1 - qk_2 a_2|.$$

<u>Second case</u>: $k_2 a_2 \geq k_1 a_1$

$$\Rightarrow a_1 = -q(k_1 a_1 - k_2 a_2) + r$$

$$r = (qk_1 + 1)a_1 - qk_2 a_2$$

$$= |\text{step}(a_1, a_2, k_1, k_2, q)a_1 - qk_2 a_2|.$$

For $q = 0$, we got $\text{step}(a_1, a_2, k_1, k_2, 0) = 1$ and the lemma is true with $|a_1| = r$. $\qquad\square$

Extracted term via functional ("Dialectica") interpretation:

The term etd is extracted via Gödel's functional ("Dialectica") interpretation from a formalization of the informal proof:

```
[n0,n1]
 [let pf 712
  ((Rec nat ⇒ nat©©nat ⇒ nat©©nat) ([p3]0©0)
   ([n3,pf4,p5]
    [if(0<Lin n0 n1 p5 impb
         Rem n0 (Lin n0 n1 p5)=0 impb
         Rem n1 (Lin n0 n1 p5)=0 impb False)
     (pf4
     [let p6
      (Step n0 n1 p5 (Quot n0 (Lin n0 n1 p5))©
                  Quot n0 (Lin n0 n1 p5)* right p5)
      [if (Lin n0 n1 p6< n3 impb 0<Lin n0 n1 p6 impb False)
        (Quot n1 (Lin n0 n1 p5)* left p5©
        Step n1 n0 (right p5©left p5)(Quot n1 (Lin n0 n1 p5)))
        p6]])
```

```
      [p5]))
n1)
[let p2
 [if (0<n1 impb Rem n0 n1=0 impb False)
  (pf712(Step n0 n1 (0©1)(Quot n0 n1)©Quot n0 n1))
  (0©1)]
 [if(0<Lin n0 n1 p2 impb
     Rem n0(Lin n0 n1 p2)=0 impb
     Rem n1(Lin n0 n1 p2)=0 impb False)
  (pf712(0©[if(0<n1)0 2]))
  p2]]]
```

Finally, application of term-to-expr to etd can be evaluated with
functions |Step| and |Lin| in the underlying programming language
(compare the symbolic of LISP):

```
define |Step|
  (lambda(a1)
   (lambda(a2)
    (lambda(p)
     (lambda(q)
      (if(and(<(*(cdr p)a2)(*(car p)a1))(< 0 q))
         (−(*q(car p))1)
         (+(*q(car p))1)))))))
(define |Lin|
  (lambda(a1)
   (lambda(a2)
    (lambda(p)
     n(abs(−(*(car p)a1)(*(cdr p)a2)))))))
```

The result for (((ev (term-to-expr etd))66)27) is (16 .
39), and |16 * 66−39 * 27 |=3 is the gcd of 27 and 66.

Applications to more ambitious mathematical proofs (e.g., in
numeric functional analysis) use a combination of functional interpretation with majorizability constructions which we already applied
previously. Analogously to the monotone modified realizability
interpretation, a monotone functional interpretation extracts terms

which majorize some functionals realizing the usual functional interpretation:

Definition of monotone functional ("Dialectica") interpretation MD. A suitable closed term t^* satisfies the monotone functional interpretation of A with

$$\exists x(t^* \ maj \ \underline{x} \wedge \forall \underline{a}, \underline{y} A_D(\underline{x}(\underline{a}), \underline{y}, \underline{a})).$$

In this case, term t^* is said to satisfy the monotone functional ("Dialectica") interpretation MD of A.

Again, we are interested in the soundness for the monotone functional interpretation MD.

For types and formulas, we use the following abbreviations:

$\underline{b} \leq_\sigma \underline{r} \ \underline{a} := \wedge_{i=1}^k (b_i \leq_{\sigma_i} r_i \underline{a})$
$\Delta :=$ set of all sentences $\forall \underline{a}^\delta \exists \underline{b} \leq_\sigma \underline{r} \underline{a} \forall \underline{c}^\gamma B_0(\underline{a}, \underline{b}, \underline{c})$ with B_0 QF
$\tilde{\Delta} :=$ set of the Skolem normal forms of the sentences in Δ
$\quad =$ set of all sentences $\exists \underline{B} \leq \underline{r} \forall \underline{a}, \underline{c} B_0(\underline{a}, \underline{B} \ \underline{a}, \underline{c})$ of $\forall \underline{a}^\delta \exists \underline{b} \leq_\sigma \underline{r} \ \underline{a}$
$\quad \quad \forall \underline{c}^\gamma B_0(\underline{a}, \underline{b}, \underline{c})$ in Δ.

Soundness theorem for monotone functional interpretation MD. *Let Δ be a set of sentences $\forall \underline{a}^\delta \exists \underline{b} <_\delta \underline{r} \underline{a} \forall \underline{c}^\gamma B_0(\underline{a}, \underline{b}, \underline{c})$ with B_0 QF and $\tilde{\Delta}$ the corresponding set of Skolem normal forms $\exists \underline{B} \leq \underline{r} \forall \underline{a}, \underline{c} B_0(\underline{a}, \underline{B} \ \underline{a}, \underline{c})$.*

The following rule holds:

$$\text{WE-HA}^\omega + \text{AC} + \text{IP}_\forall^\omega + \text{M}^\omega + \Delta \vdash A(\underline{a})$$
$$\Rightarrow \text{WE-HA}^\omega + \tilde{\Delta} \vdash \exists x(t^* maj \ \underline{x} \wedge \forall \underline{a}, \underline{y} A_D(\underline{x}\underline{a}, \underline{y}, \underline{a}))$$

With t^* suitable tuple of closed terms of WE-HA^ω which can be extracted from a given proof of the assumption.

$\underline{A}(\underline{a})$ is a formula of $\mathcal{L}(\text{WE-HA}^\omega)$ containing only $\underline{a}$ free.

The proof is by induction on the length of the proof and similar to previous soundness proofs with majorization arguments (Howard, 1973; Kohlenbach, 2008, 142 f.).

The soundness theorem of MD-interpretation allows a bound extraction from a given proof.

Theorem of bound extraction by monotone functional interpretation MD. *Let $A(x^1, y^\rho, z^\tau)$ be an any formula containing only x, y, z free, $\deg(\tau) \leq 2$, and $s^{\rho(1)}$ a closed term of WE-HA$^\omega$. The following rule holds:*

$$\text{WE-HA}^\omega + \text{AC} + \text{IP}^\omega_\forall + \text{M}^\omega + \Delta \vdash \forall x^1 \forall y \leq_\rho sx \exists z^\tau A(x,y,z)$$
$$\Rightarrow \text{WE-HA}^\omega + \text{AC} + \text{IP}^\omega_\forall + \text{M}^\omega + \Delta \vdash \forall x^1 \forall y \leq_\rho sx \exists z^\tau \leq_\tau tx A(x,y,z)$$

with t as a suitable closed term of WE-HA$^\omega$ which can be extracted from a given proof of the assumption.

Proof. For $\tau = 2$ (without any loss of generality), we assume

$$\text{WE-HA}^\omega + \text{AC} + \text{IP}^\omega_\forall + \text{M}^\omega + \Delta \vdash \forall x \forall y \leq_\rho sx \exists z A(x,y,z)$$
$$\Rightarrow \text{WE-HA}^\omega + \text{AC} + \text{IP}^\omega_\forall + \text{M}^\omega + \Delta \vdash \forall x, y \exists z (y \leq sx \rightarrow A(x,y,z))$$
$$\text{IP}^\omega_\forall$$
$$\Rightarrow \text{WE-HA}^\omega + \text{AC} + \text{IP}^\omega_\forall + \text{M}^\omega + \tilde{\Delta} \vdash \exists Z(t^* \text{maj } Z \wedge \forall x, y(y \leq sx \rightarrow$$
$$A(x,\underline{y},Zxy))^D) \qquad \text{soundness theorem}$$
$$\text{WE-HA}^\omega + \text{AC} + \text{IP}^\omega_\forall + \text{M}^\omega \vdash G^D \leftrightarrow G$$
$$\text{characterization theorem (for all } G)$$
$$\Rightarrow \text{WE-HA}^\omega + \text{AC} + \text{IP}^\omega_\forall + \text{M}^\omega + \tilde{\Delta} \vdash \exists Z \forall x \forall y \leq sx$$
$$(\lambda w^1 \cdot t^* x^M (s^* x^M) w^M \geq_2 Zxy \wedge A(x,y,Zxy))$$
$$\text{extraction theorem of functional bound}$$
$$\text{We define } tx := \lambda w^1 . t^* x^M (s^* x^M) w^M$$
$$\Rightarrow \text{WE-HA}^\omega + \text{AC} + \text{IP}^\omega_\forall + \text{M}^\omega + \tilde{\Delta} \vdash \exists Z \forall x \forall y \leq sx (Zxy \leq_2 tx$$
$$\wedge A(x,y,Zxy))$$
$$\Rightarrow \text{WE-HA}^\omega + \text{AC} + \text{IP}^\omega_\forall + \text{M}^\omega + \tilde{\Delta} \vdash \forall x \forall y \leq sx \exists z \leq_2 tx A(x,y,z)$$
$$\text{WE-HA}^\omega + \text{AC} + \text{IP}^\omega_\forall + \text{M}^\omega + \Delta \vdash \tilde{\Delta}$$
$$\text{AC}$$
$$\Rightarrow \text{WE-HA}^\omega + \text{AC} + \text{IP}^\omega_\forall + \text{M}^\omega + \Delta \vdash \forall x \forall y \leq sx \exists z \leq_2 tx A(x,y,z)$$

$\square$

Example of application. Proof mining aims at the extraction of effective procedures or, at least, bounds from given proofs. We consider different interpretations and extraction theorems which are suitable for different mathematical principles. A remarkable example is the application of monotone functional interpretation to weak König's lemma (WKL) (Troelstra, 1974). A weak form of König's lemma states that every infinite binary tree has an infinite branch. Several theorems of analysis turn out to be equivalent to WKL with respect to the second order fragment RCA_0 of recursive arithmetic. Therefore, WKL is important in the foundational program of reverse mathematics which we will consider in Chapter 8.

Weak König's lemma.

$$\forall f^1(T(f) \wedge \forall x^0 \exists n^0(lth\, n = x \wedge fn = 0)$$
$$\rightarrow \exists b \leq_1 \lambda k.1 \forall x^0(f(\bar{b}x) = 0)),$$

where $T(f) :\equiv \forall n, m(f(n * m) = 0 \rightarrow fn = 0) \wedge \forall n, x(f(n * \langle x \rangle) = 0 \rightarrow x \leq 1)$. $T(f)$ *asserts that f represents a $0, 1$ tree.*

The full König's lemma (KL) assumes an infinitely branching tree instead of a binary one. In reverse mathematics, it turns out to be equivalent to the arithmetical comprehension axiom which founded classical analysis in the tradition of Hermann Weyl and Paul Lorenzen (cf. Chapter 8). Thus, from a classical point of view, KL has a much higher potentiality of proofs than WKL. WKL can be derived from KL if we assume the AC.

König's lemma.

$$\forall f^1(\tilde{T}(f) \wedge \forall x^0 \exists n^0(lth\, n = x \wedge fn = 0) \rightarrow \exists b^1 \forall x^0(f(\bar{b}x) = 0)),$$

where $\tilde{T}(f) :\equiv \forall n, m(f(n * m) = 0 \rightarrow fn = 0) \wedge \forall n \exists m \forall x(f(n * \langle x \rangle) = 0 \rightarrow x \leq m)$.

The proof in second-order arithmetic is not accepted in intuitionistic mathematics, because it uses arguments by contradiction to guarantee that there exists an adjacent vertex followed by infinitely many other vertices. Recursively computable arithmetic extended by

WKL allows us to derive many classical theorems of mathematics and conversely many of those theorems imply WKL. Although WKL is an ineffective principle, monotone functional interpretation can be used to extract bounds for arbitrary formulas. Actually, it can be proven that WKL is an example of an axiom of the form of sentences in set Δ which was defined previously.

Another interpretation of proof mining combines Gödel's functional ("Dialectica") interpretation with the so-called negative translation (Gödel, 1933) which is abbreviated by ND. Negative or double-negation interpretation allows to embed formulas A of classical logic and theories into their intuitionistic variant A'. All these translations $A \mapsto A'$ state that A' is (or is intuitionistically equivalent to) a negative formula (Kuroda, 1951).

Definition of negative translation. Let A be a formula in a theory based on the language $\mathcal{L}(\mathrm{IL}^{\omega}_{-=})$ of the many-sorted version $\mathrm{IL}^{\omega}_{-=}$ of intuitionistic logic $\mathrm{IT}_{-=}$(without equality $=$). A' is defined by induction on the logical structure of A:

(i) $A^* :\equiv A$, if A is a prime formula.
(ii) $(A\Box B)^* :\equiv (A^*\Box B^*)$ with $\Box \in \{\wedge, \vee, \rightarrow t\}$.
(iii) $(\exists x^\rho A)^* :\equiv \exists x^\rho A^*$.
(iv) $(\forall x^\rho A)^* :\equiv \forall x^\rho \neg\neg A^*$.

For soundness of interpretation ND, program extraction by ND, and characterization by ND, we need a restriction scheme of AC to QF formulas (Kohlenbach, 2008, 165 f.).

Axiom of choice restricted to quantifier-free formulas QF-AC.

$\forall \underline{x}\exists \underline{y}A_0(\underline{x}, \underline{y}) \rightarrow \exists \underline{Y}\forall \underline{x}A_0(\underline{x}, \underline{Y}\underline{x})$ with A_0 QF and $\underline{x}, \underline{y}$ tuples of variables of arbitrary types.

In the following, WE-PA$^\omega$ means a WE variant of m-sorted Peano arithmetic E-PA$^\omega$ where the extensionality axioms are restricted to a QF rule of extensionality.

ND-soundness theorem. Let $\mathcal{P}$ be an arbitrary set of universal sentences $\forall \underline{z}^{\sigma} B_0(\underline{z})$ with B_0 QF of $\mathcal{L}(\text{WE-PA}^{\omega})$ containing only $\underline{a}$ free. Then the following rule holds:

$$\text{WE-PA}^{\omega} + \text{QF-AC} + \mathcal{P} \vdash A(\underline{a})$$
$$\Rightarrow \text{WE-HA}^{\omega} + \mathcal{P} \vdash \forall \underline{y}(A')_D(\underline{ta}, \underline{y}, \underline{a})$$

with extracted closed terms $\underline{t}$ of WE-HA^{ω}.

ND-program extraction theorem. Let $\mathcal{P}$ be an arbitrary set of universal sentences $\forall \underline{z}^{\sigma} B_0(\underline{z})$ with B_0 QF of $\mathcal{L}(\text{WE-PA}^{\omega})$ and $A_0(x^{\rho}, y^{\tau})$ be a (QF) formula of $\mathcal{L}(\text{WE-PA}^{\omega})$ which only contains x^{ρ}, y^{τ} as free variables. Then it holds:

$$\text{WE-PA}^{\omega} + \text{QF-AC} + \mathcal{P} \vdash \forall x^{\rho} \exists y^{\tau} A_0(x, y)$$
$$\Rightarrow \text{WE-HA}^{\omega} + \mathcal{P} \vdash \forall x A_0(x, tx)$$

with extracted closed term t of WE-HA^{ω}.

The following characterization theorem of ND was proved in Kreisel (1959).

ND-characterization theorem. Let A be a formula of language $\mathcal{L}(\text{WE-PA}^{\omega})$. Then

$$\text{WE-PA}^{\omega} + \text{QF-AC} \vdash A \leftrightarrow (A')^D.$$

We also mention the combination of negative translation with monotone functional interpretation which is abbreviated by NMD (Kohlenbach, 2008, p. 174). As previously, Δ denotes a set of sentences $\forall \underline{a}^{\delta} \exists \underline{b} <_{\sigma} \underline{ra} \forall \underline{c}^{\gamma} B_0(\underline{a}, \underline{b}, \underline{c})$ and $\tilde{\Delta}$ the corresponding set of their Skolem normal forms $\exists \underline{B} \leq \underline{r} \forall \underline{a}, \underline{c} B_0(\underline{a}, \underline{Ba}, \underline{c})$.

NMD-soundness theorem.

$$\text{WE-PA}^{\omega} + \text{QF-AC} + \Delta \vdash A(\underline{a})$$
$$\Rightarrow \text{WE-HA}^{\omega} + \tilde{\Delta} \vdash \exists \underline{x}(\underline{t}^* \ maj \ \underline{x} \wedge \forall \underline{a}, \underline{y}(A')_D(\underline{xa}, \underline{y}, \underline{a}))$$

with t^* as suitable tuple of closed terms of WE-HA^{ω} which can be extracted by NMD from a given proof of the assumption.

As in the previous case of majorization, we can get a uniform bound extraction by NMD.

Theorem of bound extraction by monotone functional interpretation combined with negative translation NMD. *Let $A_0(x^1, y^\rho, z^\tau)$ be a QF formula of $\mathcal{L}(\text{WE-PA}^\omega)$ containing only x, y, z as free variables, $\deg(\tau) \leq 2$, and s be a closed term. The following rule holds*:

$$\text{WE-PA}^\omega + \text{QF-AC} + \Delta \vdash \forall x^1 \forall y \leq_\rho sx \exists z^\tau A_0(x, y, z)$$
$$\Rightarrow \text{WE-HA}^\omega + \tilde{\Delta} \vdash \forall x^1 \forall y \leq_\rho sx \exists z^\tau \leq_\tau tx A_0(x, y, z)$$

with t a suitable closed term of WE-HA^ω *which can be extracted from a given proof of the assumption.*

Example of application. We remember the application of monotone functional interpretation MD to WKL. Analogously, there is a remarkable application of monotone functional interpretation combined with negative translation NMD to the uniform weak König's lemma (UWKL). In this case, we consider an infinite binary tree f which is defined by

$$T^\infty(f) :\equiv T(f) \wedge \forall x^0 \exists n^0 (lthn = x \wedge fn = 0)$$
$$\Rightarrow \text{WKL} \equiv \forall f^1 (T^\infty(f) \to \exists b \leq_1 \lambda k.1 \forall x^0 (f(\overline{b}x) = 0)).$$

Uniform weak König's lemma.

$$\exists \Phi \leq_{1(1)} 1 \forall f^1 (T^\infty(f) \to \forall x^0 (f((\overline{\Phi f})x) = 0))$$

Analogously, it can be proven that UWKL is an example of an axiom of the form of sentences in set Δ which was defined previously. Over the weakly extensional systems WE-PA^ω, UWKL is as weak as WKL with respect to the recursive functionals of type ≤ 2. But the situation changes over the extensional intuitionistic systems E-HA^ω. In this case, UWKL is stronger than WKL.

Definition of ENMD-interpretation. NMD-interpretation can be combined with the elimination of the extensionality axiom (E) from proofs of theorems with appropriate type restrictions of variables. This interpretation is called ENMD. In this case, quantifiers are restricted to the so-called hereditarily extensional functionals x^ρ (Luckhardt, 1973):

Definition of hereditarily extensional equality. Between functionals x_1^ρ, x_2^ρ of type ρ the following relation holds by induction on ρ:

$$x_1 \approx_0 x_2 :\equiv (x_1 =_0 x_2)$$

$$x_1 \approx_{\tau\rho} x_2 :\equiv \forall y_1^\rho, y_2^\rho (y_1 \approx_\rho y_2 \to x_1 y_1 \approx_\tau x_2 y_2)$$

A strong variant of $\approx_\rho$ ensures that $x \approx y$ implies $x \approx x$:

$$x =_0^e y :\equiv (x =_0 y)$$

$$x =_{\tau\rho}^e y :\equiv \forall u^\rho, v^\rho (u =_\rho^e v \to xu =_\tau^e xv \wedge xu =_\tau^e yv)$$

By a well-defined translation A_e, all quantifiers can be relativized to hereditarily extensional functionals with respect to $=_\rho^e$.

Relativization of all quantifier to hereditarily extensional functionals.

(i) $A_e :\equiv A$, if A is a prime formula.
(ii) $(A \Box B)_e :\equiv (A_e \Box B_e)$, where $\Box \in \{\wedge, \vee, \to\}$.
(iii) $(\exists x^\rho A)_e :\equiv \exists x^\rho (x =_\rho^e x \wedge A_e)$.
(iv) $(\forall x^\rho A)_e :\equiv \forall x^\rho (x =_\rho^e x \to A_e)$.

Combination of elimination of extensionality, negative translation and monotone functional interpretation (ENMD). Let Δ be a set of sentences $\forall \underline{a}^\delta \exists \underline{b} \leq_\sigma \underline{r}\underline{a} \forall \underline{c}^\gamma B_0(\underline{a}, \underline{b}, \underline{c})$ with $\deg(\underline{\sigma}) \leq 1$ and $\deg(\gamma) \leq 2$. $\tilde{\Delta}_\varepsilon$ denotes the ε-weakining of the Skolem normal form of $\tilde{\Delta}$ with $\forall c^\gamma \exists \underline{B} \leq_{\sigma\delta} \underline{r} \forall \underline{a} \forall \tilde{\underline{c}} \leq_\gamma \underline{c} B_0(\underline{a}, \underline{B}\underline{a}, \tilde{\underline{c}})$. Then, we unwind a uniform bound from a given proof (Kohlenbach, 2008, p. 184).

Theorem on uniform bound extraction by ENMD-Interpretation. *Let $A_0(x^1, y^1, z^\tau)$ be a QF formula of $\mathcal{L}(\text{E-PA}^\omega)$ containing only x, y, z as free variables, $\deg(\tau) \leq 2$ and s be a closed* *term. The following rule holds*:

$$\text{E-PA}^\omega + \text{QF-AC}^{1,0} + \text{QF-AC}^{0,1} + \Delta + \text{WKL}$$
$$\vdash \forall x^1 \forall y \leq_1 sx \exists z^\tau A_0(x, y, z)$$
$$\Rightarrow \text{WE-HA}^\omega + \tilde{\Delta}_e \vdash \forall x^1 \forall y \leq_1 sx \exists u \leq_\tau tx A_0(x, y, z),$$

with t as closed term of WE-HA$^\omega$ *which can be extracted from a given proof of the assumption.*

A really fundamental result was Spector's functional interpretation of full classical analysis (Spector, 1962). For that, we consider the following principle.

Comprehension principle over numbers CA^0.
$\exists f^1 \forall x^0 (f(x) =_0 0 \leftrightarrow A(x))$ for any formula $A(x)$ of $\mathcal{L}(\text{E-PA}^\omega)$ (not containing f free).

Function f can be interpreted as the characteristic function of a set indicating if x is an element of the set defined by formula A. Formula $A(x)$ must not contain f in order to prevent impredicative conclusions and contradictions. Spector proved that the negative translation of CA^0 can be functionally interpreted by the so-called bar recursive functionals. Together with the soundness of functional D-interpretation, this yields (via negative translation) a functional interpretation of axiom system WE-PA$^\omega$ + QF-AC + CA^0 and (with elimination of extensionality) of E-PA$^\omega$+QF-AC^{1,0}+QF-AC^{0,1}+CA^0 which contains full second order arithmetic and allows to formalize most parts of classical analysis.

With classical logic and QF-AC, it can be proved that comprehension principle CA^0 is equivalent to the following axiom of choice.

Axiom of choice AC^{0,0}. $\forall x^0 \exists y^0 A(x, y) \rightarrow \exists f^1 \forall x^0 A(x, f(x))$ for any formula $A(x, y)$ (not containing f free)

Proposition. *In* WE-PA$^\omega$+QF-AC^{1,0}, *the comprehension principle* CA^0 *is equivalent to the axiom of choice* AC^{0,0}.

Proof.

$$
\begin{array}{lll}
\text{``}\Rightarrow\text{''}: & \mathrm{CA}^0 & \text{assumption} \\
& \Rightarrow \forall x^0, y^0(g(x,y) =_0 0 \leftrightarrow A(x,y)) & \text{with function } g \\
& \forall x \exists y A(xy) & \text{assumption of} \\
& & \mathrm{AC}^{0,0} \text{ for any } A \\
\\
& \Rightarrow \forall x \exists y(g(x,y) = 0) & \\
& \Rightarrow \exists f^1 \forall x^0 g(x, f(x)) = 0 & \text{QF-AC}^{1,0} \\
& \Rightarrow \exists f^1 \forall x^0 A(x, f(x)). & \\
\text{``}\Leftarrow\text{''} & 0 \neq 1 & \\
& \Rightarrow \forall x^0 \exists n^0(n =_0 0 \leftrightarrow A(x)) & \text{PEM} \\
& \Rightarrow \exists f^1 \forall x^0(f(x) =_0 0 \leftrightarrow A(x)) & \mathrm{AC}^{0,0} \qquad \square
\end{array}
$$

A stronger statement than $\mathrm{AC}^{0,0}$ is the following.

Axiom of countable choice $\mathbf{AC}^0$. $\mathrm{AC}^0 := \bigcup_\rho \{\mathrm{AC}^{0,\rho}\}$ with $\mathrm{AC}^{0,\rho} :\equiv \forall x^0 \exists y^\rho A(x,y) \to \exists f^{\rho(0)} \forall x^0 A(x, f(x))$.

Spector even proved that the negative translation of AC^0 has a functional interpretation in the bar recursion functionals. According to the previous definition of negative translation, the negative translation $(\mathrm{AC}^0)'$ of AC^0 is equivalent to

$$
\forall x^0 \neg\neg\exists y^\rho A^*(x,y) \to \neg\neg\exists f^{\rho(0)} \forall x^0 \neg\neg A^*(x, f(x)),
$$

which can be proven in WE-HA$^\omega$ + AC^0 + DNS with the (intuitionistically underivable) double-negation shift (DNS) schema

$$
\mathrm{DNS} :\equiv \forall x^0 \neg\neg A(x) \to \neg\neg\forall x^0 A(x).
$$

Therefore, the functional interpretation of $(\mathrm{AC}^0)'$ reduces to the functional interpretation DNS^D of DNS.

Functional interpretation $\mathbf{DNS}^D$ of DNS. Functional D-interpretation of $A(x)$ is $\exists \underline{a} \forall \underline{b} A_D(x, \underline{a}, \underline{b})$:

$$
\begin{aligned}
(\mathrm{DNS})^D &\equiv (\forall x^0 \neg\neg\exists \underline{a} \forall \underline{b} A_D(x, \underline{a}, \underline{b}) \to \neg\neg\forall y^0 \exists \underline{u} \forall \underline{v} A_D(y, \underline{u}, \underline{v})^D, \\
&\equiv (\forall x^0 \exists \underline{A} \forall \underline{B} \neg\neg A_D(x\underline{AB}, \underline{B}(\underline{AB}) \\
&\qquad \to \neg\neg\exists \underline{u} \forall y^0, v A_D(y, uy, v))^D,
\end{aligned}
$$

$$\equiv (\exists \underline{A} \forall x^0, \underline{B} \neg\neg A_D(x, \underline{A}x\underline{B}, B(AxB)) \to \exists \underline{U} \forall Y, \underline{V}$$
$$\neg\neg A_D(Y(\underline{U}Y\underline{V}), \underline{U}Y\underline{V}(Y(\underline{U}Y\underline{V})), \underline{V}(\underline{U}Y\underline{V})))^D,$$
$$\equiv (\forall \underline{A} \exists \underline{U} \forall Y, \underline{V} \exists x, \underline{B}(\neg\neg A_D(x, \underline{A}x\underline{B}, \underline{B}(\underline{A}x\underline{B}))$$
$$\to \neg\neg A_D(Y(\underline{U}Y\underline{V}), \underline{U}Y\underline{V}(Y(\underline{U}Y\underline{V})), \underline{V}(\underline{U}Y\underline{V}))))^D.$$

Solution of functional interpretation DNS^D of DNS. According to (DNS)^D, one has to construct terms $t_x, t_{\underline{U}}$, and $t_{\underline{B}}$ (containing A, Y, V free) satisfying

$$\forall \underline{A}, Y, \underline{V}(\neg\neg A_D(t_x, A(t_x, t_{\underline{B}}), t_{\underline{B}}(A(t_x, t_{\underline{B}})))$$
$$\to \neg\neg A_D(Y(t_{\underline{U}}), t_{\underline{U}}(Y(t_{\underline{U}})), V(t_{\underline{U}}))).$$

These terms are determined by the following system of equations

$$t_x = Y(t_{\underline{U}}),$$
$$\underline{A}(t_{\underline{U}}, t_{\underline{B}}) = t_{\underline{U}}(Y(t_{\underline{U}})),$$
$$t_{\underline{B}}(\underline{A}(t_x, t_{\underline{B}})) = \underline{V}(t_{\underline{U}}).$$

These equations are solved by bar recursive functionals which are formalized by new constants $\underline{B}^{\rho,\tau}$ of WE-HA$^\omega$. They are defined by the following.

Axioms of bar recursion (BR$_{\rho,\tau}$).

$$y(\overline{\underline{x}n^0}) <_0 n \to B_i^{\rho,\tau}\underline{y}\underline{z}\ \underline{u}n\underline{x} =_{\tau_i} z_i n(\overline{\underline{x},n})$$
$$y(\overline{\underline{x},n}) \geq_0 n \to B_i^{\rho,\tau}\underline{y}\underline{z}\ \underline{u}n\underline{x} =_{\tau_i} u_i(\lambda D^{\underline{\rho}}.B^{\rho,\tau}\underline{y}\underline{z}\ \underline{u}(n+1)$$
$$(\overline{\underline{x},n} * D))n(\overline{\underline{x},n})$$

for $i = 1, \ldots, k$ with

$$(\overline{\underline{x},n})_j(k^0) =_{\rho_j} \begin{cases} x_j(k), & \text{if } k < n, \\ 0^{\rho_j}, & \text{otherwise.} \end{cases}$$

$$(\overline{\underline{x},n} * \underline{D})_j(k^0) =_{\rho_j} \begin{cases} x_j(k), & \text{if } k < n, \\ D_j, & \text{if } k = n, \\ 0^{\rho_j}, & \text{otherwise.} \end{cases}$$

$\mathrm{BR} := \bigcup_{\underline{\rho},\underline{\tau}\epsilon\mathrm{T}}\{\mathrm{BR}_{\underline{\rho},\underline{\tau}}\}$ is called bar recursion. Proof mining aims at the soundness of functional interpretation of the axiom of countable choice AC^0 with Bar recursion BR.

Soundness of functional interpretation of AC^0 with BR. *Let $A(\underline{a})$ be an arbitrary formula of $\mathcal{L}(\text{WE-PA}^\omega)$ containing only the free variables $\underline{a}$. Then the following rule holds:*

$$\text{WE-PA}^\omega + \text{QF-AC} + \text{AC}^0 \vdash A(\underline{a})$$
$$\Rightarrow \ \text{WE-HA}^\omega + \text{BR} \vdash \forall \underline{y}(A')_D(\underline{ta}, \underline{y}, \underline{a})$$

with $\underline{t}$ as suitable tuple of closed terms of $\text{WE-HA}^\omega + \text{BR}$ which can be extracted from a given proof of the assumption and A' denotes the negative translation of A.

Bar recursion can also be used for the functional interpretation of the negative translation of the axiom of dependent choice.

Axiom of dependent choice DC. $\mathrm{DC} := \bigcup_{\underline{\rho}}\{\mathrm{DC}^{\underline{\rho}}\}$ with

$$\mathrm{DC}^{\underline{\rho}} :\equiv \forall x^0, \underline{y}^{\underline{\rho}} \exists \underline{z}^{\underline{\rho}} A(x, \underline{y}, \underline{z}) \to \exists \underline{f}^{\underline{\rho}(0)} \forall x^0 A(x, \underline{f}(x), \underline{f}(S(x))).$$

Again, a soundness theorem can be proved for the functional interpretation with bar recursion (Howard, 1968; Luckhardt, 1973).

Soundness theorem with BR. Let $A(\underline{a})$ be an arbitrary formula of $\mathcal{L}(\text{WE-PA}^\omega)$ containing only the free variables $\underline{a}$. Then the following rule holds:

$$\text{WE-PA}^\omega + \text{QF-DC} + \text{AC}^0 \vdash A(\underline{a})$$
$$\Rightarrow \ \text{WE-HA}^\omega + \text{BR} \vdash \forall \underline{y}(A')_D(\underline{t}, \underline{a}, \underline{y}, \underline{a})$$

with $\underline{t}$ as suitable tuple of closed term of $\text{WE-HA}^\omega + \text{BR}$ which can be extracted from a given proof of the assumption, and A' denotes the negative translation of A.

Another important example of functional interpretation is the application to the axiom of arithmetical comprehension (Kohlenbach, 2008, chapter 11.3). Actually, it is a special case of full comprehension over numbers.

Axiom of arithmetical comprehension $\mathbf{CA}^0_{ar}$.

$\exists f^1 \forall x^0(f(x) =_0 0 \leftrightarrow A_{ar}(x))$ with arithmetical formula $A_{ar}(x)$ of $\mathcal{L}(\text{WE-PA}^\omega)$, i.e., $A_{ar} \in \Pi^0_n$ (arithmetical hierarchy) with only quantifiers for variables of type 0, but parameters of higher types.

The axiom of arithmetical comprehension CA^0_{ar} is remarkable because it allows to prove an important part of classical mathematics. Terms and formulas are defined by inductive definitions of terms and formulas over numbers in the arithmetical hierarchy. Therefore, impredicative concepts and proofs are avoided. In 1918, Hermann Weyl introduced basic ideas of this approach as constructive answer to Brouwer's intuitionistic critics of classical mathematics (cf. Chapter 6). In his book *Differential und Integral* (1965), Lorenzen elaborated Weyl's approach up to classical analysis.

A weaker form of CA^0_{ar} seems to be the following.

Axiom of arithmetical comprehension $\mathbf{\Pi^0_1}$-CA.

$$\forall f^{0(0)(0)} \exists g^1 \forall x^0(g(x) =_0 0 \leftrightarrow \forall y^0(f(x,y) =_0 0))$$

But, actually, CA^0_{ar} and Π^0_1-CA are equivalent in WE-PA$^\omega$. The proof direction "$\text{CA}^0_{ar} \Rightarrow \Pi^0_1$-CA" is obvious because Π^0_1-CA is a special case of CA^0_{ar}. For the converse case "Π^0_1-CA $\Rightarrow \text{CA}^0_{ar}$", one has to prove the comprehension principle for all $A_{ar} \in \Pi^0_n$. By iterated application of Π^0_1-CA, one can climb up the steps of arithmetical hierarchy to prove CA^0_{ar} for all arithmetical predicates.

Just as in the case of full comprehension, the axiom of arithmetical comprehension Π^0_1-CA (and hence CA^0_{ar}) is equivalent to a special scheme of choice.

Axiom of choice $\mathbf{\Pi^0_1}$-AC.

$$\forall f^{0(0)(0)(0)}(\forall x^0 \forall y^0 \forall z^0 f(x,y,z) =_0 0 \rightarrow \exists g^1 \forall x^0, z^0 f(x,g(x),z) =_0 0).$$

With the axiom of choice Π^0_1-AC, we can prove the principle of number choice in WE-PA$^\omega$:

Axiom of number choice $\mathbf{AC}_{ar}^{0,0}$.

$$\forall x^0 \exists y^0 A_{ar}(x,y) \to \exists f^1 \forall x^0 A_{ar}(x, f(x)) \text{ with } A_{ar} \in \Pi_n^0$$

The soundness of the functional interpretation of the axiom of number choice can be proven by a special case of Spector's bar recursion. $\mathrm{BR}_{0,1}$ denotes the restriction of BR to the bar recursor constant $\mathrm{BR}_{0,1}$.

Soundness theorem of functional Interpretation of $\mathbf{AC}_{ar}^{0,0}$ with $\mathbf{BR_{0,1}}$. *Let $A(\underline{a})$ be an arbitrary formula of $\mathcal{L}(\text{WE-PA}^\omega)$ with only $\underline{a}$ free variables. Then the following rule holds:*

$$\text{WE-PA}^\omega + \text{QF-AC} + \text{AC}_{ar}^{0,0} \vdash A(\underline{a})$$
$$\Rightarrow \text{WE-HA}^\omega + \text{BR}_{0,1} \vdash \forall \underline{y}(A')_D(\underline{ta}, \underline{y}, \underline{a})$$

with $\underline{t}$ as suitable tuple of closed terms of WE-HA$^\omega$+BR$_{0,1}$ *which can be extracted from a given proof of the assumption, and A' negative translation of A.*

In intuitionistic mathematics (cf. Chapter 6), bar induction and bar theorem are based on the fan principle which yields classically inconsistent results, e.g., that all functions $f : 2^{\mathbb{N}} \to \mathbb{N}$ are uniformly continuous. Kohlenbach (1999) suggested a version of the fan principle restricted to Σ_1^0-formulas which avoids classical inconsistencies.

Definition of the uniform $\mathbf{\Sigma_1^0}$-boundedness principle $\mathbf{\Sigma_1^0}$-UB.

$$\forall y^{1(0)}(\forall k^0 \forall x \leq_1 yk \exists z^0 A(x,y,k,z) \to \exists \chi^1 \forall k^0 \forall x \leq_1 yk \exists z \leq_0 \chi k A(x,y,k,z)$$ with $A \equiv \exists \underline{l} A_o(\underline{l})$, $\underline{l}$ tupel of variables of type 0, and A_o quantifier-free.

The uniform Σ_1^0-boundedness principle allows to extract a bound from a given proof.

$\mathbf{\Sigma_1^0}$-UB program extraction theorem. *Let $A_0(x^1, y^1, z^0, v^\tau) \in \mathcal{L}(\text{E-PA}^\omega)$ be a QF formula containing only x, y, z, v as free variables*

(where τ is an arbitrary type) and s a closed term of E-PA$^\omega$. *Then the following rule holds*:

$$\text{E-PA}^\omega + \text{QF-AC}^{1,0} + \text{QF-AC}^{0,1} + \Sigma_1^0\text{-UB} \;\vdash\; \forall x^1 \forall y \;\leq_1$$
$$sx \exists z^0, v^\tau A_0(x,y,z,v)$$
$$\Rightarrow \text{WE-HA}^\omega \vdash \forall x^1 \forall y \leq_1 sx \exists z \leq_0 \Psi x \exists v^\tau A_0(x,y,z,v)$$

with a closed term Ψ which can be extracted in E-PA$^\omega$ *from a proof of the assumption.*

Σ_1^0-UB as a kind of fan principle allows convenient proofs of important theorems in analysis (e.g., Heine–Borel, Bolzano–Weierstrass) which would demand much more efforts with WKL. Thus, the uniform Σ_1^0-boundedness principle is a nice example of Ockham's principle of parsimony.

Before considering further applications of proof mining in mathematics, we remind the reader of the representation of real numbers by Cauchy sequences of rational numbers which is fundamental for classical, intuitionistic, constructive, and computational analysis (Bishop, 1967; Lorenzen, 1965; Kleene, 1974; Troelstra and van Dalen, 1988; Schwichtenberg and Wainer, 2012).

Representation of rational numbers within WE-HA$^\omega$. Rational numbers are represented as codes $j(n,m)$ of (n,m) of natural numbers with the surjective Cantor pairing function

$$j(x^0, y^0) := \begin{cases} \min u \leq_0 (x+y)^2 + 3x + y \\ \quad [2u =_0 (x+y)^2 + 3x + y], & \text{if existent,} \\ 0^0, & \text{otherwise} \end{cases}$$

$$j_1 z := \min x \leq_0 z[\exists y \leq z(j(x,y) = z)],$$

$$j_2 z := \min y \leq_0 z[\exists x \leq z(j(x,y) = z)].$$

$j(n,m)$ represents the rational number $\frac{\frac{n}{2}}{m+1}$ if n is even, and the negative rational number $-\frac{\frac{n+1}{2}}{m+1}$ otherwise.

Each natural number can be conceived as the code of a uniquely determined rational number.

The representations of rational numbers can algebraically be characterized as ordered field (Ebbinghaus *et al.*, 1991).

Representation of the ordered field of rational numbers. On the representation of the rational numbers, an equivalence relation is defined by

$$n_1 =_{\mathbb{Q}} n_2 :\equiv \frac{\frac{j_1 n_1}{2}}{j_2 n_1 + 1} = \frac{\frac{j_1 n_2}{2}}{j_2 n_2 + 1}$$

for j_1, n_1, j_1, n_2 even (analogue in the other cases).

For $b, d > 0 : \dfrac{a}{b} = \dfrac{c}{d} \equiv ad =_0 cb$ when $bd > 0$.

<u>Abbreviation</u>: $n =_{\mathbb{Q}} \langle r \rangle$ means that n represents the rational r.

On $\mathbb{N}$, primitive recursive operations of adddition $+_{\mathbb{Q}}$, multiplication $\cdot_{\mathbb{Q}}$, and relations $<_{\mathbb{Q}}, \leq_{\mathbb{Q}}$ can be defined: e.g.,

$$\langle r_1 \rangle +_{\mathbb{Q}} \langle r_2 \rangle =_{\mathbb{Q}} \langle r_3 \rangle \equiv r_1 + r_2 = r_3 \quad \text{for rational numbers.}$$

$\mathbb{N}$ is embedded into $\mathbb{Q}$ by

$$n \mapsto \langle n \rangle := j(2n, 0); 0_{\mathbb{Q}} := \langle 0 \rangle; 1_{\mathbb{Q}} := \langle 1 \rangle.$$

The ordered field $(\mathbb{Q}, +, \cdot, 0, 1, <)$ is represented by $(\mathbb{N}, +_{\mathbb{Q}} \cdot_{\mathbb{Q}}, 0_{\mathbb{Q}}, 1_{\mathbb{Q}} <_{\mathbb{Q}})$ in formal language $\widehat{\text{WE-HA}}^{\omega} \upharpoonright$ (fragment of WE-HA with only recursor for type 0-recursion and quantifier free-induction scheme).

Representation of real numbers within $\widehat{\text{WE-HA}}^{\omega}$. Each function $f^1 : \mathbb{N} \to \mathbb{N}$ (i.e., functional of type 1) can be considered an infinite sequence of codes of rationals. Real numbers can be represented by functionals f^1 satisfying the Cauchy principle:

$$\forall n (|f(n) -_{\mathbb{Q}} f(n+1)|_{\mathbb{Q}} <_{\mathbb{Q}} \langle 2^{-n-1} \rangle).$$

The following functional $f \mapsto \hat{f}$ is primitive recursive, satisfies the Cauchy principle, and can be carried out in $\widehat{\text{WA-HA}}^{\omega}$:

$$\hat{f}(n) := \begin{cases} f(n), & \text{if } \forall k < n(|f(k) -_{\mathbb{Q}} f(k+1)|_{\mathbb{Q}} <_{\mathbb{Q}} \langle 2^{-k-1}\rangle), \\ f(k), & \text{for the least } k < n \text{ with} |f(k) -_{\mathbb{Q}} f(k+1)|_{\mathbb{Q}} \\ & \geq_{\mathbb{Q}} \langle 2^{-k-1}\rangle \text{ otherwise.} \end{cases}$$

If the Cauchy principle is already valid for f, then $\forall n(f(n) =_0 \hat{f}(n))$

The representations of rational numbers can algebraically be characterized as Archimedian ordered field (Ebbinghaus *et al.*, 1991).

Representation of the Archimedian ordered field of real numbers. On the representation of real numbers f_1, f_2, an equivalence relation $=_{\mathbb{R}}$ can be defined by

$$f_1 =_{\mathbb{R}} f_2 :\equiv \forall n(|\hat{f}_1(n+1) -_{\mathbb{Q}} \hat{f}_2(n+1)|_{\mathbb{Q}} <_{\mathbb{Q}} \langle 2^{-n}\rangle).$$

In contrast to $=_{\mathbb{Q}}$, the relation $=_{\mathbb{R}}$ is not decidable but in $\prod_0^1$:

$$f_1 <_{\mathbb{R}} f_2 :\equiv \exists n(\hat{f}_2(n+1) -_{\mathbb{Q}} \hat{f}_1(n+1) \geq_{\mathbb{Q}} \langle 2^{-n}\rangle) \in \Sigma_1^0$$

$$f_1 \leq_{\mathbb{R}} f_2 :\equiv \neg(f_2 <_{\mathbb{R}} f_1) \in \Pi_1^0$$

Functionals $+_{\mathbb{R}}, -_{\mathbb{R}}, \cdot_{\mathbb{R}}$ etc. can be defined on codes of real numbers, representing elementary operations $+, -, \cdot$, etc. on $\mathbb{R}$: e.g.,

$$(f_1 +_{\mathbb{R}} f_2)(k) := \hat{f}_1(k+1) +_{\mathbb{Q}} \hat{f}_2(k+1).$$

If $n = \langle r \rangle$ codes the rational number r, then $\lambda k.n$ represents r as real number, $0_{\mathbb{R}} := \lambda k.0_{\mathbb{Q}}$, $1_{\mathbb{R}} := \lambda k.1_{\mathbb{Q}}$.

The Archimedian ordered field $(\mathbb{R}, +, \cdot, 0, 1, <)$ is represented by $(\mathbb{N}^{\mathbb{N}}, +_{\mathbb{R}}, \cdot_{\mathbb{R}}, 0_{\mathbb{R}}, 1_{\mathbb{R}} <_{\mathbb{R}})$ in $\widehat{\text{WE-HA}}^{\omega} \upharpoonright$.

$\widehat{\text{WE-HA}}^{\omega} \upharpoonright$ is the fragment of WE-HA$^{\omega}$ with only recursor R_0 for type-0 recursion and the induction scheme restricted to QF induction.

The Archimedian ordered field of real numbers is an example of a complete metric space (Beeson, 1985, Chapter 1; Kohlenbach, 1993, Section 3). In general, complete separable metric spaces are represented as completions $(\hat{X}, \hat{d})$ of countable metric spaces (X, d) with elements of X coded by natural numbers and f_X primitive recursive enumeration of set $\langle X \rangle$ of all codes. Metric d is represented by pseudo metric d_X on $\mathbb{N}$ (i.e., a functional of type $1(0)(0)$) with $d_X(n, m) =_\mathbb{R} \langle d(x, y) \rangle$ with codes $f_X(n)$ and $f_X(m)$ of $x, y \in X$ and $\langle d(x, y) \rangle$ representative of real number $d(x, y)$. But, it is a pseudo metric, because $d_X(n, m) = 0$ does not imply $n =_0 m$. It is assumed that $d_X(n, m) =_1 \widehat{d_X(n, m)}$.

The completion $(\hat{X}, \hat{d})$ of (X, d) is represented as the completion of $(\mathbb{N}, d_X)$. An element of this completion is given by a function h with

$$\forall n (d_X(h(n), h(n + 1)) <_\mathbb{R} \langle 7 \cdot 2^{-n-1} \rangle),$$

satisfying the Cauchy principle in $(\mathbb{N}, d_X)$.

If $\hat{h}$ is defined like $\hat{f}$ for the representation of $\mathbb{R}$, then this operation would not be computable in h since $<_\mathbb{R}$ is not decidable (in contrast to $<_\mathbb{Q}$). Therefore, we modify the Cauchy principle with function h as

$$\forall n ((d_X(h(n), h(n + 1)))(n + 1) <_\mathbb{Q} \langle 6 \cdot 2^{-n-1} \rangle).$$

$\hat{h}$ can be defined as functional in h in WE-HA$^\omega$ by

$$\hat{h}(n) := \begin{cases} h(n), & \text{if } \forall k < n((d_X(h(k), h(k+1))) \\ & \quad (k+1) <_\mathbb{Q} \langle 6 \cdot 2^{-k-1} \rangle), \\ h(k) \text{ for min } k < n: \\ \quad (d_X(h(k), h(k+1))) \\ \quad (k+1) \geq_\mathbb{Q} (6 \cdot 2^{-k-1}), & \text{otherwise.} \end{cases}$$

The pseudo metric d_X is extended to a pseudo metric $\hat{d}_X$ on $\mathbb{N}^\mathbb{N}$:

$$\hat{d}_X^{1(1)(1)}(h_1, h_2)(n) :=_0 (d_X(\hat{h}_1(n + 5), \hat{h}_2(n + 5)))(n + 5)$$

with summand $+5$ ensuring the rate of convergence of $\hat{d}_X(h_1, h_2)$ with

$$\forall k (|\hat{d}_X(h_1, h_2)(k) -_\mathbb{Q} \hat{d}_X(h_1, h_2)(k + 1)| <_\mathbb{Q} (2^{-k-1})).$$

Definition of standard representation of complete separable metric spaces. $(\mathbb{N}^{\mathbb{N}}, \hat{d}_X)$ is called standard representation of $(\hat{X}, \hat{d})$ with equivalence relation

$$h_1 =_{\hat{X}} h_2 :\equiv (\hat{d}_X(h_1, h_2) =_{\mathbb{R}} 0_{\mathbb{R}}).$$

Example 1. $\mathbb{R}^n$ resp. $[0,1]^n$ with the Euclidean metric

$$\hat{d}_E((x_1, \ldots, x_n), (y_1, \ldots, y_n)) := \left(\sum_{j=1}^{n} |x_i - y_i|^2\right)^{\frac{1}{2}}$$

is the completion of the space $\mathbb{Q}^n$ resp. $[0,1]^n \cap \mathbb{Q}^n$ with the metric

$$\hat{d}_E((r_1, \ldots, r_n), (\tilde{r}_1, \ldots, \tilde{r}_n)) := \left(\sum_{j=1}^{n} |r_i - \tilde{r}_i|^2\right)^{\frac{1}{2}} \text{ on } \mathbb{Q}^n \text{ resp. } [0,1]^n \cap \mathbb{Q}^n.$$

Example 2. The space C[0,1] of all continuous functions $f : [0,1] \to \mathbb{R}$ with the maximum metric d_∞ is the completion of (A, d_∞) with set A of all finite tuples of rational numbers and

$$d_\infty((r_0, \ldots, r_m), (\tilde{r}_0, \ldots, \tilde{r}_n))$$
$$:= \sup_{x \in [0,1]} |(r_m x^m + \cdots + r_1 x + r_0) - (\tilde{r}_n x^n + \cdots + \tilde{r}_1 x + \tilde{r}_0)|.$$

Coming back to proof mining, we want to extract the constructive content involved in the Bolzano–Weierstrass principle for bounded sequences in $\mathbb{R}^d$ (Kohlenbach, 2008, p. 269). According to the Bolzano–Weierstrass principle, any sequence (x_n) bounded in $[-1, 1]^d \subset \mathbb{R}^d$ possesses a limit point x such that

$$\forall (x_n) \subset [-1, 1]^d \exists x \in [-1, 1]^d \forall k^0, m^0 \exists n >_0 m(\|x - x_n\|_{\max} \le 2^{-k})$$

with respect to the maximum norm $\| \cdot \|_{\max}$. In order to apply the tools of proof mining, we must represent the Bolzanos–Weierstrass principle in $\widehat{\text{WA-HA}}^\omega \upharpoonright$.

The Bolzano–Weierstrass principle represented in $\widehat{\text{WA-HA}}^\omega$. Using the standard representation of $[-1, 1]$, BW has the form

$$\forall x_1^{1(0)}, \ldots, x_d^{1(0)} \mathrm{BW}(x_1, \ldots, x_d)$$

with

$$\mathrm{BW}(\underline{x}) :\equiv \exists a_1, \ldots, a_d \leq_1 M \forall k^0, m^0 \exists n >_0 m$$
$$\wedge_{i=1}^d (|\tilde{a}_i -_{\mathbb{R}} \widetilde{x_i n}| \leq_{\mathbb{R}} 2^{-k}),$$

where M and $y^1 \mapsto \tilde{y}$ are representations of [-1,1] in $\widehat{\mathrm{WA\text{-}HA}}^\omega \upharpoonright$.

Program extraction of the Bolzano–Weierstrass principle.
Let Δ be a set of sentences $\forall x^\delta \exists y \leq_\rho sx \forall z^\eta G_0(x, y, z)$ and $\tilde{\Delta}$ the corresponding set of Skolem normal forms $\exists y \leq_{\rho\delta} s \forall x^\delta, z^\eta G_0(x, Yx, z)$.

Let $B_0(u^1, v^\tau, w^\gamma) \in \mathcal{L}(\mathrm{G}_\infty \mathrm{A}^\omega)$ be a QF formula containing only u^1, v^τ, w^γ free with $\gamma \leq 2$ and $\underline{\xi}, t \in \mathrm{G}_\infty \mathrm{R}^\omega$. The following rule holds:

$$\mathrm{G}_\infty \mathrm{A}^\omega + \Delta + \mathrm{QF\text{-}AC} \vdash \forall u_1 \forall v \leq_\tau tu(\mathrm{BW}(\underline{\xi}uv) \to \exists w^\gamma B_0(u, v, w))$$
$$\Rightarrow \widehat{\mathrm{WA\text{-}HA}}^\omega + \tilde{\Delta} \vdash \forall u^1 \forall v \leq_\tau tu \exists w \leq_\gamma \Phi(u) B_0(u, v, w)$$

with Φ closed term extracted from the proof of the assumption.

In the representation of complete separable metric spaces, we can derive general metatheorems on the extractability of effective uniform bounds from proofs in analysis (Gerhardy and Kohlenbach, 2008). In this context, "uniform" means that the bounds are independent of parameters in (compact) metric spaces. Metatheorems open new avenues to general tools of proof mining which can be applied not only to single principles but also to whole theories and disciplines of mathematics. The separation of mathematics and logic has been an unfortunate development during the last decades. Mathematicians are often not interested in logic and proof theory because these fields seem to have no consequences for their research fields in, e.g., numerical or functional analysis. Vice versa, logicians and proof theorists are operating on an abstract and formal level which seems to be unrealizable in advanced mathematical research. Actually, metatheorems are practical links between logic and proof theory with mathematics.

For example, in numerical and functional analysis, many problems demand the construction of solutions $x \in K$ for equation

$$A(x) :\equiv (F(x) = 0)$$

with a compact metric space K and $F : K \to \mathbb{R}$ continuous function. The solution involves two steps:

(1) Construct approximate solutions $x_n \in K$ satisfying $A_n(x_n) :\equiv (|F(x_n)| < 2^{-n})$.
(2) Conclude that either $(x_n)_{n \in \mathbb{N}}$ itself or some subsequence converges to a solutionof $A(x)$, using the compactness of K and the continuity of F.

Metatheorems of proof theory can deliver general tools of solutions if we distinguish the following cases:

(a) If F has exactly one root $\hat{x} \in K$, proof-theoretic analysis of a given proof of uniqueness of $\hat{x}$ can be applied to extract an effective rate of convergence under quite general circumstances.
(b) If the solution $\hat{x}$ is not necessarily unique, one often cannot effectively obtain a solution but weaker tasks like obtaining effective rates of asymptotic regularity might be solvable.

In approximation theory, we are interested in the extraction of rates of convergence towards a unique solution. The following metatheorem delivers a general tool of proof mining to extract a bound from a given proof for complete separable spaces in the compact case (Kohlenbach, 1993).

Metatheorem on proof mining for complete separable metric spaces.

(1) *Let X be a complete separable metric ("Polish") space, K a compact metric space and $A_1(n^0, x^1, y^1, m^0)$ a purely existential formula of $\mathcal{L}(\text{E-PA}^\omega)$ with degree of types of existential quantifier ≤ 1 having only n, x, y, m as free variables.*

(2) *Let X, K be explicitly representable in* E-PA$^\omega$.

(3) *Let A_1 be provably extensional in $x \in X$, $y \in K$ i.e.,* E-PA$^\omega$ +
QF-AC^{1,0} + QF-AC^{0,1} + WKL $\vdash \forall n^0, m^0, x_1^1, x_2^1, y_1^1, y_2^1(x_1 =_x$
$x_2 \wedge y_1 =_K y_2 \wedge A_1(n, x_1, y_1, m) \to A_1(n, x_2, y_2, m))$

Then the following rule holds:

$$\text{E-PA}^\omega + \text{QF-AC}^{1,0} + \text{QF-AC}^{0,1} + \text{WKL}$$

$$\vdash \forall n \in \mathbb{N} \forall x \in X \forall y \in K \exists m \in \mathbb{N} A_1(n, x, y, m)$$

$$\Rightarrow \text{WE-HA}^\omega \quad \vdash \forall n \in \mathbb{N} \forall x \in X \forall y \in K \exists m \leq \Phi(n, x) A_1(n, x, y, m)$$

with closed term Φ of E-PA$^\omega$ *which can be extracted from a proof of
the assumption depending on $x \in X$ via a representation $f_x \in \mathbb{N}^{\mathbb{N}}$.*

In general, the used extraction algorithm is a combination of negative translation and monotone functional interpretation which was already discussed before. In special applications, it may be simplified and optimized.

Application of metatheorem to uniqueness proofs

Let X be a complete separable metric space, K a compact metric space and $F : X \times K \to \mathbb{R}$ a continuous function, representable in E-PA$^\omega$ + QF-AC^{1,0} + QF-AC^{0,1} + WKL.

We assume a proof in E-PA$^\omega$ + QF-AC^{1,0} + QF-AC^{0,1} + WKL that for every $x \in X$, equation $F(x, \cdot)$ has at most one root in K:

$$\text{E-PA}^\omega + \text{QF-AC}^{1,0} + \text{QF-AC}^{0,1} + \text{WKL}$$

$$\vdash \forall x \in X \forall y_1, y_2 \in K(\wedge_{i=1}^2 F(x, y_i) = 0 \to y_1 = y_2).$$

With metric d_K on K, this statement can be rewritten as

$$\text{E-PA}^\omega + \text{QF-AC}^{1,0} + \text{QF-AC}^{0,1} + \text{WKL}$$

$$\vdash \forall x \in X \forall y_1, y_2 \in K \exists l \in \mathbb{N}(\wedge_{i=1}^2 (|F(x, y_i)| \leq_{\mathbb{R}} 2^{-l})$$

$$\to d_K(y_1, y_2) <_{\mathbb{R}} 2^{-l}).$$

If real numbers are represented as Cauchy sequences of rational numbers with fixed rate of convergence, we get

$$(\wedge_{i=1}^2(|F(x,y_i)| \leq_{\mathbb{R}} 2^{-l}) \to d_K(y_1,y_2) <_{\mathbb{R}} 2^{-l}) \in \textstyle\sum_1^0, \text{ since } \leq_{\mathbb{R}} \in \textstyle\prod_1^0$$
and $<_{\mathbb{R}} \in \Sigma_1^0$.

By the general metatheorem, we can now extract an explicit bound $\Phi(x,k)$ from such a proof. The bound is given by a closed term of language $\mathcal{L}(\text{E-PA}^\omega)$ of the arithmetical system E-PA$^\omega$, depending on representation f_x^1 of $x \in X$ such that

$$\text{WE-PA}^\omega \vdash \forall x \in X \forall k \in \mathbb{N} \forall y_1, y_2 \in K$$

$$(\wedge_{i=1}^2(|F(x,y_i)| < 2^{-\Phi(x,k)}) \to d_K(y_1,y_2) < 2^{-k}).$$

Application of metatheorem to monotone convergence theorems.

Let X be a complete separable metric space, K a compact metric space and $F : X \times K \times \mathbb{N} \to \mathbb{R}_+$ a function (definable in E-PA$^\omega$ + QF-AC^{1,0} + QF-AC^{0,1} + WKL) such that for any $x \in X$ and $y \in K$ the sequence $(F(x,y,n))_{n\in\mathbb{N}}$ is non-increasing. We assume a proof in E-PA$^\omega$+QF-AC^{1,0}+QF-AC^{0,1}+WKL that $(F(x,y,n))_{n\in\mathbb{N}}$ converges to zero:

$$\text{E-PA}^\omega + \text{QF-AC}^{1,0} + \text{QF-AC}^{0,1} + \text{WKL}$$

$$\vdash \forall x \in X \forall y \in K \forall k \in \mathbb{N} \exists n \in \mathbb{N} \forall m \geq n (F(x,y,m) <_{\mathbb{R}} 2^{-k}).$$

$$\Rightarrow \text{E-PA}^\omega + \text{QF-AC}^{1,0} + \text{QF-AC}^{0,1} + \text{WKL}$$

$$\vdash \forall x \in X \forall y \in K \forall k \in \mathbb{N} \exists n \in \mathbb{N}(F(x,y,n) <_{\mathbb{R}} 2^{-k}).$$

By the general metatheorem, we can now extract an explicit modulus $\delta(x,k)$ depending on representation f_x^1 of $x \in X$:

$$\text{WE-HA}^\omega \vdash \forall x \in X \forall y \in K \forall k \in \mathbb{N} \; \exists n \leq \delta(x,k)(F(x,y,n) <_{\mathbb{R}} 2^{-k})$$
$$\Rightarrow \text{WE-HA}^\omega \vdash \forall x \in X \forall y \in K \forall k \in \mathbb{N} \; \forall n \geq \delta(x,k)(F(x,y,n) <_{\mathbb{R}} 2^{-k}),$$

because sequence $(F(x,y,n))_{n\in\mathbb{N}}$ is assumed to be non-increasing: Obviously, the metatheorem extracts rates of K-uniform convergence from proofs of the pointwise convergence of K-families of monotone sequences of real numbers.

The previous general metatheorem proves the extractability of effective uniform bounds which only depend on representatives of elements in complete separable metric spaces X. The total boundedness and the completeness of K are necessary for this result. In the next step of generalization, bounds can be obtained only under the assumption of abstract metric spaces. The bounds are independent even from non-compact but only metrically bounded spaces. Separability of the spaces must not be assumed. The proofs are given in (Kohlenbach, 2005) and (Gerhardy and Kohlenbach, 2008).

Extension of formal systems for abstract metric spaces

In order to represent abstract metric spaces, axioms of classes of metric spaces (X, d) must be added to systems $\mathcal{A}^\omega := \text{WE-PA}^\omega + \text{QF-AC} + \text{DC}$ with axiom schema

$$DC := \{DC^{\underline{\rho}} | \underline{\rho} \in \mathrm{T}\} \text{ of dependent choice with}$$

$$DC^{\underline{\rho}} :\equiv \forall x^0, \underline{y}^{\underline{\rho}} \exists \underline{z}^{\underline{\rho}} A(x, \underline{y}, \underline{z}) \to \exists \underline{f}^{\underline{\rho}(0)} \forall x^0 A(x, \underline{f}(x), \underline{f}(st(x))).$$

The set of all finite types T^x over the ground types 0 (for $\mathbb{N}$) and X is defined inductively by

(i) $0, X \in \mathrm{T}^X$,
(ii) $\rho, \tau \in \mathrm{T}^X \Rightarrow \tau(\rho) \in \mathrm{T}^X$.

The language $\mathcal{L}(\mathcal{A}^\omega[X, d])$ of $\mathcal{A}^\omega[X, d]$ results from $\mathcal{L}(\mathcal{A}^\omega)$ by extending it to all the types in the set T^X.

All axioms and rules of $\mathcal{A}^\omega$ are extended to the new set of types T^X.

The new constants d_X and b_X of metric spaces are defined by additional axioms:

(1) $\forall x^X (d_X(x, x) =_{\mathbb{R}} 0_{\mathbb{R}})$
(2) $\forall x^X, y^X (d_X(x, y) =_{\mathbb{R}} d_X(y, x))$

(3) $\forall x^X, y^X, z^X (d_X(x, z) \leq_{\mathbb{R}} d_X(x, y) +_{\mathbb{R}} d_X(y, z))$

(4) $\forall x^X, y^X (d_X(x, y) \leq_{\mathbb{R}} (b_X) \; \mathbb{R} := \lambda k^0 \cdot j(2b_X, 0^0))$.

Again, real numbers are represented by Cauchy sequences.

In this abstract framework of mathematical theories, we can still apply proof mining, in order to obtain a general metatheorem of effective bounds.

Metatheorem on proof mining for abstract metric spaces. Let ρ, σ be of degree ≤ 1 and τ be a type of degree $(1, X)$. Let $s^{\rho(\sigma)}$ be a closed term of $\mathcal{A}^\omega[X, d]$ and let $B_\forall(x^\sigma, y^\rho, z^\tau, u^0)$(resp. $C_\exists(x^\sigma, y^\rho, z^\tau, u^0))$ be a $\forall$-formula that contains only $x^\sigma, y^\rho, z^\tau, u^0$ free (resp. a $\exists$-formula that contains only $x^\sigma, y^\rho, z^\tau, u^0$ free).

Then the following rule holds:

$$\mathcal{A}^\omega[X, d] \vdash \forall x^\sigma \forall y \leq_\rho s(x) \forall z^\tau (\forall u^0 B_\forall(x, y, z, u) \to$$
$$\exists v^0 C_\exists(x, y, z, v))$$
$$\Rightarrow \mathcal{A}^\omega[X, d] \vdash \forall y \leq_\rho s(x) \forall z^\tau (\forall u \leq \Phi(x, b) B_\forall(x, y, z, u) \to$$
$$\exists v \leq \Phi(x, b) C_\exists(x, y, z, v)$$

with computable functional $\Phi : S_\sigma \times \mathbb{N} \to \mathbb{N}$ for all $x \in S_\sigma$ and $b \in \mathbb{N}$ in any metric space (X, d) with a metric bounded by $b \in \mathbb{N}$.

In the next step, the Σ_1^0-uniform boundedness (UB) principle is generalized to bounded metric spaces.

Generalized uniform boundedness principle $\exists$-UB^X. A generalized $\exists$-formula has the form $\exists \underline{a}^{\underline{\sigma}} F_{qf}(\underline{a})$ with F_{qf} QF and $\underline{\sigma} \in \mathrm{T}^X$.

Definition of degrees for types $\rho \in \mathrm{T}^X$.

(1) A type $\rho \in T^X$ is of degree $(\cdot, 0), (\text{resp.}(\cdot, X))$ iff it has the form $0(\rho_k) \ldots (\rho_1)(\text{resp.} X(\rho_k) \ldots (\rho_1)$ with $\rho_1, \ldots, \rho_k \in \mathrm{T}^X$ including $0 \; (\text{resp.} X)$ as special case.

(2) If ρ is a type of degree $(\cdot, 0)$, then $\min_\rho(x^\rho, y^\rho) := \lambda \underline{v}.\min_0(x\underline{v}, y\underline{v})$.

Definition of the uniform boundedness scheme $\exists\text{-}UB^X$ for generalized $\exists$-formulas.

$$\forall y^{\alpha(0)} (\forall k^0, x^\alpha, \underline{z}^{\underline{\beta}} \exists n^0 A_\exists (y, k \, \min_\alpha(x, yk), \underline{z}, n)$$

$$\to \exists \chi^1 \forall k^0, x^\alpha, \underline{z}^{\underline{\beta}} \exists n \leq_0 \chi k A_\exists (y, k \, \min_\alpha(x, yk), \underline{z}, n)).$$

The metatheorem on proof mining for abstract metric spaces remains valid if system $\mathcal{A}^\omega[X, d]$ is extended by the $\exists\text{-}UB^X$ principle.

Generalized $\exists\text{-}UB^X$ extraction theorem. *Let σ, ρ be types of degree ≤ 1 and τ be a type of degree $(1, X)$.*
Let $s^{\rho(\sigma)}$ be a closed term of $\mathcal{A}^\omega[X, d]$ and $B_\forall(x^\sigma, y^\rho, z^\tau, u^0)$ (resp.$C_\exists(x^\sigma, y^\rho, z^\tau, v^0)$) be a $\forall$-formula containing only x, y, z, u free (resp. an $\exists$-formula containing only x, y, z, v free).
Then the following rule holds:

$$\mathcal{A}^\omega[X, d] + \exists\text{-}UB^X \vdash \forall x^\sigma \forall y \leq_\rho s(x) \forall z^\tau (\forall u^0 B_\forall(x, y, z, u)$$
$$\to \exists v^0 C_\exists(x, y, z, v))$$
$$\Rightarrow \quad \mathcal{A}^\omega[X, d] + \exists\text{-}UB^X \vdash \forall y \leq_\rho s(x) \forall z^\tau (\forall u \leq \Phi(x, b) B_\forall(x, y, z, u)$$
$$\to \exists v \leq \Phi(x, b) C_\exists(x, y, z, v))$$

with a computable functional $\Phi : S_\sigma \times \mathbb{N} \to \mathbb{N}$ extracted from a proof of the assumption for all $x \in S_\sigma$ and $b \in \mathbb{N}$ in any metric space (X, d) with a metric bounded by $b \in \mathbb{N}$.

Applications of the $\exists\text{-}UB^X$ principle yields several remarkable properties of metric spaces (Kohlenbach, 2008, 442 f.).

First application of $\exists\text{-}UB^X$ principle. $\mathcal{A}^\omega[X, d] + \exists\text{-}UB^X$ proves that metric space (X, d) is complete.

Proof. We prove that the $\exists\text{-}UB^X$ principle reduces the assumption of completeness to falsity:

(X, d) is not complete $\hspace{3cm}$ assumption

$\Rightarrow \exists(x_n)$ Cauchy sequence $\forall x \in X$

$\quad ((x_n)$ does not converge to $x)$, i.e.,

$\quad \forall x \in X \neg(\lim x_n = x).$

$(*)\forall k \forall m, n \geq k(d(x_m, x_n) <_\mathbb{R} 2^{-k}$ assumption by axiom of choice $\mathrm{AC}^{0,0}$ (in $\mathcal{A}^\omega[X, d]$)

$\Rightarrow \ \forall x^X \exists n^0(d_X(x_n, x) >_\mathbb{R} 2^{-n+1})$ because of $\neg(\lim x_n = x)$

$\Rightarrow \ \exists n^0 \forall x^X \exists \tilde{n} \leq n(d_X(x_n, x) >_\mathbb{R} 2^{-\tilde{n}+1})$ $\exists\text{-UB}^X$ principle

$\Rightarrow \ \exists n^0 \forall x^X(d_X(x_n, x) >_\mathbb{R} 2^{-n})$ assumption $(*)$

$\Rightarrow$ contradiction for $x := x_n$

$\square$

Under the $\exists\text{-UB}^X$ principle, the assumption of separability of a bounded metric space (X, d) leads to its total boundednesss. Let (X, d) be a totally bounded metric space. A function $\alpha : \mathbb{N} \to \mathbb{N}$ satisfying $\exists(a_n)_n$ in $X \forall k \in \mathbb{N} \forall x \in X \exists n \leq \alpha(k)(d(x, a_n) < 2^{-k})$ is called a modulus to total boundedness.

Second application of $\exists\text{-UB}^X$ principle. If (X, d) is separable, then (X, d) is totally bounded and has a modulus of total boundedness α, i.e.,

$$\mathcal{A}^\omega[X, d] + \exists\text{-UB}^X \vdash \forall f^{X(0)}(\forall k^0, x^X \exists n^0(d_X(f(n), x) <_\mathbb{R} 2^{-k})$$
$$\to \exists \alpha^1 \forall k^0, x^X \exists n \leq \alpha(k)(d_X(f(n), x) <_\mathbb{R} 2^{-k})).$$

Proof.

$\forall k^0, x^X \exists n^0(d_X(f(n), x) <_\mathbb{R} 2^{-k})$ assumption

$\Rightarrow \ \exists \alpha^1 \forall k^0, x^X \exists n^0 \leq \alpha(k)(\underbrace{d_X(f(n), x) <_\mathbb{R} 2^{-k}})$ $\exists\text{-UB}^X$ principle
$\qquad\qquad\qquad\qquad\qquad\qquad \exists\text{-formula}$

$\square$

Third application of $\exists\text{-UB}^X$ principle. If $\Phi : X \to \mathbb{R}$ in X has approximate solutions, then there exists a real solution, i.e., $\mathcal{A}^\omega[X, d] + \exists\text{-UB}^X \vdash \forall \Phi^{1(\beta)}(\forall k^0 \exists y^\beta(|\Phi(y)|_\mathbb{R} <_\mathbb{R} 2^{-k}) \to \exists y^\beta(\Phi(y) =_\mathbb{R} 0))$.

Proof.

$\forall y^\beta(\Phi(y) \neq_\mathbb{R} 0)$ assumption of negation

$\Rightarrow \ \forall y^\beta \exists k^0(|\Phi(y)|_\mathbb{R} >_\mathbb{R} 2^{-k})$

$\Rightarrow \ \exists k^0 \forall y^\beta(|\Phi(y)|_\mathbb{R} >_\mathbb{R} 2^{-k})$ $\exists\text{-UB}^X$ principle

$\Rightarrow$ contradiction to assumption

$\square$

We end this chapter with an application of proof mining to ergodic theory (Kohlenbach, 2008, p. 499) which has special interest in statistical physics and dynamical systems theory (cf. Chapter 13; Mainzer, 2007a,b). Ergodic theory studies the behavior of a dynamical system running for a long time. Poincaré's recurrence theorem demands that certain systems, after long but finite time, return to a state very close to the initial state. More precise information is provided by ergodic theorems. They claim that, under certain conditions, the time average of a function along the trajectories exists almost everywhere and is related to the space average.

The ergodic theorems of Birkhoff (1931) and von Neumann (1932) assert the existence of a time average along each trajectory. For ergodic systems, the time average is the same for all initial points: The system evolving for a long time "forgets" its initial state.

Ergodic theory studies the asymptotic behavior of the averaging operator defined by

$$A_n(x) := \frac{1}{n} S_n(x) \text{ with } S_n(x) := \sum_{i=0}^{n-1} f^i(x).$$

A self-mapping $f : X \to X$ of metric space (X, d) is called non-expansive iff

$$\forall x, y \in X (d(f(x), f(y)) \le d(x, y)).$$

In general, fixed points of non-expansive self-mappings of complete metric spaces do not exist. A classical result for Hilbert's spaces was proven by von Neumann (1932).

von Neumann's mean ergodic theorem. Let X be a Hilbert space and $f : X \to X$ a non-expansive linear operator.

Then for any point $x \in X$ the sequence $(A_n(X))_n$ converges (in the Hilbert's space norm).

The proof of von Neumann's mean ergodic theorem delivers the no-counterexample version of the Cauchy property

$$\forall g : \mathbb{N} \to \mathbb{N} \forall \varepsilon > 0 \forall x \in X$$

$$\exists n \forall i, j \in [n; n + g(n)](\|A_i(x) - A_j(x)\| < \varepsilon).$$

Then, one can extract a computable bound $\Phi(d, \varepsilon, g)$ on $\exists n$ that only depends on a norm upper bound $d \geq \|x\|, \varepsilon$ and g (Avigad *et al.*, 2007).

Computable bound extraction of ergodic theorem.

$$\forall g : \mathbb{N} \to \mathbb{N} \forall \epsilon > 0 \forall x \in X \exists n \leq \Phi$$

$$\forall i, j \in [n; n + g(n)](\|A_i(x) - A_j(x)\| < \varepsilon)$$

with $\Phi = h^{(k)}(0)$, $\rho := \left\lceil \frac{\|x\|}{\varepsilon} \right\rceil$, $k := 2^9 \rho^2$, $h(n) := n + 2^{13} \rho^4 \tilde{g}((n + 1)$
$\tilde{g}(2n\rho)\rho^2)$ *and* $\tilde{g}(n) := \max\{i + g(i) : i \leq n\}$.

We will discuss dynamical systems in Chapter 14 of complexity and information dynamics.

Chapter 8

Reverse Mathematics Bridging Logic, Mathematics, and Computer Science

Since Euclid (mid-fourth century–mid-third century BA), mathematicians used axioms to deduce theorems. But the "forward" procedure from axioms to theorems is not always obvious. How can we find appropriate axioms for a proof starting with a given theorem in a "backward" (reverse) procedure?

Pappos of Alexandria (290–350 AD) understood the "forward" procedure as "synthesis" (Greek for the Latin word "constructio" or English translation "construction") because, in Euclid's geometry, logical deductions from axioms to theorems are connected with geometric constructions of corresponding figures (Pappos 1876–1878 II, 634 ff.). The reverse search procedure of axioms for a given theorem was called "analysis" which decomposes a theorem in its necessary and sufficient conditions and the corresponding geometric figures in their elementary building blocks (e.g., circle and straight line) (Hintikka and Remes, 1974; Mainzer, 1994, 120 ff.).

In modern times, reverse mathematics is a modern research program to determine the minimal axiomatic system required to prove theorems (Friedman, 1975; Friedman and Simpson, 2000). In general, it is not possible to start from a theorem τ to prove a whole axiomatic subsystem T_1. A weak base theory T_2 is required to supplement τ:

Definition of equivalent theories. If $T_2 + \tau$ can prove T_1, this proof is called a reversal. If T_1 proves τ and $T_2 + \tau$ is a reversal, then T_1 and τ are said to be equivalent over T_2.

Reverse mathematics enable us to determine the proof-theoretic strength resp. complexity of theorems by classifying them with respect to equivalent theorems and proofs. Many theorems of classical mathematics can be classified by subsystems of second-order arithmetic $\mathcal{Z}_2$ with variables of natural numbers and variables of sets of natural numbers (Simpson, 1999, 2005).

Definition of second-order arithmetics $\mathcal{Z}_2$. $\mathcal{Z}_2$ is a formal system consisting of language $\mathcal{L}_2$ and axioms:

$\mathcal{L}_2$:

- variables (two-sorted) of natural numbers $x, y, \dots$ and sets of natural numbers $A, B, \dots$
- constants 0 and 1,
- binary operations,
- atomic formulas $x = y, x < y, x \in A$,
- formulas $\varphi, \psi, \dots$ with propositional connectives $\wedge, \vee, \neg, \leftrightarrow$ and quantifiers $\forall, \exists$.

Axioms:

- axioms of Peano arithmetic,
- inductive axiom: $(0 \in X \wedge \forall n(n \in X \to n+1 \in X)) \to \forall n(n \in X)$,
- comprehension scheme: $\exists X \forall n(n \in X \leftrightarrow \varphi(n))$ for any formula $\varphi(n)$ of $\mathcal{L}_2$ where X does not occur freely.

Arithmetical formulas can be classified according to the arithmetical hierarchy $\sum_n^0$, $\prod_n^0$, and Δ_n^0. We can distinguish $\sum_n^0$, $\prod_n^0$, and Δ_n^0-schemes of induction and comprehension. For example, in the induction axiom, let X be the set that exists due to the comprehension axiom with formula $\varphi(n)$. Then, the Σ_1^0-induction scheme is the restriction of the induction axiom to Σ_1^0-formulas $\varphi(n)$. A set X is the comprehension of a Σ_1^0-formula iff it is recursively enumerable.

The Δ_1^0-comprehension scheme is the restriction of the comprehension axiom to Δ_1^0-formulas $\varphi(n)$. A set X is the comprehension of a Δ_1^0-formula iff it is recursively computable. Thus, Δ_1^0-comprehension means recursive comprehension (RC) (Soare, 2016).

In the following, we distinguish subsystems of $\mathcal{Z}_2$ with different comprehension schemes. The arithmetical and analytical hierarchies yield classifications of axiomatic subsystems of $\mathcal{Z}_2$ with increasing proof-theoretic power and corresponding structures of $\mathcal{Z}_2$-models. We start with the weak $\mathcal{Z}_2$-subsystem RCA_0 consisting of the axioms of Peano arithmetic, the Σ_1^0-induction scheme, and the Δ_1^0-comprehension scheme.

Definition of $\mathcal{Z}_2$-subsystem RCA_0.

$$\mathrm{RCA}_0 := \text{Peano axioms} + \Sigma_1^0\text{-induction} + \Delta_1^0\text{-comprehension}.$$

A structure of an (arithmetical) set M defines its variables and non-logical symbols (constants, operations) satisfying relations between variables: e.g., $\mathbb{Q} = (M, +_\mathbb{Q}, -_\mathbb{Q}, \cdot_\mathbb{Q}, 0_\mathbb{Q}, I_\mathbb{Q}, <_\mathbb{Q}, =_\mathbb{Q})$ is the structure of rational numbers.

A model of a set of (arithmetical) formulas is a structure with the same non-logical symbols and all formulas in the set are in the model as well. An ω-model refers its number variables to the natural numbers (Soare, 2016).

Definition of an ω-model of RCA_0:

An ω-model S satisfies

(1) $S \neq \emptyset$
(2) $A \in S$ and $B \in S$ imply $A \oplus B \in S$
(3) $A \in S$ and $B \leq_T A$ (B Turing reducible to A) imply $B \in S$.

The minimum ω-model of RCA_0 is the set of computable sets. Obviously, RCA_0 cannot prove the existence of non-computable sets. Examples of theorems provable in RCA_0 are the intermediate value theorem, soundness theorem, and Gödel's completeness theorem. Besides basic properties of numbers, even theorems in calculus, algebra, analysis, and topology (e.g., Baire's category theorem) can be proved in the subsystem RCA_0. In the sense of reverse mathematics, these parts of mathematics are equivalent to the recursively computable subsystem RCA_0. Their proof-theoretic strength has the degree of recursive computability.

In a next step of distinguished $\mathcal{Z}_2$-subsystems, we consider WKL. König's lemma states that every infinite, finitely branching tree has an infinite long path. WKL adds the condition of binary trees. The $\mathcal{Z}_2$-subsystem WKL_0 consists of RCA_0 extended by WKL.

Definition of $\mathcal{Z}_2$-subsystem WKL_0.

$$WKL_0 := RCA_0 + WKL$$

In the sense of reverse mathematics, WKL_0 is equivalent (over RCA_0) to, e.g.,

- Σ_1^0-separation,
- Heine–Borel theorem for $[0,1]$ and compact metric spaces,
- separable Hahn–Banach theorem,
- Brouwer–Schauder fixed point theorems.

Σ_1^0-separation means that given two Σ_1^0-formulas of number variable n which are exclusive, there exists a set containing all numbers n satisfying one formula and none satisfying the other.

Definition of Σ_1^0-separation.

$$\neg \exists n(\varphi_0(n) \wedge \varphi_1(n))$$
$$\Rightarrow \exists X \, \forall n((\varphi_0(n) \to n \in X) \wedge (\varphi_1(n) \to n \notin X))$$

for Σ_1^0-formulas $\varphi_0(n)$ and $\varphi_1(n)$.

Actually, WKL proves Σ_1^0-separation and Σ_1^0-separation implies WKL_0 over RCA_0 which is the reversal.

In the next step of distinguished $\mathcal{Z}_2$-subsystems, we consider a comprehension scheme for all arithmetical formulas with no set quantifiers. The $\mathcal{Z}_2$-subsystem ACA_0 consists of RCA_0 extended by the arithmetical comprehension of all arithmetical formulas of the arithmetical hierarchy.

Definition of $\mathcal{Z}_2$-subsystem $\mathrm{ACA_0}$.

$$\mathrm{ACA_0} := \mathrm{RCA_0} + \text{arithmetical comprehension}$$

In the sense of reverse mathematics, $\mathrm{ACA_0}$ is equivalent (over $\mathrm{RCA_0}$) to, e.g.,

- sequential compactness of $[0,1]$ and compact metric spaces,
- existence of the strong algebraic closure of a countable field,
- König's lemma for subtrees of $\mathbb{N}^{\mathbb{N}}$,
- least upper bound principle for sequences of real numbers.

It is remarkable that the strength of arithmetical comprehension can be reduced to Σ_1^0-comprehension.

Theorem. *ACA_0 is equivalent to Σ_1^0-comprehension over* $\mathrm{RCA_0}$.

$(``\Rightarrow")$

    arithmetical comprehension

$\Rightarrow \Sigma_1^0$-comprehension   definition of arithmetical hierarchy

$(``\Leftarrow")$

Every arithmetical formula consists of alternating $\forall$ and $\exists$ quantifiers followed by a recursive formula. If the formula starts with a $\forall$ quantifier, then, an existential quantifier with a dummy variable can be put at the beginning of the formula. In this way, each arithmetical formula can be written as a Σ_n^0-formula. Therefore, it is sufficient to prove by induction on n that Σ_1^0-comprehension implies Σ_n^0-comprehension:

$n = 0, 1$: trivial

$n \Rightarrow n + 1$: Σ_n^0-comprehension $\Rightarrow \Sigma_{n+1}^0$-comprehension:

    $\varphi(n)$ Σ_{n+1}^0-formula (with $n \geq 1$)   assumption

$\Rightarrow \varphi(n) \equiv \exists k \psi(n, k)$ (with $\psi(n, k)$

    Π_n^0-formula)   definition of Σ_{n+1}^0-formula

$$\Rightarrow Y := \{(n,k)|\neg\psi(n,k)\} \text{ exists} \qquad \Sigma_n^0\text{-comprehension}$$
$$X := \{n|\exists k(n,k) \notin Y\} \text{ exists} \qquad \Sigma_1^0\text{-comprehension}$$
$$\Rightarrow (n \in X \Leftrightarrow \underbrace{\exists k\psi(n,k)}_{\varphi(n)}) \qquad \text{definition of } \varphi(n)$$

$\square$

Reverse mathematics can be extended to the analytical hierarchy $\Sigma_n^1, \prod_n^1$, and Δ_n^1. A first example is the Σ_1^1-weak choice axiom:

$$\forall n \exists! x \varphi(n,x) \rightarrow \exists(y_n)\forall n\varphi(n,y_n)$$

with φ arithmetical formula and unique existence (!) of x.

Definition of $\mathcal{Z}_2$-subsystem Σ_1^1-WC.

$$\Sigma_1^1 - \text{WC} := \text{RCA}_0 + \Sigma_1^1 - \text{weak choice axiom}$$

The Σ_1^1-choice axiom is like the Σ_1^1-weak choice axiom without assuming uniqueness.

Definition of $\mathcal{Z}_2$-subsystem Σ_1^1-C.

$$\Sigma_1^1 - \text{C} := \text{RCA}_0 + \Sigma_1^1 - \text{choice axiom}$$

The Δ_1^1-comprehension axiom reduces the comprehension of sets to Δ_1^1-formulas, i.e., hyperarithmetic theories. From a proof-theoretic point of view, hyperarithmetic theories are distinguished by predicative proofs with an ordinal depth below ω_1^{CK} (compare Chapter 4).

Definition of $\mathcal{Z}_2$-subsystem Δ_1^1-CA$_0$.

$$\Delta_1^1\text{-CA}_0 := \text{RCA}_0 + \Delta_1^1 - \text{comprehension axiom}$$

Arithmetic transfinite recursion means that the arithmetic comprehension scheme can be iterated transfinitely.

Definition of $\mathcal{Z}_2$-subsystem ATR_0.

$$\mathrm{ATR}_0 := \mathrm{RCA}_0$$

$$+ \text{arithmetic comprehension axiom (iterated transfinitely)}$$

We can also consider extended systems with Π_1^1- and Π_2^1-comprehension axioms which are no longer predicative. Thus, along the complexity degrees of arithmetic and analytic hierarchies, we can distinguish main systems of reverse mathematics like RCA_0 with Δ_1^0-comprehension, WKL_0 with WKL, ACA_0 with arithmetic comprehension, ATR_0 with arithmetic comprehension iterated transfinitely, and Π_1^1-comprehension axiom. Σ_1^1-WC with Σ_1^1-weak choice axiom, Δ_1^1-CA_0 with Δ_1^1-comprehension, and Σ_1^1-C with Σ_1^1-choice axiom are systems of hyperarithmetic analysis.

Hierarchy of subsystems of analysis

RCA_0 with Δ_1^0-comprehension

$\downarrow$

WKL_0 with weak König's lemma

$\downarrow$

ACA_0 with arithmetic comprehension

$\downarrow$

Σ_1^1-WC with Σ_1^1-weak choice axiom

$\downarrow$

Δ_1^1-CA_0 with Δ_1^1-comprehension

$\downarrow$

Σ_1^1-C with Σ_1^1- choice axiom

$\downarrow$

ATR_0 with arithmetic comprehension iterated transfinitely

$\downarrow$

Π-CA_0 with Π_1^1-comprehension

The five most commonly used Z_2-subsystems in reverse mathematics correspond to philosophical programs in foundations of mathematics with increasing proof-theoretic power starting with the weakest RCA_0-subsystem.

Z_2-Subsystems and philosophical research programs

RCA_0 corresponds to the research program of computable mathematics. It consists of arithmetic axioms with induction principle reduced to Σ_1^0-formulas and the comprehension principle for Δ_1^0-formulas. Any set of natural number which is provable in RCA_0 is computable. Any theorem on non-computable sets is not provable in RCA_0. Therefore, RCA_0 is a constructive system, but not in the sense of intuitionism because it accepts classical logic with PEM. It is close to Bishop's constructivism (Bishop, 1967), but with classical logic.

Many classical theorems of arithmetic and analysis can be proven in RCA_0. For example, the set of rational numbers satisfy the axioms of an ordered field. If the real numbers are defined by Cauchy sequences of rational numbers, they can be proven to satisfy the axioms of an Archimedian ordered field. Karl T. W. Weierstrass (1815–1897) defined real numbers by nested sequences of closed intervals. Any nested sequence of closed intervals whose length tends to zero can be proven to have a single point in its intersection (Ebbinghaus *et al.*, 1991). In RCA_0, we can also prove the existence and uniqueness of the real closure of a countable ordered field.

WKL_0 is a stronger research program than RCA_0, because it implies that non-computable sets exist. For example, in WKL_0, we can prove that separating sets for effectively inseparable recursively enumerable sets exist. It can be considered a partial realization of Hilbert's finitistic program. Remarkable classical theorems are provable in WKL_0, but they do not follow from RCA_0. They are equivalent to WKL and, therefore, to WKL_0 over RCA_0. These second-order statements are, for example, the Heine–Borel theorem for the closed unit real interval, the boundedness of a continuous real function on the closed unit interval, the uniform continuity of a continuous real function on the closed unit interval or the Jordan curve theorem.

ACA_0 with the comprehension scheme of arithmetical formulas corresponds to the research program of Weyl's (1918) and Lorenzen's (1955, 1965) predicativism which was followed by Feferman (1968a). Nevertheless, there are predicatively provable theorems that are not provable in ACA_0. Its research program is stronger than WKL_0, because it exibits a model of WKL_0 which does not contain all arithmetical sets. Famous theorems are equivalent to ACA_0 over RCA_0 — Bolzano–Weierstrass theorem as well as König's lemma for arbitrary finitely branching trees.

ATR_0 proves the consistency of ACA_0. Therefore, by Gödel's theorem, it is stronger than ACA_0. It has the proof-theoretic ordinal Γ_0 which is the supremum of ordinals of predicative systems. $\Delta_1^1\text{-}CA_0$ yields systems of hyperarithmetic analysis with Δ_1^1-predicativism.

$\Pi_1^1\text{-}CA_0$ which consists of RCA_0 with the comprehension axiom for Π_1^1-formulas is stronger than arithmetical transfinite recursion. Therefore, it is impredicative. The $\Pi_1^1\text{-}CA_0$ comprehension principle is equivalent to theorems of set theory with impredicative proofs. The equivalence proves that the impredicative proofs cannot be reduced to predicative ones. Therefore, they are essentially impredicative.

Classical reverse mathematics uses classical logic and classification of proof-theoretic strength with RCA_0 (Δ_1^0-recursive comprehension) as basic subsystem. The research program of reverse classical mathematics tries to clarify which set existence axioms are necessary to prove the theorems of ordinary mathematics. Constructive reverse mathematics (Ishihara, 2005, 2006) uses intuitionistic logic and Bishop's constructive mathematics BISH as basic subsystem of a constructive classification with Brower's intuitionistic mathematics, Markov's constructive recursive mathematics RUSS ("Russian"), and Hilbert's classical mathematics CLASS.

Hierarchy of constructive reverse mathematics

$$\text{BISH} = \mathcal{Z}_2 + \text{intuitionistic logic}$$
$$+ \text{axioms of countable, dependent, and unique choice}$$
$$\downarrow$$
$$\text{INT} = \text{BISH} + \text{axiom of continuous choice} + \text{Fan theorem}$$
$$\downarrow$$
$$\text{RUSS} = \text{BISH} + \text{MP} + \text{Church's thesis}$$
$$\downarrow$$
$$\text{CLASS} = \text{BISH} + \text{principle of excluded middle} + \text{full AC}$$

Bishop's constructive mathematics (BISH)

BISH is an informal mathematics with intuitionistic logic and function existence axioms.

Axiom of countable choice:

$$\forall n \in \mathbb{N}\, \exists x \in X\, A(n, x) \to \exists f \in X^{\mathbb{N}}\forall n \in \mathbb{N}\, A(n, f(n))$$

Axiom of dependent choice:

$$\forall x \in X\, \exists y \in X\, A(x, y)$$
$$\to \forall x \in X\, \exists f \in X^{\mathbb{N}}(f(0) = x \wedge \forall n \in \mathbb{N}\, A(f(n), f(n+1)))$$

Axiom of unique choice:

$$\forall x \in X\, \exists!\, y \in Y\, A(x, y) \to \exists f \in Y^X \forall x \in X\, A(x, f(x))$$

Bishop's constructive (forward) mathematics (BISH) intends to find a constructive substitute A' for a classical theorem A such that

$$\text{BISH} \vdash A' \text{ and } \text{CLASS} \vdash A \leftrightarrow A'$$

When A and A' are not equivalent in BISH, A does not admit a constructive proof with "Brouwerian counterexamples" P to A such that

$$\text{BISH} \vdash A \to P \text{ and } \text{BISH} \nvdash P.$$

The following principles are equivalent in BISH. Limited principle of omniscience (LPO) is an instance of the principle of excluded middle (PEM) $A \vee \neg A$:

(1) LPO: $\forall \alpha \in \mathbb{N}^{\mathbb{N}}(\exists n(\alpha(n) \neq 0) \vee \neg\exists n(\alpha(n) \neq 0))$
(2) $\forall x \in \mathbb{R}(0 < x \vee \neg(0 < x))$
(3) Monotone convergence theorem (Mandelkern, 1988): Every bounded monotone sequence of real numbers converges.
(4) Bolzano–Weierstrass theorem (Mandelkern, 1988): Every bounded sequence of real numbers has a convergent subsequence.
(5) Sequential compactness theorem (Ishahara and Schuster, 2002): Every compact metric space is sequentially compact.
(6) Cantor intersection theorem: Every sequence of closed sets of a compact metric space with the finite intersection property has a non-empty intersection.

LPO is false both in INT and RUSS (Troelstra and van Dalen, 1988, pp. 194, 204). The following principles are also equivalent in BISH. Lesser limited principle of omniscience (LLPO) is an instance of De Morgan's law $\neg(A \wedge B) \rightarrow \neg A \vee \neg B$:

(1) LLPO:
   $\forall \alpha, \beta \in \mathbb{N}^{\mathbb{N}}(\neg(\exists n\, \alpha(n) \neq 0 \wedge \exists n\, \beta(n) \neq 0) \rightarrow \neg\exists n\, \alpha(n) \neq 0 \vee \neg\exists n\, \beta(n) \neq 0)$
(2) $\forall x \in \mathbb{R}(\neg(0 < x) \vee \neg(x < 0))$
(3) $\forall x, y \in \mathbb{R}(xy = 0 \rightarrow x = 0 \vee y = 0)$
(4) WKL: Every infinite free has an infinite path.

LLPO is false both in INT and RUSS (Troelstra and van Dalen, 1988, pp. 195, 210).

Markov's constructive recursive mathematics (RUSS)

RUSS refers to "Russia" because of the Russian mathematician Markov and his school of constructive mathematics in the sense of Turing computability. Therefore, RUSS is Bishop's constructive mathematics (BISH) with MP and Church's thesis. MP is an instance of the double negation elimination $\neg\neg A \to A$. The following statements are equivalent in BISH:

(1) MP:
$\forall \alpha \in \mathbb{N}^{\mathbb{N}}(\neg\neg\exists n(\alpha(n) \neq 0) \to \exists n(\alpha(n) \neq 0))$
(2) $\forall x \in \mathbb{R}(\neg\neg(0 < x) \to 0 < x$

MP is an instance of the double negation elimination, false in INT, but accepted in RUSS.

MP is weaker than LPO. The following principles are equivalent with the weak Markov's principle (WMP) which holds in INT as well as in RUSS (Ishahara, 1992; Mandelkern, 1988b):

(1) WMP:
$\forall \alpha \in \mathbb{N}^{\mathbb{N}}(\forall \beta \in \mathbb{N}^{\mathbb{N}}(\neg\neg\exists n\beta(n) \neq 0) \vee \neg\neg\exists n(\alpha(n) \neq 0 \wedge \beta(n) \neq 0)) \to \exists n\, \alpha(n) \neq 0)$
(2) $\forall x \in \mathbb{R}(\forall y \in \mathbb{R}(\neg\neg(0 < y) \vee \neg\neg(y < x)) \to 0 < x)$

Intuitionistic mathematics (INT)

INT strongly depends on the acceptance of the fan theorem (FAN) (cf. Chapter 6) which has immense consequences for constructive analysis (Berger and Ishahara, 2005):

- A binary fan is the collection of all (finite) sequences of 0s and 1s (including empty sequence).
- A bar of a binary fan is a set of cut-offs such that for each infinite path (sequence) α through the fan there exists a natural number n such that $\overline{\alpha}(n)$ is in the bar.
- A bar is uniform if there exists a natural number k such that for each infinite path α through the fan, there is some $i \leq k$ such that $\overline{\alpha}(i)$ is in the bar.

Brouwer's FAN for detachable bar is stated as FAN_Δ which is a contrapositive form of WKL. FAN_Δ is weaker than LLPO (Ishahara,

1990) and WKL, accepted in INT, but false in RUSS (Troelstra and van Dalen, 1988, p. 220). In order to characterize FAN_Δ with compactness and uniform continuity, we introduce the following definitions.

Definition. A pointwise continuous function f from the Cantor space $2^{\mathbb{N}}$ into $\mathbb{N}$ is representable if there exists $\gamma : \{0,1\}^* \to \mathbb{N}$ such that $\forall \alpha \in 2^{\mathbb{N}} \exists n (\forall k < n \gamma(\overline{\alpha}(k)) = 0 \wedge \gamma(\overline{\alpha}(n)) = f(\alpha) + 1)$.

A subset S of a metric space X is a cozero set if there is a pointwise continuous function $f : X \to \mathbb{R}$ such that $s = \{x \in X | f(x) \neq 0\}$.

The following principles are equivalent in BISH:

(1) FAN_Δ.
(2) Uniform continuity theorem for representable functions (Loeb, 2005): *Every representable pointwise continuous function from $2^{\mathbb{N}}$ to $\mathbb{N}$ is uniformly continuous.*
(3) Heine–Borel theorem for cozero sets (Ishahara and Schuster, 2004): *Every cover of compact metric space by a sequence of cozero sets has a finite cover.*

Finally, we introduce a boundedness principle which is weaker than LPO and provable both in INT and RUSS.

Boundedness principle BD-N. Every countable pseudo-bounded subset of $\mathbb{N}$ is bounded, where a subset A of $\mathbb{N}$ is said to be pseudo-bounded if for each sequence (a_n) in A, $a_n < n$ for all sufficiently large n.

BD-N is weaker than LPO and provable in INT as well as in RUSS (Ishahara, 1992). There are remarkable equivalents of BD-N, for example, with the following.

Pointwise continuity theorem. *Every sequentially continuous mapping from a separable metric space into a metric space is pointwise continuous.*

There are useful relations of principles which can be proven in constructive reverse mathematics:

BISH $\vdash$ LPO $\leftrightarrow$ WLPO + MP
BISH $\vdash$ WLPO $\rightarrow$ LLPO
BISH $\vdash$ LLPO $\rightarrow$ FAN$_\Delta$ (Ishahara, 1990)
BISH $\vdash$ LPO $\rightarrow$ BD-N

The following principle is a weak form of continuous choice in intuitionism.

Weak continuity for numbers WC-N.

$$\forall \alpha \in \mathbb{N}^{\mathbb{N}} \exists n A(\alpha, n)$$
$$\rightarrow \forall \alpha \in \mathbb{N}^{\mathbb{N}} \exists m, n \forall \beta \in \mathbb{N}^{\mathbb{N}} (\overline{\alpha}(m) = \overline{\beta}(m) \rightarrow A(\beta, n))$$

There are remarkable relations with other principles in BISH:

BISH + WC-N $\vdash \neg$ LLPO (Troelstra and van Dalen, 1988, p. 210),
BISH + WC-N $\vdash$ WMP (Ishahara, 1992),
BISH + WC-N $\vdash$ BD-N (Ishahara, 1992).

Church's thesis is crucial for the research program RUSS.

Church's thesis CT$_0$.

$\forall n \exists m A(n, m) \rightarrow \exists k \forall n \exists m (A(n, U(m)) \wedge T(k, n, m))$ with Kleene's T-predicate and the result-extracting function U.

CT$_0$ has the following relations to other principles in constructive reverse mathematics:

BISH + CT$_0$ $\vdash \neg$ FAN$_\Delta$ (Troelstra and van Dalen, 1988, p. 221),
BISH + CT$_0$ $\vdash$ WMP (Ishahara, 1993),
BISH + CT$_0$ + MP $\vdash$ BD-N (Ishahara, 1992).

If BISH + CT$_0$ + MP is consistent, then

BISH $\nvdash$ MP $\rightarrow$ FAN$_\Delta$,
BISH $\nvdash$ BD-N $\rightarrow$ FAN$_\Delta$.

From a philosophical point of view, constructive reverse mathematics strictly follows Ockham's principle of parsimony: It classifies mathematical principles according to their most efficient costs avoiding unnecessary abstractions. In the hierarchy of BISH, INT, RUSS, and CLASS, we can determine how far mathematical principles are away from constructive and computational proofs. In this sense, constructive reverse mathematics bridges logic, mathematics, and computer science.

Chapter 9

From Intuitionistic to Homotopy Type Theory — Bridging Logic, Mathematics, and Computer Science

In computer programs, a data type is a classification of data which determines how the compiler or interpreter should use the data. Programming languages support various types of data (e.g., Boolean, integer, real). A data type provides a set of values which are related to a formal expression (e.g., variable, function). The type defines the operations that can be applied to the data, the meaning and store of the data. Thus, data types are necessary to reduce software bugs. Historically, data types in computer languages are rooted in logical type theory which was introduced to avoid logical paradoxes (e.g., Russell). In this case, every term of a formal language has a type. Operations are restricted to terms of a certain type. Since Church's typed λ-calculus, type theory has been extended to represent more and more abstract mathematical objects from numbers up to functions, functionals, functional of functionals, etc. Thus, type theories are a bridge between logic, mathematics, and computer science.

Martin-Löf's intuitionistic type theory offers both a philosophical foundation of constructive mathematics as well as a proof assistant in computer science (e.g., Coq). Recently, homotopy type theory extends formalization from set-theoretical objects up to categories related to more and more inaccessible cardinals (e.g., Grothendieck universe in algebraic geometry). Homotopy type theory (HoTT) tried to develop a universal ("univalent") foundation of mathematics as well as computer languages with respect to proof assistants for advanced mathematical proofs. Behind all that, there is the fundamental epistemic question how far even abstract mathematical objects such as sets, structures, and categories can be reduced to the digital world of computers (represented by data types of computer languages). We must not forget the Platonic belief that mathematical universes with structures and categories are ontologically real. Following this line, physical universes are only parts of the mathematical world and the digital world is only their formal representation in computer languages (compare Chapter 15).

Curry–Howard correspondence

The BHK-interpretation uses "construction" as an intuitive notion which can be interpreted as an effective function (compare Chapter 4). In Kleene's and Kreisel's approach (compare Chapter 7), proofs are realized by machines resp. computational functionals. Haskell Curry related proofs in formal Hilbert-style systems to computational methods of combinatory logic. In 1969, the logician William A. Howard observed that Gentzen's proof system of natural deduction can be directly interpreted in its intuitionistic version as a typed variant of the mode of computation known as lambda calculus (Howard, 1969).

According to Church, $\lambda a.b$ means a function mapping an element a onto the function value b with $\lambda a.b[a] = b$. In the following, proofs are represented by terms $a, b, c, \ldots$; propositions are represented by $A, B, C, \ldots.$

Examples.

$$
\begin{array}{cc}
 & [A] \\
\lambda a(\lambda b.a) \; \vdots & \\
\underline{\quad B \to A \quad} & \\
(\to \text{I}) \quad A \to (B \to A) &
\end{array}
\qquad
\begin{array}{cc}
 & [A] \\
\lambda a.b \quad \vdots & \\
\underline{\quad B \quad} & \\
(\to \text{I}) \quad A \to B &
\end{array}
$$

In general, the Curry–Howard correspondence means a correspondence between formal proofs and computational type systems.

Natural deduction and Lambda calculus

In natural deductions, Γ, Γ_1 and Γ_2 denote ordered sequences of formulas, while in Lambda-calculus type rules, they denote sequences of typed formulas:

<table>
<tr><td align="center">Intuitionistic implicational
natural deduction</td><td align="center">Lambda calculus type
assignment rules</td></tr>
</table>

$$\frac{}{\Gamma_1, \alpha, \Gamma_2 \vdash \alpha}\,\text{Ax} \qquad\qquad \frac{}{\Gamma_1, x : \alpha,\ \Gamma_2 \vdash x : \alpha}$$

$$\frac{\Gamma, \alpha \vdash \beta}{\Gamma \vdash \alpha \to \beta}\to I \qquad\qquad \frac{\Gamma, x : \alpha \vdash t : \beta}{\Gamma \vdash \lambda x.t : \alpha \to \beta}$$

$$\frac{\Gamma \vdash \alpha \to \beta \quad \Gamma \vdash \alpha}{\Gamma \vdash \beta}\to E \qquad\qquad \frac{\Gamma \vdash t : \alpha \to \beta \quad \Gamma \vdash u : \alpha}{\Gamma \vdash tu : \beta}$$

A proof $\Gamma \vdash \alpha$ means a program that, given values with the types listed in Γ, works on an object of type α. An axiom corresponds to the introduction of a new variable with an unconstrained type, the $\to I$ rule corresponds to function abstraction, and the $\to E$ rule corresponds to function application. In Curry–Howard correspondence, $t : \alpha$ means "t proves α" as well as "t is of type α".

Propositions as types in intuitionistic type theory. In Curry–Howard correspondence, propositions are understood as types and predicates are dependent types. For example, the predicate $prime(x)$ is interpreted as type A depending on x (Dybjer and Palmgren, 2016).

$$\bot = \emptyset$$
$$\top = 1$$
$$A \vee B = A + B$$
$$A \wedge B = A \times B$$
$$A \supset B = A \to B$$

$$\exists x : A.B = \Sigma x : A.B$$

$$\forall x : A.B = \Pi x : A.B$$

$A + B$ is the disjoint sum of A and B, and $\Sigma x : A.B$ is the disjoint sum of the A-indexed family of type B. The elements of $\Sigma x : A.B$ are pairs (a, b) such that $a : A$ and $b : B[x := a]$ (the type obtained by substituting all free occurrences of x in B by a). $A \times B$ is the Cartesian product. The elements of $\Pi x : A.B$ are computable functions f such that $fa : B[x := a]$, whenever $a : A$.

Theorem of prime numbers under Curry–Howard interpretation. Euclid's theorem expresses that there are arbitrarily large primes:

$$\forall m : \mathrm{N}.\exists n : \mathrm{N}.m < n \wedge \mathrm{prime}(n).$$

Under the Curry–Howard interpretation this becomes the type of functions which map a number m to a triple $(n(p, q))$, where n is a number, p is a proof that $m < n$ and q is a proof that n is prime:

$$\Pi m : \mathrm{N}\Sigma n : \mathrm{N}.m < n \times \mathrm{prime}(n).$$

The proofs-as-programs principle demands a constructive proof that there are arbitrarily large primes. In this case, the constructive proof is understood as a program which given any number produces a larger prime together with proofs that it indeed is larger and indeed is prime.

Martin-Löf's intuitionistic type theory

In addition to the type formers of the Curry–Howard interpretation, the logician and philosopher Per Martin-Löf extended the basic intuitionistic type theory (containing Heyting's arithmetic of higher types HA^{ω} and Gödel's system T of primitive recursive functions of higher type) with primitive identity types, well founded tree types, universe hierarchies and general notions of inductive and inductive–recursive definitions. His extension increases the proof-theoretic strength of type theory and its application to programming as well as to formalization of mathematics (Martin-Löf, 1998). We start with the transformation of intuitionistic predicate logic into type theory.

Intuitionistic type predicate logic

We only consider the Π-type former for the $\forall$-quantifier, in order to illustrate the principle of type formation, introduction, elimination, and equality:

Π-formation	Π-introduction	Π-elimination
$\dfrac{\Gamma \vdash A \quad \Gamma, x : A \vdash B}{\Gamma \vdash \Pi x : A.B}$	$\dfrac{\Gamma, x : A \vdash b : \beta}{\Gamma \vdash \lambda x.b : \Pi x : A.B}$	$\dfrac{\Gamma \vdash f : \Pi x : A.B \quad \Gamma \vdash a : A}{\Gamma \vdash fa : B[x := a]}$

Π-equality is introduced by β-conversion and η-conversion:

β-conversion	η-conversion
$\dfrac{\Gamma, x : A \vdash b : B \quad \Gamma \vdash a : A}{\Gamma \vdash (\lambda x.b)a = b[x := a] : B[x := a]}$	$\dfrac{\Gamma \vdash f : \Pi x : A.B}{\Gamma \vdash \lambda x.fx = f : \Pi x : A.B}$

Congruence rules preserve equality: e.g.,

Congruence rule
$\dfrac{\Gamma \vdash A = A' \quad \Gamma, x : A \vdash B = B'}{\Gamma \vdash \Pi x : A.B = \Pi x : A'.B'}$

Intuitionistic type arithmetic

As in Peano arithmetic, the natural numbers are generated by 0 and the successor operation s. Again, we consider N-formation, N-introduction, N-elimination, and N-equality:

N-formation	N-introduction	
$\Gamma \vdash \mathrm{N}$	$\Gamma \vdash 0 : \mathrm{N}$	$\dfrac{\Gamma \vdash 0 : \mathrm{N}}{\Gamma \vdash s(a) : N}$

The elimination rule states that these are the only ways to generate a natural number (Dybjer and Palmgren, 2016). The function $f(c) = R(c, d, xy.e)$ is defined be primitive recursion on the natural number c with base d and step function $xy.e$ (or $\lambda xy, e$) which maps the value y for the previous number $x : \mathrm{N}$ to the value for $s(x)$:

N-elimination

$$\frac{\Gamma, x : \mathrm{N} \vdash C \quad \Gamma \vdash c : \mathrm{N} \quad \Gamma \vdash d : C[x := 0] \quad \Gamma, y : \mathrm{N}, z : C[x := y] \vdash e : C[x := s(y)]}{\Gamma \vdash R(c, d, yz.e) : C[x := c]}$$

N-elimination is the type of a function which is defined by primitive recursion. Under the Curry–Howard interpretation, it also expresses the rule of mathematical induction.

N-equality (under appropriate premises)

$$R(0, d, yz.e) = d : C[x := 0]$$
$$R(s(a), d, yz.e) = e[:= a, z := R(a, d, yz.e)] : C[x := s(a)]$$

The universe of small types

In his first version of type theory (unpublished, 1971), Martin-Löf used an axiom assuming a type of all types (in analogy to the set of all sets in set theory) which was obviously an impredicative conception and inconsistent. In analogy to set-theoretic universes with impredicative concepts, Martin-Löf restricted types to a universe U *of* small types closed under all type formers of the theory, except that it does not contain itself (Martin-Löf, 1975):

U-formation

$$\Gamma \vdash \mathrm{U}$$

U-introduction

$$\Gamma \vdash \varnothing : U \qquad\qquad \Gamma \vdash 1 : U$$

$$\frac{\Gamma \vdash A : U \quad \Gamma \vdash B : U}{\Gamma \vdash A + B : U} \qquad \frac{\Gamma \vdash A : U \quad \Gamma \vdash B : U}{\Gamma \vdash A \times B : U}$$

$$\frac{\Gamma \vdash A : U \quad \Gamma \vdash B : U}{\Gamma \vdash A \to B : U}$$

$$\frac{\Gamma \vdash A : U \quad \Gamma, x : A \vdash B : U}{\Gamma \vdash \Sigma x : A.B : U} \qquad \frac{\Gamma \vdash A : U \quad \Gamma, x : A \vdash B : U}{\Gamma \vdash \Pi x : A.B : U}$$

$$\Gamma \vdash N : U$$

U-elimination

$$\frac{\Gamma \vdash A : U}{\Gamma \vdash A}$$

Type-theoretic universe U and the Grothendieck universe

The type-theoretic universe U is analogous to a Grothendieck universe in set theory which is a set of sets closed under all the methods sets can be constructed in Zermelo-Fraenkel (ZF) set theory. This is a reminder of a Grothendieck universe U:

1. $x \in U$, $y \in x \Rightarrow y \in U$ (transitivity),
2. $x, y \in U \Rightarrow \{x, y\} \in U$,
3. $x \in U \Rightarrow \mathcal{P}(x) \in U$ (power set),
4. $\{x_\alpha\}_{\alpha \in I}$ family of elements of U, $I \in U \Rightarrow \cup_{\alpha \in I} x_\alpha \in U$.

Alexander Grothendieck (1928–2014) used his universe as a way of avoiding proper classes in algebraic geometry (Grothendieck, 1971–1977). Its existence goes beyond the usual axioms of ZF set theory and implies the existence of strongly inaccessible cardinals. Tarski–Grothendieck set theory is an axiomatic treatment of set theory, used in some automatic proof systems, in which every set belongs to a Grothendieck universe. The concept of a Grothendieck universe can also be defined in a topos (category theory). Thus, Martin-Löf's type-theoretic universe is an enormous extension of intuitionistic type theory to grasp abstract mathematical objects.

The axiom of choice in intuitionistic type theory. In intuitionistic type theory, the axiom of choice is a provable theorem. This result is an immediate consequence of the BHK-interpretation of the intuitionistic quantifiers:

Theorem. $(\Pi x : A.\Sigma y : B.C) \to \Sigma f : (\Pi x : A.B)C[y := fx]$

Proof. The proof simply follows from the definitions of the Π- and Σ-type constructors:

$\Pi x : A.\Sigma y : B.C$ is the type of functions which map elements $x : A$ to pairs (y, z) with $y : B$ and $z : C$.

$\Rightarrow$ The choice function f is obtained by returning the first component $y : B$ of this pair. $\qquad\qquad\qquad\qquad\qquad\qquad\qquad\qquad\qquad\square$

In set theory, the axiom of choice is in general not constructive. In intuitionistic type theory, we consider an axiom of choice for types. Types are not in general appropriate constructive approximations of sets in set theory (Martin-Löf, 2009).

General identity type former

Identity of natural numbers can be defined by the rule of primitive recursion. In 1975, Martin-Löf extended intuitionistic type theory with an identity type former I for all types (Martin-Löf, 1975). The rules for I express that the identity relation is inductively generated by the proof of reflexivity (constant r):

I-formation	I-introduction
$\dfrac{\Gamma \vdash A \quad \Gamma \vdash a : A \quad \Gamma \vdash a' : A}{\Gamma \vdash I(A, a, a')}$	$\dfrac{\Gamma \vdash A \quad \Gamma \vdash a : A}{\Gamma \vdash r : I(A, a, a)}$

The elimination rule for the identity type is a generalization of identity elimination in predicate logic (with elimination constant J):

I-elimination

$$\frac{\Gamma, x : A, y : \mathrm{I}(A, a, x) \vdash C \quad \Gamma \vdash b : A \quad \Gamma \vdash c : \mathrm{I}(A, a, b) \quad \Gamma \vdash d : C[x := a, y := r]}{\Gamma \vdash \mathrm{J}(c, d) : C[x := b, y := c]}$$

I-equality (under appropriate assumptions)

$$\mathrm{J}(r, d) = d$$

Trees are important data structures in programming languages (as well as in intuitionistic mathematics). In 1982, Martin-Löf extended his intuitionistic type theory to grasp these formal objects (Martin-Löf, 1982).

Well-founded trees and iterative sets in intuitionistic type theory

Elements of the type $\mathrm{W}x : A.B$ of well-founded trees are varying and arbitrarily branching. Therefore, the type is given by a generalized inductive definition (e.g., W-formation and W-introduction, analogously W-elimination and W-equality). Each $a : A$ represents a term constructor sup a with (possibly infinite) arity $B[x := a]$:

W-formation   W-introduction

$$\frac{\Gamma \vdash A \quad \Gamma, x : A \vdash B}{\Gamma \vdash \mathrm{W}x : A.B} \qquad \frac{\Gamma \vdash a : A \quad \Gamma, y : B[x := a] \vdash b : \mathrm{W}x : A.B}{\Gamma \vdash \mathrm{sup}(a, y, b) : \mathrm{W}x : A.B}$$

Well-founded trees can be applied to a type-theoretic model of constructive ZF set theory (Aczel, 1978). The proof-theoretic strength of intuitionistic type theory is extended to constructive ZF set theory by the type of iterative sets:

$$V = \mathbf{W}x : \mathbf{U}.x$$

If $a : \mathbf{U}$ is a small type and $x : A \vdash M$ an indexed family of iterative sets, then $\{M | x : A\}$ is an iterative set. An iterative set is a data structure in functional programming. Type V can also be interpreted as a constructive type theoretic model of the cumulative hierarchy in ZF set theory.

The proof-theoretic strength of intuitionistic type theory can be extended to increasing set-theoretical universes in analogy to increasing large cardinal numbers. At first, we remind the reader of some basics of cardinal numbers (Conway and Guy, 1996):

Reminder of cardinal numbers

Cardinality measures the size of sets. Two sets have the same cardinality iff there is a one-to- one mapping (bijection) between the elements of the two sets. The cardinality of set X is denoted by $|X|$. The sequence of cardinal numbers starts with the natural numbers as finite cardinals which are followed by infinite cardinal numbers of well-ordered sets (aleph numbers) which are indexed by ordinal numbers:

$$0, 1, 2, 3, \ldots, n, \ldots; \quad \aleph_0, \aleph_1, \aleph_2, \ldots, \aleph_\alpha, \ldots$$

According to the index of ordinal numbers, we distinguish successor and limit cardinals. In general, a number can describe the size of a set, or its position in a sequence (counting). Cardinal numbers relate to the size of sets and ordinal numbers to the position aspect. For finite sets, the aspects of size and position cannot be distinguished in the natural numbers. But for infinite sets we get different notions. Under the axiom of choice, there is a relation between cardinal and ordinal numbers: The cardinality of a set X is the least ordinal number α such that there is a bijection between X and α.

There is a smallest transfinite cardinal number $\aleph_0$ which is the size of the set of all natural numbers and corresponds to the ordinal number ω. For every cardinal number there is a next-larger cardinal. Cantor's continuum hypothesis states that the cardinality c of all real numbers (continuum) is the same as $\aleph_1$. The cardinality c of the continuum is set-theoretically the cardinality of the power set $2^{\aleph_0} := \mathcal{P}(\aleph_0)$ of $\aleph_0$. Thus, the continuum hypothesis states $\aleph_1 = 2^{\aleph_0}$. The continuum hypothesis is independent of the standard axioms of set theory which means that it can neither be proved nor disproved from these axioms. The generalized continuum hypothesis states that for every infinite set X, there are no cardinals between $|X|$ and $2^{|X|}$.

Roughly speaking, "large" cardinal numbers relate to certain set-theoretical constructions which are not sufficient to "access" these numbers (Drake, 1974). Examples are the so-called inaccessible cardinals.

A cardinal ν is called inaccessible iff

(a) ν is a limit cardinal larger than $\aleph_0$;
(b) the union of less than ν sets with cardinalities smaller than ν has a cardinality smaller than ν;
(c) If the cardinality of a set is smaller than ν, then its power set is also smaller than ν.

The existence of inaccessible cardinals is equivalent to the existence of a set-theoretical universe which is closed with respect to certain set-theoretical constructions:

A class U is called a (set-theoretic) universe iff

(a) $\aleph_0 \in U$ (infinity),
(b) $x \in M \in U \Rightarrow x \in U$ (transitivity),
(c) $M \in U \Rightarrow \mathcal{P}(M)$ (power set),
(d) $\mathfrak{M}$ set of sets, $\mathfrak{M} \in U \Rightarrow \cup \, \mathfrak{M} \in U$ (union),
(e) $f : M \to N$ function, $M \in U \Rightarrow f(M) \in U$.

It is well-known that there are even "larger" cardinals than inaccessible cardinals with respect to certain set-theoretical conditions which must be demanded by new axioms.

Large cardinal numbers and set-theoretic universes in intuitionistic type theory

Universes	Large cardinals
Set-theoretic universes (Palmgren, 1998)	inaccessible cardinals
superuniverses (Rathjen *et al.*, 1998)	hyperinaccessible cardinals
Mahlo universes (Setzer, 2000)	Mahlo cardinals
$\vdots$	$\vdots$

These universe constructions can be characterized by types if the schema of inductive-definitions in intuitionistic type theory is extended by inductive recursive definitions. In this case, we generate simultaneously inductively a type and a function from that type defined by recursion.

In analogy to Martin-Löf intuitionistic type theory, another higher-order typed lambda calculus was introduced by Tierry Coquand (Coquand and Huet, 1988). His calculus of constructions (CoC) became the type-theoretic fundament of the proof assistant Coq:

Calculus of constructions (CoC)

CoC is a type theory which can serve as a typed programming language as well as a constructive foundation of mathematics. With inductive types, the calculus of inductive constructions (CiC) removes impredicativity. It extends the Curry–Howard isomorphism to proofs in the full intuitionistic predicate calculus.

Terms in CoC are constructed inductively:

(1) T is a term (called **Type**),
P is a term (called **Prop** which is the type of all propositions),
Variables $x, y, \ldots$ are terms.
(2) A, B terms $\Rightarrow (AB)$ term,
A, B terms, X variable $\Rightarrow (\lambda X : A.B), (\forall X : A.B)$ term.

Objects in CoC are distinguished in five kinds:

(1) proofs (terms whose types are propositions),
(2) propositions (small types),
(3) predicates (functions which return propositions),
(4) large types (types of predicates, e.g., P),
(5) T (type of larger types).

CoC proves typed judgements of the form $x_1 : A_1, x_2 : A_2, \ldots \vdash t : B$ ("If variables $x_1, x_2, \ldots$ have types $A_1, A_2, \ldots$, the term t has type B"). Judgements are derived from the inference rules of CoC with the following abbreviations Γ (sequence $x_1 : A_1, x_2 : A_2, \ldots$), K (either P or T), and $A : B : C$ (A has type B, and B has type C):

$$(1) \quad \frac{}{\Gamma \vdash P : T}$$

$$(2) \quad \frac{\Gamma \vdash A : K}{\Gamma, x : A \vdash x : A}$$

$$(3) \quad \frac{\Gamma, x : A \vdash t : B : K}{\Gamma \vdash (\lambda x : A.t) : (\forall x : A.B) : K}$$

$$(4) \quad \frac{\Gamma \vdash M : (\forall x : A.B) \quad \Gamma \vdash N : A}{\Gamma \vdash MN : B[x := N]}$$

$$(5) \quad \frac{\Gamma \vdash M : A \quad A =_\beta B \quad B : K}{\Gamma \vdash M : B}$$

CoC only needs the basic operator $\forall$ to define the logical operators:

$$A \Rightarrow B \equiv \forall x : A.B \ (x \notin B)$$

$$A \wedge B \equiv \forall C : P.(A \Rightarrow B \Rightarrow C) \Rightarrow C$$

$$A \vee B \equiv \forall C : P.(A \Rightarrow C) \Rightarrow (B \Rightarrow C) \Rightarrow C$$

$$\neg A \equiv \forall C : P.(A \Rightarrow C)$$

$$\exists x : A.B \equiv \forall C : P.(\forall x : A(B \Rightarrow C)) \Rightarrow C$$

The basic operator $\forall$ is also used to define the data types:

Booleans:	$\forall A : P.A \Rightarrow A \Rightarrow A$
naturals:	$\forall A : P.(A \Rightarrow A) \Rightarrow (A \Rightarrow A)$
product $A \times B$:	$A \wedge B$
disjoint union $A + B$:	$A \vee B$

A proof assistant or interactive theorem prover is a software tool which supports the process of formal proving by human-machine interaction. We already discussed the example of Minlog in previous chapters. The type theory CoC is the basis of the proof assistant Coq.

The proof assistant Coq

Coq implements a program specification which is based on the Calculus of Inductive Constructions (CiC) combining both a higher-order logic and a richly-typed functional language (Bertot and Castran, 2004).

The <u>commands</u> of Coq allow:

— to define functions or predicates (that can be evaluated efficiently).
— to state mathematical theorems and software specifications.
— to interactively develop formal proofs of these theorems.
— to machine-check these proofs by a relatively small certification (kernel).
— to extract certified programs to languages (e.g., Objective Caml, Haskell, Scheme).

Coq provides interactive proof methods, decision and semi-decision algorithms. Connections with external theorem provers are available. Coq is a platform for the formalization of mathematics as well as the development of programs.

In Coq, proof checking is reduced to type-checking in type theories such as CoC resp. CiC. Thus, the heart of Coq is a type-checking algorithm in the language of CiC: Coq objects are sorted into the `Prop` sort and the `Type` sort:

— Prop is the sort of propositions (with type `Prop`):

e.g., $\forall$ A B : Prop, A $\bigwedge$ B $\rightarrow$ A $\bigvee$ B

New predicates can be defined either inductively or by abstracting over existing propositions:

e.g., `Definition divide (x y:N)` := $\exists z, x * z = y$

— `Type` is the sort for datatypes and mathematical structures (with type `Type`):

Types can be inductive structures or types for tuples or a form of subsets (Σ-types):

e.g., the type of even natural numbers $\{n : N | \text{even } n\}$

Coq implements a functional programming language supporting several types:

e.g., the pairing function of type $z \rightarrow z * z$ is written fun $x \Rightarrow (x, x)$

Further details of Coq can be found in several tutorials. In the meantime, Coq and CiC were applied in advanced difficult proofs (e.g., the four-color-theorem (Gonthier, 2008)).

Homotopy type theory

Since their very beginning, data types have played a crucial role in computer languages: How far can mathematical objects be represented with types of computer languages? Homotopy theory is an outgrowth of algebraic topology (Spanier, 1966) and homological algebra (Eilenberg and Cartan, 1956) with relationships to higher category theory which can be considered as fundamental concepts of mathematics. Type theory is a branch of mathematical logic and theoretical computer science. Homotopy type theory (HoTT) interprets types as objects of abstract homotopy theory. Therefore, HoTT tried to develop a universal ("univalent") foundation of mathematics as well as computer language with respect to the proof assistant Coq.

Reminder: Homotopy theory

In topology, two continuous functions from one topological space to another are called homotopic (Greek: *homós* "same, similar" and *tópos* "place") if they can be continuously deformed one into the other (Fig. 8).

Formally, a homotopy between two continuous functions f and g from a topological space X to a topological space Y is defined to be a continuous function $H : X \times [0,1] \to Y$ from the product of the space X with the unit interval $[0,1]$ to Y such that, if $x \in X$ then $H(x,0) = f(x)$ and $H(x,1) = g(x)$. In this case, f and g are said to be homotopic (Hazewinkel, 2001).

Two spaces X and Y are homotopy equivalent $(A \simeq B)$ if there exist continuous maps $f : X \to Y$ and $g : Y \to X$ such that $g \circ f$ is homotopic to the identity map id_X and $f \circ g$ is homotopic to id_Y.

The homotopy category is the category whose objects are topological spaces, and whose morphisms are homotopy equivalence classes of continuous maps. Two topological spaces X and Y are isomorphic in this category if and only if they are homotopy-equivalent.

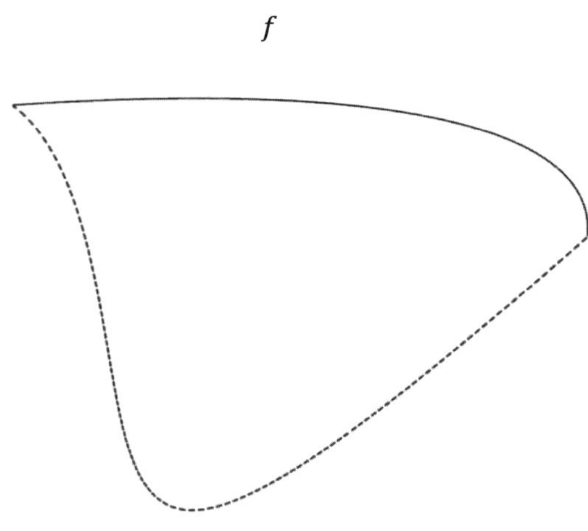

Fig. 8. Homotopic curves (Wikipedia Homotopy [https://en.wikipedia.org/wiki/Homotopy]).

In HoTT, every type A is interpreted as a space. The elements of the space correspond to the terms of type A. The identity type $a = b$ (better: $\text{id}_A(a, b)$ to indicate the type A of a and b) is the type of all paths from the point a to b in space space A (Awodey and Warren, 2009). Obviously, a proof that a point a equals a point b can be interpreted as a path from the point a to b. The fact that, for every point a, there is a path of type $a = a$, corresponds to the reflexivity of equality. The symmetry of equality corresponds to a path of type $a = b$ which can be inverted as a path of type $b = a$. The fact that two paths of type $a = b$ and $b = c$ can be connected as path of type $a = c$ corresponds to the transitivity of equality.

The homotopy interpretation of types leads to important differences in the meaning of substitution of equality. What does substitution of equality mean? In classical algebra, the equality of two values a and b means that a and b can be used interchangeably without any distinction. In the homotypy interpretation, a path $p : a = b$ and a proof of property $P(a)$ are given. In this case, the proof can be transported along the path to a proof of property $P(b)$. But there may be different paths $a = b$ which provide different results. Therefore, the path which is actually used must be indicated.

Intuitionistic type theory and homotopy theory

Intuitionistic Type Theory	Homotypy Theory
types A	spaces A
terms a	points a
$a : A$	$a \in A$
dependent type $x : A \vdash B(x)$	fibration $B \to A$
identity type $\text{id}_A(a, b)$	space of paths from a to b
$p : \text{id}_A(a, b)$	path $p : a \mapsto b$
$\alpha : \text{id}_{\text{Id}_A(a,b)}(p, q)$	homotopy $\alpha : p \Rightarrow q$

In intuitionistic type theory, a term $a : A$ can be understood as an element of the type A or a proof of the proposition A or, in homotopy type theory, as a point of the space A. Proofs p of identity between two elements a, b of a type A are geometrically illustrated as paths connecting the corresponding points. All proofs of an identity are not forced to be equal (Hofmann and Streicher, 1998): This was shown by a model where each type is interpreted as a groupoid (i.e., a category, in which each morphism is invertible resp. an isomorphism). A groupoid does not only contain paths between points (1-dimensional paths), but also paths between paths between points (2-dimensional paths) and so forth up to a tower of n-dimensional paths.

Category theory has become a remarkable bridge between contemporary mathematics and theoretical computer science. Mathematical categories yield semantical meaning of highly advanced type theories with respect to constructive mathematics. This is the reason why we consider categorical constructions in more details. Historically, the concepts of categories and functors were introduced by Samuel Eilenberg and Saunders Mac Lane, when they tried to understand how mathematical structures are preserved in algebraic topology (Eilenberg and MacLane, 1942). Therefore, structures must be collect in new units ("categories") in order to study their common properties. The preservation of mathematical structures was already a key idea of 20th century mathematics since Dedekind, Noether, Bourbaki, and others, which was extended by increasing abstractions in category theory (Feferman, 1977). Further on, category theory allows constructions of new mathematical structures out of given ones which opens new avenues of mathematical research (Baez and May, 2010).

Definition of category. A category C consists of objects and morphisms which satisfy the following conditions:

- <u>Morphism</u>: For objects X and Y, there is a set $\mathrm{Hom}(X, Y)$ of morphisms $f : X \to Y$ from X to Y in C.
- <u>Identity</u>: For every object X, there exists a morphism id_X in $\mathrm{Hom}(XX)$ (called the identity on X).
- <u>Composition</u>: For objects X, Y and Z, there exists a partial binary operation from $\mathrm{Hom}(X, Y) \times \mathrm{Hom}(Y, Z)$ to $\mathrm{Hom}(X, Z)$ (called the

composition of morphisms in C). If $f : X \to Y$ and $g : Y \to Z$, the composition of f and g is denoted by $g \circ f : X \to Z$.

Morphisms, identity, and composition satisfy the following axioms:

(1) Associativity: $h \circ (g \circ f) = (h \circ g) \circ f$ for all $f : X \to Y, g : Y \to Z$, and $h : Z \to W$.
(2) Identity: $(\mathrm{id}_Y \circ f) = f$ and $(f \circ \mathrm{id}_X) = f$ for all $f : X \to Y$.

Well-known structures with structure-preserving maps can be collected in categories (Marquis, 2014). Examples are the category SET with sets as objects and functions as morphisms, category TOP with topological spaces as objects and continuous functions as morphisms, category DIFF with differential manifolds as objects and smooth maps as morphisms, BOOL with Boolean algebras as objects and structure-preserving homomorphisms ($\bot, \top, \wedge, \vee$) as morphisms, category HEYT with (intuitionistic) Heyting algebras and ($\bot, \top, \wedge, \vee$) morphisms, etc. The category $\overline{\mathrm{TOP}}$ consists of topological spaces as objects and equivalence classes of homotopic functions as morphisms which are not structure-preserving functions.

From a proof-theoretical point of view, it is remarkable that categories can be understood as deductive systems. A graph consists of the classes of arrows and objects. The arrows are called oriented edges and the objects are nodes. We distinguish source and target mapping which assigns an arrow to a source node resp. a target node. A deductive system is a graph with an arrow $\mathrm{id}_X : X \to X$ and a binary composition $g \circ f : X \to Z$ of arrows $f : X \to Y$ and $g : Y \to Z$. In logics, the objects of a deductive system are formulas, the arrows are deductions or proofs. Operations on arrows are rules of inference. The following definition of categories as deductive systems was introduced in (Lambek, 1968–1972).

Definition of category as deductive system. A category is a deductive system in which the following equations hold between proofs for all $f : X \to Y$, $g : Y \to Z$, and $h : Z \to W$:

$$(\mathrm{id}_Y \circ f) = f, \quad (f \circ \mathrm{id}_X) = f, \quad h \circ (g \circ f) = (h \circ g) \circ f.$$

Foundations of mathematics can be related to higher category theory. In the homotopical model of type theory, types are interpreted as certain categories. In Martin-Löf type theory and CiC, types are stratified in homotopy levels (abbreviated h-level): Types of h-level 0 are those equal to the one point type. Types of h-level 1 are those in which any two elements are equal. Types of h-level 2 are called sets. For example, the type of natural numbers has h-level 2. Categories are definied as types of h-level 3. Higher category theory yields models of Martin-Löf type theory and CiC which is also the basis of the proof assistant Coq. Thus, homotopy type theory (HoTT) bridges categorical foundations of mathematics with proof theory and proof assistants (The Univalent Foundations Program, 2013).

HoTT connects type theory with proof theory, set theory, and homotopy theory

Types	Proofs	Sets	Homotopy
A	proposition	set	space
$a : A$	proof	element	point
$B(x)$	predicate	family of sets	fibration
$b(x) : B(x)$	conditional proof	family of elements	section
$0, 1$	$\bot, \top$	$\varnothing, \{\varnothing\}$	$\varnothing, *$
$A + B$	$A \vee B$	disjoint union	coproduct
$A \times B$	$A \wedge B$	set of pairs	product space
$A \to B$	$A \Rightarrow B$	set of functions	function space
$\Sigma_{(x:A)} B(x)$	$\exists_{x:A} B(x)$	disjoint sum	total space
$\Pi_{(x:A)} B(x)$	$\forall_{x:A} B(x)$	product	space of sections
Id_A	Equality $=$	$\{(x, x) \mid x \in A\}$	path space A^I

According to HoTT, Vladimir Voevodsky suggested a formalization of modern mathematics which became known as "univalent foundation". It is a version of Martin-Löf type theory extended by a new axiom considering a universal property of identity and equivalence of mathematical structures.

The univalence axiom (UA). Let U be a universe with small types which are themselves terms $A : U$. In homotopy theory, U is a space with spaces (types) as points. There is a canonical function to turn identity of these spaces (types) into their equivalence:

UA: A function of the type $(A = B) \to (A \simeq B)$ is itself an equivalence:

$$(A = B) \simeq (A \simeq B)$$

For points A and B of the universe U (i.e., small types), UA identifies the following three notions (Awodey, 2014):

- (logical) an identification $p : A = B$ of A and B
- (topological) a path $p : A \mapsto B$ from A to B in U
- (homotopical) an equivalence $p : A \simeq B$ between A and B.

Equivalence ($\simeq$) refers to higher dimensional objects from set theory up to category theory (e.g., equal elements, isomorphic sets, equivalent groupoids): UA expresses that "everything" is preserved by equivalence. In particular, equivalent types (e.g., isomorphic structures) are identical.

Provability, constructivity, and computability in HoTT

HoTT allows mathematical proofs to be translated into a computer programming language for computer proof assistants (e.g., Coq, CiC) even for advanced mathematical categories with "isomorphism as equality" (UA). The connection of univalent foundations of mathematics and proof assistants was especially considered by Thierry Coquand, one of the founders of the Coq proof assistant. In short, an essential goal of HoTT is (Bezem *et al.*, 2014):

type checking in highly advanced type theories
$\Rightarrow$ proof checking in higher categories
(for advanced mathematical proofs)

Besides UA, HoTT is extended by higher inductively defined structures (e.g., inductively defined spaces with collections of points, paths, higher paths, etc.) which can be characterized by appropriate induction principles. HoTT is consistent with respect to a model in the category of Kan complexes (Voevodsky, 2012; Joyal, 2002). Thus, it is consistent relative to ZFC (with as many large cardinals which are necessary for nested univalent universes). But it is still an open question whether it is possible to provide a constructive justification of the univalence axiom (UA).

Chapter 10

Real Computability and Real Analysis

Reverse mathematics measures the degrees of proof-theoretical strength which distinguish mathematical principles, axioms, and theories. They are more or less constructive. Ordinary mathematics cannot completely be reduced to second-order arithmetic. From a computational point of view, it cannot completely reduced to the digital world of computers. What about mathematical thinking beyond Turing computability. What about mathematical theories based on, e.g., the fields $\mathbb{R}$ and $\mathbb{C}$ of real and complex numbers (Ebbinghaus *et al.*, 1991)? The mathematical theory of complex dynamical systems is based on continuous differential equations over $\mathbb{R}$ and $\mathbb{C}$ (Mainzer, 2007a). Most algorithms in mathematics (e.g., Newton, Euler, Gauss) are defined on $\mathbb{R}$ and $\mathbb{C}$. But the logical theory of decidability and computability is defined over digital numbers $\mathbb{Z}_2$.

The conflict of a discrete and continuous model of the world is deeply rooted in the beginning of modern science. Ancient philosophers of atomism (e.g., Democritus) believed in a world of interacting discrete particles as indivisible building blocks (atoms). They were attacked by philosophers of the continuum (Aristotle and coworkers) which cannot be explained by a chain of coupled pearls. Aristotle described change in nature by continuous dynamics. In the beginning of modern science, physicists assumed a mechanistic world of interacting atoms. But, mathematically, atoms were considered mass points determined by continuous differential equations. Leibniz as inventor of the differential calculus tried to bridge the discrete and continuous world by his philosophical concept of monads which were mathematically represented by infinitesimally small, but nevertheless non-zero quantities called differentials (Mainzer, 1981).

Scientific computing in physics and chemistry was mainly based on continuous functions. In practical cases, the solutions of their (real or complex) equations could only be approximated by algorithms of numerical analysis (e.g., Newton's method). Therefore, the procedures of numerical analysis depend strongly on the continuous concept of real numbers. But until today, numerical analysis is still a highly efficient collection of successful procedures of problem solving which must miss a strict logical foundation. The discrete theory of computability is well-founded by the discrete concept of Turing machines (Church's thesis).

Besides algorithmic procedures, there is also a long standing tradition of decidability problems in analysis and algebra. Since antiquity, decidability of geometric constructions like the squaring of the circle, trisection of angles, or Deli's problem were discussed. But, it was the transcendental real number π and Galois's proof of the unsolvability by radicals of polynomial equations of degree five and more which decided these problems (Mainzer, 1980). Thus, the foundations of algorithms with real numbers must be clarified.

John von Neumann, computer pioneer and brilliant mathematician, felt the gap between the digital world of logic and computers and the "real" world of ordinary mathematics, when he proclaimed in the Hixon Symposium Lecture (1948):

> "There exists today a very elaborate system of formal logic, and specifically, of logic as applied to mathematics. This is a discipline with many good sides, but also serious weakness.... The reason for this is that it deals with rigid, all-or-none concepts, and has very little contact with the continuous concept of the real or of the complex number, that is, with mathematical analysis. Yet analysis is the technically most successful and best-elaborated part of mathematics..." (Neumann, 1948)

Roger Penrose argued exactly on this line, when he considered chaos theory in his book *The Emperor's New Mind* (Penrose, 1991, p. 124):

> "Now we witnessed ... in certain extraordinarily complicated looking set, namely the Mandelbrot set. Although the rules which provide its definition are surprisingly simple, the set itself exhibits and endless variety of highly elaborate structures. Could this be an example of non-recursive (i.e., undecidable) set, truly exhibited before our mortal eyes?"

Penrose rejects recursive analysis, because in this case, a real (or complex) number must be input to a Turing machine bit by bit: Problems arise when one wants to decide if two numbers are equal. It is also problematic to reduce a real (resp. complex) problem to its rational "skeleton" because, e.g., the curve $x^3 + y^3 = 1$ has no rational points with both x and y positive.

Thus, we need a generalized concept of machines over a ring R (e.g., ring $\mathbb{Z}$ or fields $\mathbb{R}$ and $\mathbb{C}$) to unify the theory of complex dynamical systems with computational complexity in numerical analysis and mathematical physics. A canonical model of computation over the reals was introduced by Blum *et al.* (1989). Following their arguments, we start with the questions: How can one decide if a differential equation is chaotic? Is the Mandelbrot set decidable (Mandelbrot, 1990)?

Decision algorithm of Mandelbrot set

Consider a complex dynamical system with state function (endomorphism)

$$g_c : \mathbb{C} \to \mathbb{C}(\mathbb{C} = \mathbb{R} \times \mathbb{R}) \text{ with } g_c(z) = z^2 + c$$

The Mandelbrot set $\mathcal{M}$ consists of all complex numbers c such that the sequence ("orbit") $g_c(0) = c$, $g_c^2(0) = c^2 + c$, $g_c^3(0) = (c^2 + c)^2 + c, \ldots$ remains bounded. $\mathcal{M}$ is the set of all inputs c of the flowchart in Fig. 10 that do not halt. (If the orbit ever escapes the disk of radius 2, it will go off to infinity.) The flowchart is defined by a finite diagram with input node, computation node, branch node, and output node (Fig. 9). It has the power to accept and to compare

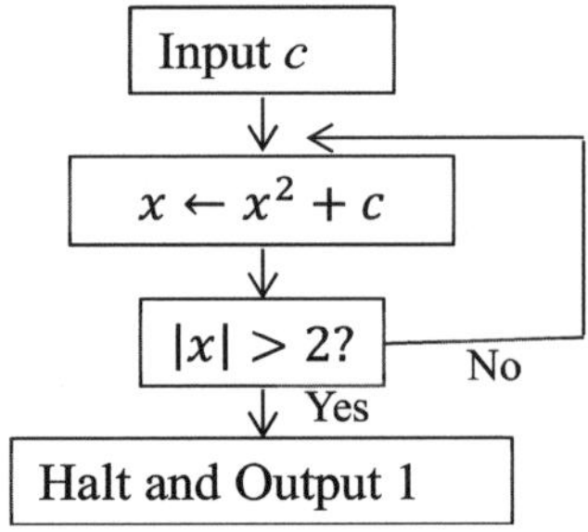

Fig. 9. Flowchart of Mandelbrot set $\mathcal{M}$.

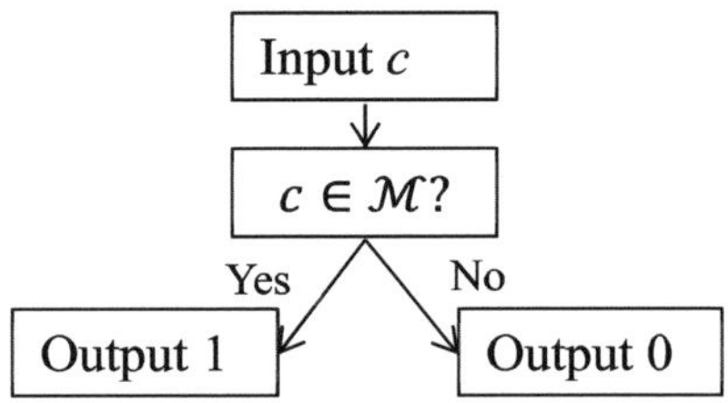

Fig. 10. Decision algorithm for $\mathcal{M}$.

complex numbers and to perform basic arithmetic operations on them. The halting set of the flowchart is the set of inputs where the computation ("orbit") halts.

The decision problem of the Mandelbrot set needs an algorithm ("machine") that, given input c (complex number), will decide in a finite number of steps whether or not c is in M (Fig. 10).

Newton's method is a typical search algorithm of numerical analysis and scientific computation (Goldstine, 1977). It is an iterative method to approximate the roots of non-linear equations. Thus, it is classical example to bridge the gap between digital algorithms of logic and computability theory and real algorithms of ordinary mathematics:

Newton's method over real numbers $\mathbb{R}$ and complex numbers $\mathbb{C}$

Given an initial approximation a to a root of the polynomial equation, $f(z) = 0$. Newton's method replaces a by the exact solution

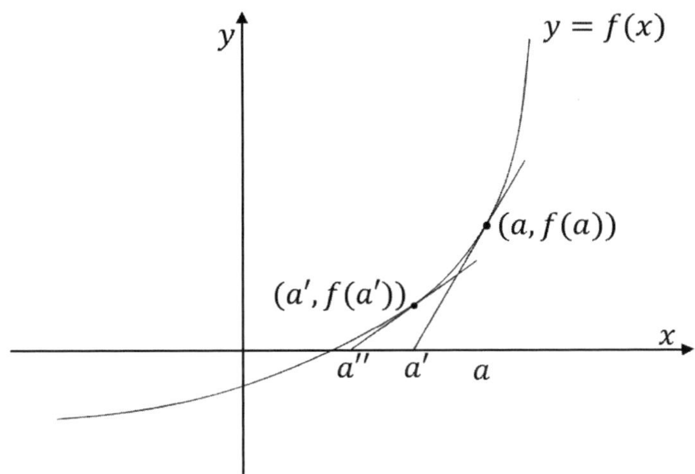

Fig. 11. Newton's method.

a' of the best linear approximation to f which is given by the tangent to the graph of f at point $(a, f(a))$ in Fig. 11. With a', the approximation is iterated to generate a'', etc.

The algorithm is defined by Newton's endomorphism $N_f : \mathbb{C} \to \mathbb{C}$ with

$$N_f(z) = z - \frac{f(z)}{f'(z)}.$$

Newton's method is not generally convergent.

Search machine of Newton's method over real numbers $\mathbb{R}$ and complex numbers $\mathbb{C}$

Newton's machine is represented by a finite directed graph in Fig. 12 with four types of nodes — input, computation, branch, and output — each with associated functions and conditions on incoming and outcoming edges.

For an input z_0, the machine generates the orbit $z_0, z_1 = N_f(z_0)$, $z_2 = N_f(z_1), \ldots, z_{k+1} = N_f(z_k) = N_f^{k+1}(z_0), \ldots$

The machine halts if $|f(z_k)| < \varepsilon$ and output is z_k. If $N_f(z_k)$ is undefined at some stage, there is no output.

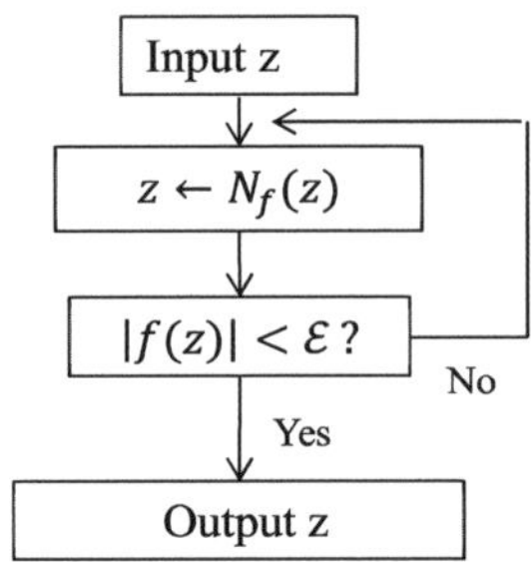

Fig. 12. Search algorithm for Newton's method.

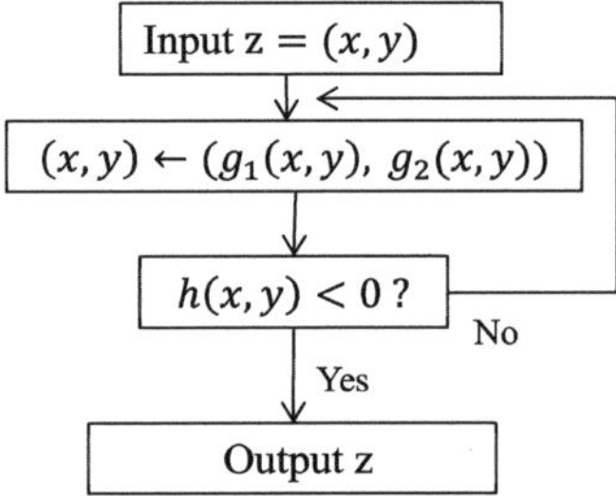

Fig. 13. Computation node of Newton's machine.

The machine will not in general halt on all inputs (e.g., infinite loops). The set $\Omega \subset \mathbb{C}$ is called the halting set of the machine iff it contains all inputs for which the machine halts with an output. The input–output map φ is defined on Ω with $\varphi : \Omega \to \mathbb{C}$.

Newton's machine (Fig. 13) is actually a machine over $\mathbb{R}$ with $\mathbb{C} = \mathbb{R}^2$ as input, output, and state space. Newton's endomorphism (computation node) is given by a rational function (quotient of two polynomials)

$$g = (g_1, g_2) : \mathbb{R}^2 \to \mathbb{R}^2 \text{ with } g_1(x, y) = \operatorname{Re} N_f(x + iy) \text{ and } g_2(x, y) = \operatorname{Im} N_f(x + iy).$$

These examples of $\mathbb{R}$- and $\mathbb{C}$-based machines can mathematically be generalized for an ordered commutative ring R. Examples are the integers $R = \mathbb{Z}$ and the real numbers $R = \mathbb{R}$. The direct sum of R with itself n times is denoted by R^n. These examples of graphic systems can be generalized in a finite-dimension machine over a ring (Blum *et al.*, 1998).

Definition of a finite-dimension graphic machine over ring R. A finite-dimension graphic machine M over R (ordered commutative ring or field with unit) consists of a finite directed graph with four types of nodes (input, computation, branch, output) and three spaces (input space $\mathcal{J}_M = R^n$, state space $S_M = R^m$, output space $\mathcal{O}_M = R^l$). Associated with each node are maps of these spaces and next nodes:

(1) <u>input node</u>: linear map $I : \mathcal{J}_M \to S_M$ and unique next node β_1.
(2) <u>computation node n</u>: polynomial (or rational) map $g_n : S_M \to S_M$ and unique next node β_n. If R is a field, g_n can be a rational map.
(3) <u>branch node n</u>: non-zero polynomial function $h_n : S_M \to R$. The next node along the "yes" outgoing edge β_n^+ is associated with condition $h_n(z) \geq 0$ and the next node along the "*no*" outgoing edge β_n^- with $h_n(z) < 0$.
(4) <u>output node n</u>: linear map $O_n : S_M \to \mathcal{O}_M$ with no next node.

A polynomial map $g : R^m \to R^m$ is given by m polynomials $g_j : R^m \to R(1 \leq j \leq m)$. A rational map (in the case R is a field) is given by a fixed pair of polynomials (p_j, q_j) with $g_j(x) = p_j(x)/q_j(x)$. If R is a ring or field without order (e.g., $R = \mathbb{C}$), in (3), β_n^+ is associated with $h_n(z) = 0$ and β_n^- with $h_n(z) \neq 0$.

Definition of the halting set of a graphic machine. If M is a finite-dimension graphic machine over R, the halting set of M is defined as the subset Ω_M of the input space $\mathcal{J}_M$ with all y where the computation of M halts.

An input–output map $\varphi_M : \Omega_M \to \mathcal{O}_M$ of halting set Ω_M in the output space $\mathcal{O}_M$ defines the computation procedure of M. For input $y \in \mathcal{J}_M$, the computation on y starts with state $I(y) \in S_M$ at node 1:

- If $\beta(1) = n$ is a computation node, the computation g_n is performed on the state $x = I(y)$ to produce $g_n(x)$ which replaces x.

- If n is a branch node and $h_n(x) < 0$, the next node is $\beta^-(n)$. Otherwise, the next node is $\beta^+(n)$. In both cases, the next state is still x.
- The computation proceeds until an output node n is reached and $O_n(x)$ for some $x \in S_M$ is computed. In this case, the computation is said to halt with $\varphi_M(x) = O_n(x)$.

If for some computation node n, the function value $g_n(x)$ is not defined because a polynomial denominator h is zero at x, we may modify the machine by changing in Fig. 14.

In an infinite-dimension machine, the input space and output space are equal to infinite countable direct sums R^∞ with infinite sequences as elements. The four nodes are the same as before in the finite-dimension case. A fifth type of nodes is supplemented in order to handle finite, but unbounded sequences.

Definition of a graphic machine over ring. A graphic machine M over R is a finite, connected, directed graph, containing four types of nodes as in the finite-dimension case (input, computation, branch, output) and a set Σ_M of shift nodes with associated maps, which can handle finite, but unbounded sequences (Blum *et al.*, 1989).

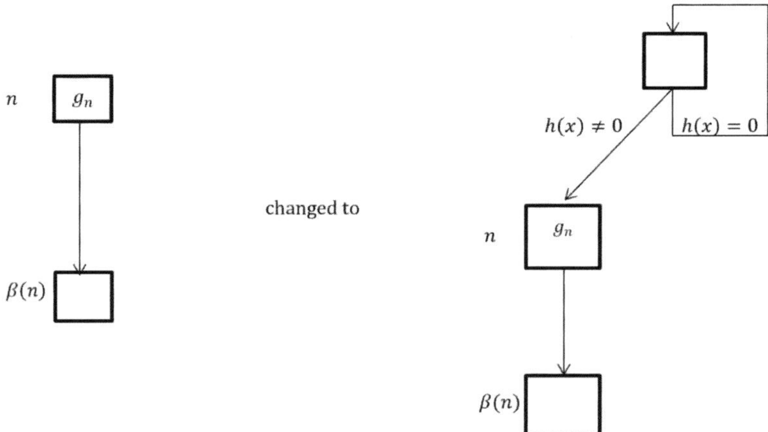

Fig. 14. Modification for undefined polynomials.

The <u>input space</u> I_M and <u>output space</u> $\mathcal{O}_M$ are the disjoint union $R^\infty = \bigcup_{n \geq 0} R^n (R^0 = \{0\})$. The <u>state space</u> S_M is the bi-infinite direct sum over R with elements $x = (\ldots, x_{-2}, x_{-1}, x_0 . x_1, x_2, \ldots)$ with x_i element of R for all integers i, $x_k = 0$ for $|k|$ sufficiently large and dot $(.)$ lying between x_0 and x_1.

Each <u>shift node</u> n is associated with a shift operation, shift left σ_l, and shift right σ_r, where $\sigma_l(x)_i = x_{i+1}$ and $\sigma_r(x)_i = x_{i-1}$.

The <u>dimension</u> K_M and degree D_M of M are the maximums of the dimensions and the degrees of all maps associated with its computation and branch nodes.

A classical Turing machine is machine over $\mathbb{Z}_2$. Thus, a graphic machine over $\mathbb{R}$ might be considered a "real" Turing machine.

The input–output map φ_M for the infinite-dimension machine M is defined as for finite-dimension machines.

Definition of computability over ring R. For $l, m \leq \infty$, a partial map $\varphi : R^l \to R^m$ is said to be computable over R iff there is a machine M over R such that the domain Ω_φ of map φ is equal to the halting set Ω_M of machine M, i.e., $\Omega_\varphi = \Omega_M$ with $\varphi_M(x) = \varphi(x)$ for all $x \in \Omega_\varphi$. In this case, machine M is said to compute map φ.

Machines M and M' are called equivalent iff they compute the same map.

Definition of recursive enumerability over ring. A subset $Y \subset R^n$ is said to be recursively enumerable over R iff it is equal to the halting set Ω_M of some machine M over R, i.e., $Y = \Omega_M$.

Definition of decidability over ring R. A subset $Y \subset R^n$ is said to be decidable iff Y and its complement are both recursively enumerable over R.

The ring $\mathbb{Z}$ of integers can be proven to be a decidable subset of R over R for any Archimedian ring R.

Definition of graphic machine code over ring R. The code $\pi(M)$ of a graphic machine M over R is a finite sequence of coded types n of nodes with numbers 1 to 5 denoting an input, computation, branch, output, and shift node, its type number of next node β_n (or a pair (β_n^+, β_n^-) for branching node), and its associated map g_n.

A polynomial of degree d and m variables can be described by the sequence $(d, m, \bar{a})$, where $\bar{a}$ denotes the sequence of its coefficients (in lexicographic order). A rational function is given by an ordered pair of polynomials. A rational map is given by a finite sequence of rational functions.

Definition of a universal graphic machine over ring R. A universal graphic machine over R is a machine U which, when input a pair $(\pi(M), x)$ with code number $\pi(M)$ and input $x \in R^\infty$ of machine M over R, will output $\phi_U(x) = \phi_M(x)$.

U systematically reads the program code $\pi(M)$ and carries out the instructions applied to x, starting with instruction labeled 1, proceeding to instruction labeled β_1, and so forth. If U ever arrives at the instruction labeled N, it halts and outputs exactly what M would output. An essential component of U is the universal polynomial evaluator, a subroutine, which takes as input (f, x) in R^∞, with $f \in R[x_1, x_2, \ldots, x_m]$ and $x \in R^m$ and outputs $f(x) \in R$.

Computing endomorphism of graphic machine M over R. Let $\mathcal{N} = \{1, \ldots, N\}$ be the set of nodes of M(with 1 the input node and N the output node) and S its state space. $\mathcal{N} \times S$ is called the full state space of the machine with the computing endomorphism

$$H : \mathcal{N} \times S \to \mathcal{N} \times S,$$

that is, H maps each node/state pair (n, x) to the unique next node/state pair (n', x') determined by the graph of M and the associated maps.

Let $\gamma(=\gamma_x)$ be the computation path $n^0 = 1, n^1, \ldots, n^k, \ldots$ and let $\gamma(k)$ be the initial computation path $(n^0, n^1, \ldots, n^k) \in \mathcal{N}^{k+1}$ of γ with length k. We call $\mathcal{V}_{\gamma(k)} = \{x' \in \mathcal{J}_M | \gamma_{x'}(k) = \gamma(k)\}$ the initial path set with all inputs whose computation paths coincide with γ for the first k steps.

Computation paths and branching conditions. Initial path sets are characterized by the branching conditions along the path

$\gamma(k)$ with two sets of left and right step-k branching functions

$$L_{\gamma(k)} = \{f_{\gamma(k')}|k' < k, k' \text{ branch step in } \gamma, \text{ and } n^{k'+1} = \beta^-(n^{k'})\},$$
$$R_{\gamma(k)} = \{f_{\gamma(k')}|k' < k, k' \text{ branch step in } \gamma, \text{ and } n^{k'+1} = \beta^+(n^{k'})\}.$$

Then, in case R is ordered, the initial path set is

$$\mathcal{V}_{\gamma(k)} = \{x \in \mathcal{J}_M|f(x) < 0, g(x) \geq 0, f \in L_{\gamma(k)}, g \in R_{\gamma(k)}\}$$

or, in case R is unordered,

$$\mathcal{V}_{\gamma(k)} = \{x \in \mathcal{J}_M|f(x) \neq 0, g(x) = 0, f \in L_{\gamma(k)}, g \in R_{\gamma(k)}\}.$$

A subset S of R^n is basic semi-algebraic over R in the ordered case (or basic quasi-algebraic, in the unordered case) if S is the set of elements in R^n that satisfy a finite system of polynomial equalities and inequalities over R. A semi-algebraic (or quasi-algebraic) set is a finite union of basic semi-algebraic (or basic quasi-algebraic) sets.

Halting sets as semi-algebraic sets.

(1) If R is an ordered ring or field, then the initial path set $\mathcal{V}_{\gamma(k)}$ is basic semi-algebraic (or basic quasi-algebraic in the unordered case).

(2) If $\gamma_1(k) \neq \gamma_2(k)$, then $\mathcal{V}_{\gamma_1(k)} \cap \mathcal{V}_{\gamma_2(k)} = \emptyset$.

Let $\Gamma_T = \{\gamma_x(T)|T_M(x) \leq T \text{ for } x \in \mathcal{J}_M\}$ be the set of time-T halting paths with halting time $T_M(x)$, i.e., the least such T with $T_M : \Omega_M \to \mathbb{Z}^+$. The set of halting paths of M is $\Gamma_M = \cup_{T<\infty}\Gamma_T$ and the set of minimal halting paths $\Gamma'_M = \{\gamma \in \Gamma_M|N \text{ occurs only once in } \gamma\}$.

For any machine M over R the halting set $\Omega_M = \cup_{\gamma \in \Gamma'_M}\mathcal{V}_\gamma$ is a countable disjoint union of basic semi-algebraic (resp. basic quasi-algebraic) sets.

With graphic machines, the concepts of computability and decidability can be generalized over a ring.

Definition of computability, enumerability, and decidability over ring R. For $l, m \leq \infty, a$ (partial) map $\varphi : R^l \to R^m$ is computable over R iff there is a machine M over R such that its halting set $\Omega_M = \Omega_\varphi$, the domain of φ, and $\phi_M(x) = \varphi(x)$ for all $x \in \Omega_\varphi$ and input–output map $\phi_M : \Omega_M \to \mathcal{O}$ (output space).

A set $Y \subset R^n$ is called recursive enumerable over R iff $Y = \Omega_M$ (halting set) for some machine M over R. (Y is not computable for $R = \mathbb{R}$.) It is said to be decidable iff it and its complement are both recursive enumerable over R.

We are now able to solve decidability problems of, e.g., the Mandelbrot set, Julia sets, and Newton's method:

Theorem. *The Mandelbrot set is not decidable over $\mathbb{R}$.*

Proof. The Mandelbrot set is not a halting set of any machine over $\mathbb{R}$. Why?

The boundary of the Mandelbrot set has Hausdorff dimension 2.

Therefore, the Mandelbrot set cannot be represented by a countable union of semi-algebraic sets over $\mathbb{R}$.

But, any halting set can be represented by a countable union of basic semi-algebraic sets over $\mathbb{R}$ because the input–output map is a piecewise polynomial or rational map. $\qquad\square$

A set $S \subset R^n$ is basic semi-algebraic over R in the ordered case (or basic quasi-algebraic in the unordered case) if S is the set of elements in R^n that satisfy a finite system of polynomial equalities and inequalities.

Theorem. *A Julia set is decidable over $\mathbb{R}$ iff it is*

(1) *a round circle,*
(2) *an arc of a round circle, or*
(3) *the whole sphere*

Most Julia sets are not decidable over $\mathbb{R}$, since most Julia sets have fractional Hausdorff dimension. But halting sets must have integral Hausdorff dimension.

Theorem. *The set of points that converge under Newton's method is generally undecidable over $\mathbb{R}$.*

Proof. It is sufficient to use the cubic $f(x) = x^3 - 2x + 2$ to obtain the undecidability result.

It is known that the set of points that do not converge to a root of f under iteration by Newton's method is exactly a Cantor set, i.e., uncountable.

A Cantor set cannot be the countable union of semi-algebraic sets. $\qquad\qquad\square$

The mathematical theory of dynamical systems is based on real and complex numbers (Ebbinghaus *et al.*, 1991). Therefore, with real and complex computing, general problems of decidability and enumerability can be solved for dynamical systems.

Recursive enumerable basin of attraction. A discrete complex dynamical system with (rational) state function $g : \mathbb{C} \to \mathbb{C}$ generates orbits of points $z_0, g(z_0), \ldots, g^k(z_0), \ldots$. Fixed points $(g(z_0) = z_0)$ or periodic points $(g^n(z_0) = z_0$ with period $n)$ can be attracting (respect. repelling) with conditions $|(g^n)'(z_0)| < 1$ (resp. $|(g^n)'(z_0)| > 1$), where $(g^n)'(z_0)$ is the derivative of the nth iterate of g at z_0:

If z_0 is attracting of period n, the derivative condition implies there is a neighborhood U of z_0 with $g^n(U) \subset U$. Thus, orbits of points in neighborhood U of z_0 with $g^n(U) \subset U$ asymptotically approach the orbit of z_0, i.e., the basin of attraction of z_0. Their union is the basin of attraction of g.

Proposition. *The basin of attraction of g is a recursive enumerable set over $\mathbb{R}$, i.e., there is a machine M whose halting set Ω_M is the basin of g.*

Proof. For rational maps, there is only a finite number of attracting periodic points.

$\Rightarrow$ There is a real polynomial h of two real variables such that $h(z) < 0$ iff z belongs to a finite union of discs around the attracting points which is contracted into itself by g.

$\Rightarrow$ A point is in the basin of g iff $h(z) < 0$ for some z in its orbit.

According to the definition of a graphic machine, we associate z with the input node, polynomial g with the computation node, and polynomial h with the branch node of a machine M which is defined by the following graphic scheme:

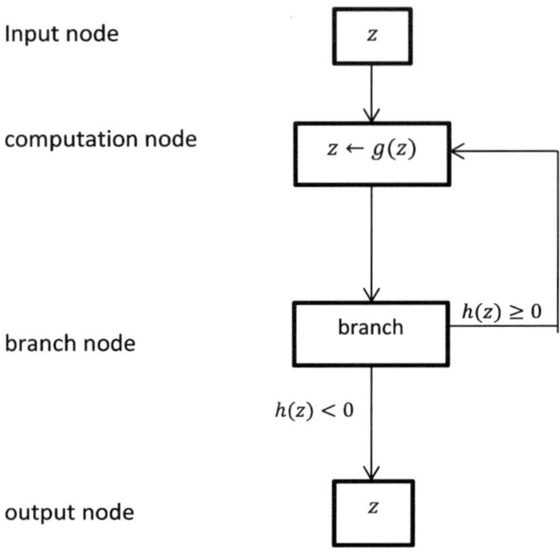

By definition of M, the halting set Ω_M is the basin of attraction of g. □

The basin of attraction of g is recursively enumerable, but not decidable. Decidability demands that the complement of the basin of attraction of g is also recursively enumerable. But the complement of the basin of attraction is the Julia set of g which, in general, is not decidable.

From a mathematical point of view, graphic machines on a ring R are convenient to illustrate algebraic properties of real computing. But, they are somewhat abstract with respect to the architectures of computers. Actually, a graphic machine on a ring R is a kind of register machine with unlimited storage which processes elements of a ring with exact arithmetic. In computer science, it is more convenient to illustrate computing processes by generalized register machines which are equivalent to graphic machines on a ring R. Analytic computations are infinite computations generating a convergent sequence of outputs.

In the following definition, we generalize register machines over rational numbers ($\mathbb{Q}$-machines) with a register δ of precision which determines the degree of rounding real numbers as input of a rational register. These analytic δ-$\mathbb{Q}$-machines can also compute real functions over $\mathbb{R}$. A register machine over $\mathbb{Q}$ can generate infinite computations of rational numbers converging to real numbers. Its set of instructions is extended by an instruction δ to compute real functions with approximations (Hotz *et al.*, 1995; Chadzelek and Hotz, 1997).

Definition of a register machine over ring R. A register machine M over ring R is defined by $M = (K,\ K_s,\ K_e,\ K_g,\ \Delta,\ R,\ \text{in},\ \text{out})$ with set K of configurations and $K_s, K_e, K_g \subset K$ for starting, ending, and goal configurations, transfer function $\Delta : K \to K$ with $\Delta|K_e = id_{K_e}$, input function in: $R^* \to K_s$, and output function out: $K \to R^*$:

$$K = \{k\,|\,k = (\alpha, \beta, \gamma, \delta, \pi, x, y, z)\},$$
$$K_s = \{k \in K\,|\,\alpha = \gamma = \delta = 0, \beta = 1, \forall_j : y_j = z_j = 0\},$$
$$K_e = \{k \in K\,|\,\pi_\beta = \text{end}\},$$
$$K_g = \{k \in K\,|\,\pi_\beta = \text{print}\}.$$

A configuration of a register machine is given by assignment $\pi_0, \ldots, \pi_N$ of program store, values $\alpha \in R$, $\beta_0, \ldots, \beta_N$ instructions,

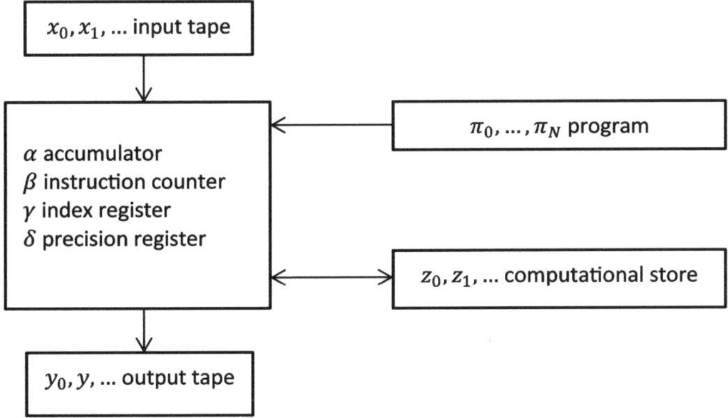

Fig. 15. Register machine over ring R.

$\gamma \in \mathbb{N}$ index register, $\delta \in \mathbb{N}$ precision register, and assignments $x_0, x_1, \ldots$ and $y_0, y, \ldots$ of input resp. output tape (see Fig. 15).

The set I of <u>instructions</u> is given by

1. <u>assignments</u>:

   (a) $\alpha := x_i$, $\alpha := z_i$, $y_i := \alpha$, $z_i := \alpha$ for $i \in \mathbb{N} \cup \{\gamma\}$,
   (b) $\alpha := r$ for $r \in \mathbb{R}$,
   (c) $\alpha := \delta$,
   (d) $\gamma := 0$.

2. <u>arithmetic</u>:

   (a) $\alpha := \alpha + z_i$, $\alpha := \alpha \cdot z_i$ for $i \in \mathbb{N} \cup \{\gamma\}$,
   (b) $\alpha := -\alpha$, $\alpha := \alpha^{-1}$,
   (c) $\gamma := \gamma + 1$, $\gamma := \gamma \dot{-} 1$.

3. <u>branches</u>:

   If $\alpha > 0$ then goto m else goto n for $m, n = 1, \ldots, N$.

4. <u>special instructions</u>:

   end, next δ, print.

Instruction $\alpha := x_i$ reads a rational approximation $q \in \mathbb{Q}$ with $|x_i - q| < 2^{-\delta}$ instead of the actual input x_i. Instruction $\alpha := r$ is also handled in that way for irrational constants r. The degree of rounding is increased step by step with each next δ instruction, the machine starts again, and δ is augmented by 1:

$$k = (\alpha, \beta, \gamma, \delta, \pi, x, y, z) \xrightarrow[\pi_\beta = \text{next } \delta]{} \Delta(k) = (0, 1, 0, \delta + 1, \pi, x, y, z)$$

In register machines over $\mathbb{R}$, real numbers are allowed on the input tape and in the program, i.e., in: $\mathbb{R}^* \to K_s, x : \mathbb{N} \to \mathbb{R}, \pi : \{1, \ldots, N\} \to I$ if limit $\lim_{j \to \infty} \text{out}(k_{i_j})$ exists. A register machine over $\mathbb{Q}$ extended with a rounding function of the instruction next δ is also called δ-$\mathbb{Q}$-machine.

A sequence $b = (k_i)_{i=0}^{\infty}$ of states (configurations) $k_i = \Delta^i(k_0)$ is called computation of M. It is called regular if it is finite with length $n, k_0 \in K_s$, $k_n \in K_e$, and out (k_n) as result. It is called analytic if $k_0 \in K_s, k_{i_j} \in K_g$ for infinite many i_j as infinite partial sequence

$(k_{i_j})_{j=0}^{\infty}$ of all goal configurations, and $\lim_{j\to\infty} \text{out}(k_{i_j})$ exists. The limit is called result and $\text{out}(k_{i_n})$ the nth approximation of the result. A register machine M defines a (partial) function $\Phi_M : R^* \to R^*$ with $\Phi_M(x) = y$ if the computation for $x \in R^*$, starting with $\text{in}(x)$ and generating the result $y \in R^*$ with $\Phi_M(x) := y$, is regular or analytic. The nth approximation of this function can be defined by $\Phi_M^{(n)}(x) = \text{out}(k_{i_n})$.

On the basis of register machines over a ring R, we can distinguish a hierarchy of classes of computable real functions.

Hierarchy of computable functions over ring R

- A function $f : R^* \supset D \to R^*$ is called analytically R-computable (or R-analytic) iff there is a R-register machine M with $f = \Phi_M | D$.
- If D is subset of the halting set Ω_M, then f is called R-computable.
- $f : R^* \supset D \to R^*$ is called δ-$\mathbb{Q}$ computable resp. δ-$\mathbb{Q}$-analytic if there is a δ-$\mathbb{Q}$ machine to compute f.
- A function $f : R^* \to R^*$ is called R^i-analytic for $i \in \mathbb{N}$ if f can be computed analytically by composition of i R-machines.
- Be $(k_{i_j})_{j=0}^{\infty}$ the sequence of all final configurations of a δ-$\mathbb{Q}$-analytic computation with infinite applications of instruction next δ_f, $\text{out}(k_{i_j}) = \left(y_1^{(j)}, \ldots, y_{n_j}^{(j)} \right) \in \mathbb{Q}^*$, and $n := \lim_{j\to\infty} n_j$. For

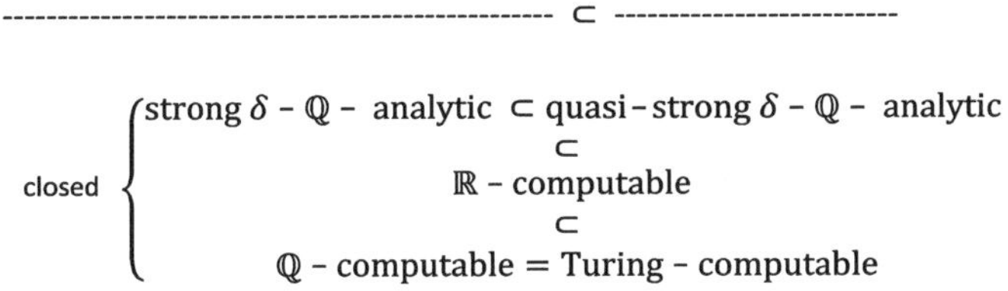

$$
\text{not closed} \left\{
\begin{array}{c}
\vdots \\
\mathbb{R}^2\text{ - analytic} \\
\subset \\
\mathbb{R}\text{ - analytic} \\
\subset \\
\delta\text{ - }\mathbb{Q}\text{ - analytic} \\
\subset \\
\text{robust } \delta\text{ - }\mathbb{Q}\text{ - analytic}
\end{array}
\right.
$$

$$\cdots\cdots\cdots\cdots \subset \cdots\cdots\cdots$$

$$
\text{closed} \left\{
\begin{array}{c}
\text{strong } \delta\text{ - }\mathbb{Q}\text{ - analytic} \subset \text{quasi-strong } \delta\text{ - }\mathbb{Q}\text{ - analytic} \\
\subset \\
\mathbb{R}\text{ - computable} \\
\subset \\
\mathbb{Q}\text{ - computable} = \text{Turing - computable}
\end{array}
\right.
$$

Fig. 16. Hierarchy of computable functions (closed and not closed with respect to composition).

$i \leq n$, the limit of the ith position of output is $y_i := \lim_{j \to \infty} y_i^{(j)}$. This computation is called quasi-strong δ-$\mathbb{Q}$-analytic if $y_1 = 0$ and $|y_i - y_i^{(j)}| \leq y_1^{(j)}$ for all $j > j_0$ and a certain j_0.

- A program is called robust iff all computations are regular resp. analytic and coincide with the result or no computation leads to a result (see Fig. 16).

Halting and convergence problems for graphic and register machines over ring R

With respect to the hierarchy of graphic resp. register machines over ring R, we can distinguish several results of the halting problem:

- The halting problem for regular $\mathbb{R}$-machines is not $\mathbb{R}$-computable.
- The halting problem for regular $\mathbb{R}$-machines is $\mathbb{R}$-analytic.
- The halting problem for regular $\mathbb{R}$-machines is not robust δ-$\mathbb{Q}$-analytic.
- The halting problem for regular $\mathbb{R}$-machines is not δ-$\mathbb{Q}$-analytic.

In analogy to the halting problem of Turing machines, we can consider the convergence problem of analytic $\mathbb{R}$-machines: Is such a machine able to generate a converging output sequence from a certain input?

- The convergence problem for analytic $\mathbb{R}$-machines is not $\mathbb{R}$-analytic.
- The convergence problem for analytic $\mathbb{R}$-machines is not $\mathbb{Q}$-analytic.
- The convergence problem for analytic $\mathbb{R}$-machines is $\mathbb{R}^3$-analytic.
- The convergence problem for analytic $\mathbb{R}^i$-machines is not $\mathbb{R}^i$-analytic.

Again (like for graphic machines), a representation theorem states that the domain of a R-computable function can be represented by a countable union of semi-algebraic sets (i.e., piece-to-piece by rational polynomials) (Hotz *et al.*, 1995, Section 4). In constructive reversible mathematics (cf. Chapter 8), the classes of δ-$\mathbb{Q}$-analytic functions are interesting, because δ-$\mathbb{Q}$-machines illustrate constructive procedures to compute real numbers. How far can real mathematics be reduced to the level of, e.g., δ-$\mathbb{Q}$-analytic functions? Let us consider applications of scientific computing with real mathematics.

Real computing and dynamical systems

Dynamical systems are fundamental modeling tools in natural, economic, and social sciences (cf. Chapter 13; Mainzer, 2007a,b). Changing states of dynamical systems (e.g., planetary systems, cellular systems, populations) are modeled by time-dependent differential equations. In general, differential equations describe functions of one (time-dependent) resp. several (spatial) variables by their ordinary resp. partial derivations of first or higher order. From a physical point of view, systems of differential equations with finitely many degrees of freedom (e.g., finitely many points of mass in Newtonian mechanics of planets) correspond to ordinary differential equations. Systems with infinitely many degrees of freedom (e.g., fluids, gases, or electromagnetic fields) correspond to partial differential equations.

Real computable functions are important for determining solutions of differential equations. Unique solutions of differential equations are computable under certain conditions. In scientific modeling, solutions of differential equations are used to explain or to predict effects of dynamical systems under certain constraints. Examples are initial value problems and stability problems of dynamical systems which we will consider in more detail (Glendinning, 1994). In initial value problems, we search for mathematical models of time-dependent processes with given initial states of dynamical systems. Mathematically, an initial value problem is an ordinary differential equation with a specified value, called the initial condition, of the unknown function at a given point in the domain of the solution.

Euler's method and initial value problems

In numeric analysis, Euler's method is used to approximate a solution of an initial value problem (Stoer and Bulirsch, 2002). In a simplified linear example, the initial value problem is given by differential equation $y' = \frac{dy}{dx} = f(x, y)$ with initial condition $y(x_0) = y_0$. Euler's method generates a numerical solution by starting on a certain interval which is divided into small subdivisions of length h. Then, using the initial condition as starting point, the rest of the solution is generated by using the iterative formulas

$$x_{n+1} = x_n + h,$$

$$y_{n+1} = y_n + h f(x_n, y_n)$$

to find the coordinates of the points in the numerical solution.

In Fig. 17, the slope of the tangent line at the point (x_0, y_0) is given by $y'(x_0) = f(x_0, y_0)$. Given the slope of the tangent line and an initial point (x_0, y_0), we want to find out the value of y_1 located at $x_1 = x_0 + h$. Obviously, by the definition of a slope, $y_1 = y_0 + h f(x_0, y_0)$. Now, we recursively continue with (x_1, y_1) to find the values of (x_2, y_2). The slope of the tangent line at (x_1, y_1) is equal to $y'(x_1) = f(x_1, y_1)$. Then using the same step as above, we get $y_2 = y_1 + h f(x_1, y_1)$. The process is iterated by the general

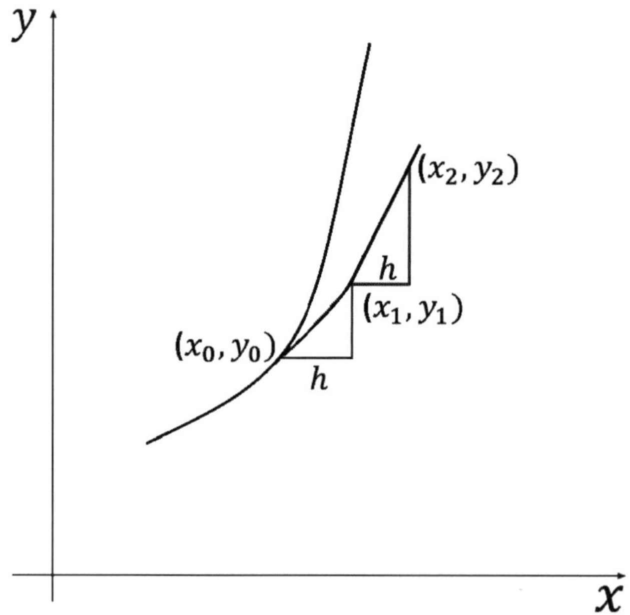

Fig. 17. Euler's method and initial value problem.

formula $y_{n+1} = y_n + hf(x_n, y_n)$, and the approximation is improved by decreasing the increment h smaller and smaller.

For real computing results, initial values problems can be generalized.

Definition of an initial value problem. Let $N \in \mathbb{N}$ be the dimension of the dynamical system, $f : \mathbb{R} \times \mathbb{R}^N \to \mathbb{R}^N$ with $(t, x) \mapsto f(t, x)$ computable over $\mathbb{R}$, and $(t_0, x_0) \in \mathbb{R} \times \mathbb{R}^N$. The initial value problem is given by $x'(t) = f(t, x(t))$ with $x(t_0) = x_0$. We search for a differentiable solving function $x : \mathbb{R} \supset I \to \mathbb{R}^N$ with $t \mapsto x(t)$ which, on an interval $I \ni t_0$, satisfies the components of equations $x'_i(t) = f_i(t, x(t))(1 \le i \le N)$, and the initial condition.

According to the representation theorem, the domain $\mathbb{R} \times \mathbb{R}^N$ of the computable function f can be represented by a countable union of separated (disjunct) semi-algebraic sets. The regions D_σ of these semi-algebraic sets are determined by the branching conditions of the program path σ which is followed by the computational machine of f. By definition, semi-algebraic sets are sets of solutions satisfying systems of polynomial inequalities with $>$ and $\ge$ in $1 + N$ variables. The path σ is separated into components i $(1 \le i \le N)$ corresponding to semi-algebraic sets D_{σ_i} which are characterized by polynomial (resp. rational) functions $f_{\sigma,i}$ Therefore, the function f can be represented by pieces of polynomial (resp. rational) functions $f_{\sigma,i}$.

Under general conditions, the search procedure of an initial value problem can be reduced to rational arithmetic. Therefore, in the sense of reverse mathematics (cf. Chapter 8), initial value problems have constructive solutions.

Theorem. *For every initial value problem with appropriate initial value (t_0, x_0), there exists a unique (maximal global) solution which is robust δ-$\mathbb{Q}$-computable and computable without division (i.e., without instruction $\alpha := \alpha^{-1}$).*

The main problem of the proof is to define a δ-$\mathbb{Q}$-machine which can evaluate function f with sufficient precision and in finite time (Chadzelek, 1998). The proof starts with the initial value (t_0, x_0) in region D_{σ_0} and uses Euler's method to approximate the solution with robust δ-$\mathbb{Q}$-computable methods. Instead of a general proof, a practical example should illustrate strong δ-$\mathbb{Q}$-computation (without division).

Example. Exponential function exp: $\mathbb{R} \to \mathbb{R}$ with $x \mapsto e^x$ is strong δ-$\mathbb{Q}$-analytic (without division).

Proof. The exponential function is given by

$$e^x = \lim_{n \to \infty} \left(1 + \frac{x}{n}\right)^n$$

with partial sequence $n = 2^k$.

$2^{-k} = \left(\frac{1}{2}\right)^k$ can easily be computed without division.

For estimating the error of approximation, we consider for all $n, m \in \mathbb{N}$:

$$\left(1 + \frac{x}{n}\right)^n \leq e^x \leq \left(1 + \frac{x}{m}\right)^{m + \lceil x \rceil}$$

with monotonically increasing left side and monotonically decreasing right side.

The right side is always at least as big as the left side and converges to the same limit (proof with Bernoulli's inequality).

For $n = m$, it follows

$$0 \leq e^x - \left(1 + \frac{x}{n}\right)^n \leq \left(1 + \frac{x}{n}\right)^n \left[\left(1 + \frac{x}{n}\right)^{\lceil x \rceil} - 1\right],$$

which is easily computable for $n = 2^k$ without division.

Obviously, this procedure can be realized by an analytic $\mathbb{R}$-machine.

For $k = \delta$, we get a strong δ-$\mathbb{Q}$-analytic algorithm:

Let $x_\delta := \rho(x, \delta)$ be the rational approximation of real input with precision δ. With $n = 2^\delta$ and because of $|x - x_\delta| < \frac{1}{n}$, we get

$$\left(1 + \frac{nx - 1}{n^2}\right)^n < \left(1 + \frac{x_\delta}{n}\right)^n < \left(1 + \frac{nx + 1}{n^2}\right)^{n + \lceil x \rceil}.$$

The left side increases monotonically, the right side decreases monotonically. Both sides are separated by e^x.

We prove that their difference converges to zero which includes that
$\lim_{\delta \to \infty} \left(1 + \frac{x_\delta}{n}\right)^n = e^x$:

$$0 < \left(\frac{n^2 + nx + 1}{n^2}\right)^{n + \lceil x \rceil} - \left(\frac{n^2 + nx - 1}{n^2}\right)^n$$

$$= \underbrace{\left(\frac{n^2 + nx + 1}{n^2}\right)^n}_{\leq e^{x+1}} \left[\underbrace{\left(\frac{n^2 + nx + 1}{n^2}\right)^{\lceil x \rceil}}_{\to 1 \text{ for } n \to \infty} - \left(\frac{n^2 + nx - 1}{n^2 + nx + 1}\right)^n\right].$$

The last term converges to 1 because it follows for large n according to Bernoulli:

$$1 \geq \left(1 - \frac{2}{n^2 + nx + 1}\right)^n \geq 1 - \frac{2}{n^2 + nx + 1} \geq 1 - \frac{2}{n + x} \xrightarrow{n \to \infty} 1.$$

For $\frac{2}{n+x}$, an upper bound of the form 2^{-l} can be found without division. Therefore, the estimation of error can also be evaluated without division. $\qquad \square$

Another aspect of stability asks whether dynamical systems remain stable after local perturbations of their initial states. In this case, the solution of an initial value problem must mathematically converge to a fixed point. If small perturbations are damped in the long run, the system is called asymptotically stable (cf. Chapter 13; Deuflhard and Bornemann, 1994).

Definition of stability.

(a) A dynamical system with differential equation $x_{f,t,x}$ and initial value (t_0, x_0) is called asymptotically stable iff

$$\exists \delta > 0 \forall (\tau, \varepsilon) \in U_\delta(t_0, x_0) \lim_{t \to \infty} |x_{f,t_0,x_0}(t) - x_{f,\tau,\varepsilon}(t)| = 0.$$

(b) A dynamical system with differential equation $x_{f,t,x}$ and initial value (t_0, x_0) is called Lyapunov-stable iff

$$\forall \varepsilon > 0 \; \exists \delta > 0 \; \forall (\tau, \varepsilon) \in U_\delta(t_0, x_0) \forall t \geq t_0$$

$$(x_{f,\tau,\varepsilon}(t) \in U_\varepsilon(x_{f,t_0,x_0}(t))).$$

For a given $\mathbb{R}$-machine $M_\pi^\mathbb{R}$ with program π, we can construct a differential equation damping (right side f_π^-) resp. enforcing (left side f_π^+) errors iff the machine does not halt (Chadzelek, 1998, p. 64). The function $f_\pi^\pm : \mathbb{R} \times \mathbb{R} \to \mathbb{R}$ is defined by

$$f_\pi^\pm(t, \xi) := \begin{cases} \pm\xi, & \text{if } t \leq t_{\max}, \\ 0, & \text{otherwise} \end{cases}$$

with length $t_{\max} \in \mathbb{N}$ of computation by $M_\pi^\mathbb{R}$ and input ε. For initial value $(0,0)$, we get $x_{f_\pi^\pm,0,0} = 0$ and in general

$$x_{f_\pi^\pm,0,x_0} = \begin{cases} e^{\pm t}x_0, & \text{if } t \leq t_{\max}, \\ e^{\pm t_{\max}}x_0, & \text{otherwise.} \end{cases}$$

Now, the undecidability of error damping and stability can easily be proven.

Theorem.

(a) *For given f, t_0, and x_0, an $\mathbb{R}$-machine cannot regularly decide if the corresponding differential equation with right side f and initial value (t_0, x_0) is asymptotically stable.*

(b) *For given f, t_0, and x_0, an $\mathbb{R}$-machine cannot regularly decide if the corresponding differential equation with right side f and initial value (t_0, x_0) is Lyapunov-stable.*

Proof.

(a) The assumption that an $\mathbb{R}$-machine M solves the problem leads to a contradiction: For each π, a machine M' can easily compute a program for the right side f_π^-. Then, error damping can be decided by machine M. We know that

$$\lim_{t \to \infty} x_{f_\pi^-,0,x_0}(t) = 0 \iff (x_0 = 0 \vee t_{\max} = \infty).$$

Therefore, the differential equation damps initial errors in the region of $(0,0)$ iff machine $M_\pi^\mathbb{R}$ does not halt with empty input. In this case, M' can decide the halting problem which is impossible.

(b) The assumption that an $\mathbb{R}$-machine M solves the problem leads to a contradiction: For each π, a machine M' can easily compute a program for the right side f_π^+. Then, error damping can be decided by machine M. For initial values in the region of $(0,0)$, the solution of the differential equation is stable iff the machine $M_\pi^\mathbb{R}$ does not halt with empty input. Otherwise, each tiny initial error is enforced above all bounds. Therefore, M' can decide the halting problem which is impossible. $\square$

Many processes in science and technology depend essentially on time. Therefore, modeling and predicting time-dependent processes are important targets of the dynamical systems approach (cf. Chapter 13; Mainzer, 2007a,b). In examples like the planetary system, processes do not depend on the initial point of time, but on the initial state of the dynamical system. Mathematically, these processes are modeled by autonomous differential equations with right side f which does not depend on time. The previous results on undecidability and stability can be transmitted to autonomous differential equations. In this case, the initial value is sufficiently determined by component x_0 and time can be chosen $t_0 = 0$.

Theorem. *It is not analytically decidable using an $\mathbb{R}$-machine whether the solution of the initial value problem $x' = f(x)$ with $x(0) = x_0$ and autonomous, $\mathbb{R}$-computable right side $f : \mathbb{R}^N \to \mathbb{R}^N (N > 1)$ with initial value $x_0 \in \mathbb{R}^N$ converges in the long run.*

Classical results on decidability and undecidability in logic have deep impact on real computing. Examples are Alfred Tarski's result that the reals are decidable, but the integers are not according to Gödel.

Definition of definability. A set $S \subset R^n$ is definable over $(R, <)$ iff S is derived from a semi-algebraic set over R by a finite sequence of projections and complements.

Proposition.

(1) *Every definable set over $\mathbb{R}$ is decidable over $\mathbb{R}$ (in bounded time) (Tarski, 1951).*
(2) *There exist definable, but undecidable sets over $\mathbb{Z}$ (implication of Gödel, 1931).*

Proof (2).

(a) For each $n \leq \infty$, there exist undecidable halting sets in $\mathbb{Z}^n$ (Cantor style diagonal argument on the halting set of a universal machine).
(b) For each $n < \infty$, time-T halting set in $\mathbb{Z}^n$ are definable over $\mathbb{Z}$ (via semi-algebraic equations of a graphic machine describing the halting computation of points (Blum *et al.*, 1998, p. 59).
(c) For each $n < \infty$, halting sets in $\mathbb{Z}^n$ are definable over $\mathbb{Z}$.

(b) $\Rightarrow$ (c)
(a), (c) $\Rightarrow$ (2) $\square$

Proposition. *Hilbert's Nullstellensatz (HN) is undecidable over $\mathbb{Z}$.*

Proof.

(c') For each $n < \infty$, halting sets in $\mathbb{Z}^n$ are diophantine, that is, projections of algebraic sets over $\mathbb{Z}$ (Matiyasevich, 1970, 1993).
(a), (c') $\Longrightarrow$ undecidability of Nullstellensatz over $\mathbb{Z}$. $\square$

If $\mathbb{Z}$ were the projection of an algebraic set over $\mathbb{Q}$, then HN would not be decidable over $\mathbb{Q}$. But that is unknown! However, $\mathbb{Z}$ is definable in $\mathbb{Q}$ (Robinson, 1949). Thus, there exist definable, but undecidable sets over $\mathbb{Q}$.

Definability, decidability, and undecidability over $\mathbb{R}$.

(a) For each $n \leq \infty$, there exist undecidable halting sets in $\mathbb{R}^n$.
(b) For each $n < \infty$, time-T halting sets in $\mathbb{R}^n$ are definable over $\mathbb{R}$.
(c) There are halting sets in $\mathbb{R}^n$ which are not definable over $\mathbb{R}$ (otherwise with Tarski's (1951) contradiction to (1)).

Proposition: undefinability of $\mathbb{Z}$ over $\mathbb{R}$. $\mathbb{Z}$ *cannot be definable over* $\mathbb{R}$.

Proof. Definable subsets of $\mathbb{R}$ are semi-algebraic (Tarski, 1951), and thus are finite unions of intervals with algebraic endpoints. $\qquad\square$

Proposition: undecidability of $\mathbb{Z}$ over $\mathbb{R}$. $\mathbb{Z}$ *is decidable over* $\mathbb{R}$.

Proof by program.

```
Input   x ∈ ℝ
z := |x|
while z ≥ 0 do z := z − 1 end do
z := z + 1
if z = 0 output 1, else output 0.
```

Each subset $S \subseteq \mathbb{Z}$ can be decided by a machine over $(\mathbb{R}, <)$ with a built-in constant c_s coding the characteristic function χ_s. Such machines over $\mathbb{R}$ can be considered as a type of oracle Turing machine. $\qquad\square$

Proposition: definability and decidability over $\mathbb{C}$.

(a) *Every definable set over* $\mathbb{C} = \mathbb{R}^2$ *is decidable over* $\mathbb{C}$.
(b) *Every decidable set in* $\mathbb{C}^n$ *is definable over* $\mathbb{C}$.
(c) $\mathbb{Z}$ *cannot be decided over* $\mathbb{C}$.

For further elaboration, cf. Cucker and Rosseló (1993) and Blum and Smale (1993). In short, with respect to definability and decidability, $\mathbb{C}$ is simpler than $\mathbb{R}$ and $\mathbb{R}$ is simpler than N.

Chapter 11

Complexity Theory of Real Computing

Theory of computational complexity over ring R

In classical computer science, an algorithm defined by a machine M is tractable (in polynomial time, or in P) iff the computation time $T(x)$ associated to input x satisfies the bound $T(x) \leq c(\text{size}(x))^q$ for all x with constants c and q (depending only on M). Time is the number of Turing machine operations and size is the number of bits. There are subtle differences between theories of P and NP over $\mathbb{R}$ and over $\mathbb{Z}$: Any NP-problem over $\mathbb{Z}$ is seen to be solvable in $2^{\text{poly}(n)}$ time by counting arguments. An analogous result over $\mathbb{R}$ is far from obvious since there is a continuum of possible guesses over $\mathbb{R}$.

The complexity of a function is measured by the optimal computing cost. The computing cost refers to the number of operations which are needed to transform an input into an output. Therefore, cost depends on input size. A machine over $\mathbb{R}$ works in polynomial time if for each input $x \in \mathbb{R}^n \subset \mathbb{R}^\infty$ the computing cost is $\text{cost}(x) \leq cn^q$ for some fixed $c, q \geq 1$, where $\text{cost}(x)$ is the number of nodes used during the computation and n is the size of x. In the digital case, size is given by the bit length. Over a ring R, size depends on a height function ht_R defined on R and on the vector length. The unit height is $\text{ht}_R(x) = 1$ for all $x \in R$. Unit height refers to algebraic complexity over $\mathbb{R}$ and $\mathbb{C}$. Over $\mathbb{Z}$ and $\mathbb{Q}$, bit complexity is defined by logarithmic or bit height:

$$\text{ht}_{\mathbb{Z}}(x) = \lceil \log(|x| + 1) \rceil \qquad \text{for all } x \in \mathbb{Z}$$

$$\text{ht}_{\mathbb{Q}}(x) \doteq \max\left(\text{ht}_{\mathbb{Z}}(p), \text{ht}_{\mathbb{Z}}(q)\right) \quad \text{for } x \in \mathbb{Q} \text{ with } x = p/q \text{ and } p$$
$$\text{and } q \text{ relatively prime integers.}$$

Definition of length and size. For $x \in \mathbb{R}^n \subset \mathbb{R}^\infty$, let the height function ht_R defined on ring R with values in the non-negative integers, i.e., $\text{ht}_R(x) = \max_i \text{ht}_R(x_i)$ for $x = (x_1, \ldots, x_n) \in \mathbb{R}^n \subset \mathbb{R}^\infty$. Then length and size of x are defined by

$$\text{length}(x) = n$$

$$\text{size}(x) = n \cdot \text{ht}_R(x).$$

Over $\mathbb{R}$, $\mathbb{C}$, and $\mathbb{Z}_2$ with unit height, the size of the input is the dimension or vector length of the input. Over $\mathbb{Z}_2$, the size is also the bit length of the string x. Over $\mathbb{Z}$ with logarithmic height, size corresponds to the bit length of the sequence x.

Definition of computing costs over a ring. Let M be a machine over a ring R with height function ht_R. For $x \in \mathbb{R}^n \subset \mathbb{R}^\infty$, the cost is

$$\text{cost}_M(x) = T(x) \times \text{ht}_{\max}(x),$$

where $T(x)$ is the halting time of M on input x and $\text{ht}_{\max}(x)$ maximum height of any element in the computation of M on input x.

Over real numbers or a ring with unit height, unit or algebraic cost is the halting time and relates to the number of algebraic operations from input to output. Over integers and rational numbers with logarithmic height, logarithmic or bit cost relates to the number of bit operations. Computational complexity must be considered with respect to these different cost measures.

P, NP, and NP-completeness can be introduced over an arbitrary ring R (Blum *et al.*, 1998, Chapter 5). In this framework, old and new NP-completeness results can be obtained.

Definition of polynomial time reductions. A decision problem S is p (polynomial)-reducible to decision problem $S'(S \hookrightarrow_p S')$ iff there is an in polynomial time computable map $\varphi : R^\infty \longrightarrow R^\infty$ such that $\varphi(S) \subseteq S'$ and $\varphi(R^\infty - S) \subseteq R^\infty - S'$, In other words, in the case of $S \hookrightarrow_p S'$, S' is at least as "hard" as S.

Examples of decision problems are

HN:	The Hilbert's Nullstellensatz
QUAD:	The feasibility of quadratic systems
4-FEAS:	The feasibility of degree-4 polynomials
TSP:	The Traveling Salesman problem
KP:	The Knapsack problem
SAT:	The Satisfiability problem

Over any ring or field R, HN $\hookrightarrow_p$ QUAD. Over any ordered ring or field R, QUAD $\hookrightarrow_p$ 4-FEAS.

Definition of class NP over R. *A decision problem* $S \subseteq R^\infty$ *is in class NP over* $R(S \in \mathrm{NP}_R)$ *if there is a machine* M *over* R *with* $\mathcal{I}_M = R^\infty \times R^\infty$ *and positive integers* c *and* q *such that*

(1) If $x \in S$, then there exists some $w \in R^\infty$ (called a guess or witness for a solution) such that if $\varphi_M(x, w) = 1$ and $\mathrm{cost}_M(x, w) \leq c(\mathrm{size}\ x)q$.
(2) *If* $x \notin S$, *then there is no* $w \in R^\infty$ *such that* $\varphi_M(x, w) = 1$.

A decision problem is said to be NP-hard over R iff every $S \in \mathrm{NP}_R$ is p-reducible to it. A decision problem is said to be NP-complete over R iff it is NP_R-hard and in class NP_R.

By definition, it is obvious that

(1) For every NP-complete problem S over R follows $S \in \mathrm{P}_R$ iff $\mathrm{P}_R = \mathrm{NP}_R$.
(2) For every NP-hard problem S' over R with $S' \hookrightarrow_p S$ follows that S is NP_R-hard.

Examples:

(a) HN and QUAD are NP-hard over $(F, =)$ for any field F (e.g., $\mathbb{C}$).
(b) HN, QUAD, and 4-FEAS are NP-hard over $(\mathbb{Z}, <)$, $(\mathbb{Q}, <)$, and $(\mathbb{R}, <)$ (or for any real closed field).

Cost measures of complexity

Over the reals or any ring with unit height, unit cost is the halting time and refers to the number of basic algebraic operations (as well as the number of shifts) from input to output. Over the integers or rationals with logarithmic height, bit cost refers to the number of bit operations. In Table 1, we distinguish unit and bit costs of computational complexity over $\mathbb{Z}$, $\mathbb{Q}$, $\mathbb{R}$, $\mathbb{C}$, and $\mathbb{Z}_2$:

In the first row, the undecidability of HN (HN $\notin$ Dec) over $\mathbb{Z}$ implies P $\neq$ NP over $(\mathbb{Z}, <)$ with respect to unit cost. The reason is that, according to Matiyasevich's theorem (Matiyasevich 1970, 1993), NP is equal to the class HALT of halting sets over $(\mathbb{Z}, <)$ which cannot be P, the class of polynomially tractable sets. EXP is the class of exponentially computable problems. With respect to unit cost, NP $\subseteq$ EXP cannot be true because the assumption of NP $\subseteq$ EXP and HN $\in$ NP leads to HN $\in$ EXP which is a contradiction to the undecidability of HN.

In the second row, with respect to bit cost over $\mathbb{Z}$, NP $\subseteq$ EXP is true, i.e., NP-problems are exponentially computable. Therefore,

Table 1. Unit and bit costs of computational complexity.

Ring	Branch	Cost	HN $\in$ Dec	HN $\in$ NP	NP $\subseteq$ EXP	P $=$ NP
$\mathbb{Z}$	$<$	unit	no	yes	no	no
$\mathbb{Z}$	$<$	bit	no	no	yes	?
$\mathbb{Z}$	$=$	unit	no	yes	no	no
$\mathbb{Z}$	$=$	bit	no	no	yes	no
$\mathbb{Q}$	$<$	unit	?	yes	no	no
$\mathbb{R}$	$<$	unit	yes	yes	yes	?
$\mathbb{C}$	$=$	unit	yes	yes	yes	?
$\mathbb{Z}_2$	$=$	unit	yes	yes	yes	?

HN $\in$ NP cannot be true because the assumption of HN $\in$ NP and NP $\subseteq$ EXP leads to HN $\in$ EXP which is a contradiction to the undecidability of HN.

In the third and fourth rows, it is P $\neq$ NP over $(\mathbb{Z},=)$ with either unit or bit cost. In the case of unit cost, P $\neq$ NP follows from the undecidability of HN over $\mathbb{Z}$. In general, according to Lagrange's theorem over $(\mathbb{Z},=)$, for every $u \geq 0$, there are numbers u_1, u_2, u_3, u_4 with $u = u_1^2 + u_2^2 + u_3^2 + u_4^2$. Therefore, $\mathbb{N}$ is in class NP with respect to unit and bit costs. It follows $\mathbb{N} \notin$ P with respect to unit cost, otherwise $\mathbb{N}$ would be decidable in constant time. $\mathbb{N} \notin$ P is also true in the case of bit cost (Shub, 1993).

In the fifth row, it is P $\neq$ NP over $(\mathbb{Q},<)$ with respect to unit cost. Actually, we do not know whether HN is undecidable over $\mathbb{Q}$. Therefore, P $\neq$ NP over $(\mathbb{Q},<)$ with respect to unit cost must be justified with different arguments: Projections of semi-algebraic sets are in NP. If we assume P $=$ NP over $(R, <)$, projections of semi-algebraic sets which are in NP must be in P and hence must be semi-algebraic. According to (the reverse of) Tarski's theorem, R must be real-closed (Macintyre *et al.*, 1983). But, $\mathbb{Q}$ is not real-closed. Thus, the assumption of $\mathbb{N} =$ NP leads to a contradiction.

Over $\mathbb{R}$ and $\mathbb{C}$, NP is decidable and NP-complete with respect to unit cost. NP $\subseteq$ EXP is obvious in this case, but $\mathbb{N} =$ NP is still unknown.

In classical complexity theory over $\mathbb{Z}$, the arithmetical and analytical hierarchies are introduced in a recursive way (cf. Chapter 3). Relative recursiveness is defined with respect to Turing's computational degrees. According to a suggestion of Turing, relative recursiveness can be illustrated by oracle machines which can be generalized over a ring R. Thus, oracle machines are also used as a tool to reduce complexity of problems.

An example is the p-reduction that is given by the NP$_\mathbb{R}$-completeness of 4-FEAS: For any set S in NP$_\mathbb{R}$, a polynomial time-computable function φ_S exists such that for every $x \in \mathbb{R}^\infty$, $x \in S$ iff $\varphi_S(x) \in$ 4-FEAS. In this case, we can imagine a machine M_S which computes an element $\varphi_S(x) \in \mathbb{R}^\infty$ for input $x \in S$ in polynomial time and queries the set 4-FEAS whether the element belongs to it. With respect to unit cost for each query, the machine works in polynomial time. Thus, if a polynomial time machine to decide 4-FEAS would be known, then this machine could also decide S. Oracle machines realize queries about a membership in a set. They are more powerful than a conventional machine because they can make not one query, but as many queries as their time bounds permit.

Definition of an oracle machine over R. An oracle machine over $\mathbb{R}$ is a machine over $\mathbb{R}$ which, in addition to its input and output spaces $\mathbb{R}^\infty$, and the state space $\mathbb{R}_\infty$, possesses an oracle space $\mathcal{R} = \mathbb{N} \times \mathbb{R}^\infty$, an oracle set $A \subseteq \mathbb{R}^\infty$, and two additional types of nodes: write and query. Denote the content of the state space by $(\ldots, x_{-1}, x_0.x_1, \ldots)$ and the content of the oracle space by $(j, y_1, y_2, \ldots)$. At the beginning of the computation, the content of the oracle space $\mathcal{R}$ is $(0, 0, \ldots)$.

Both write and query nodes have a unique next node:

Write nodes. The content of the coordinate x_0 of the state space is written in the jth coordinate of the real part of the oracle space, where j is the content of its natural number coordinate. Then the natural number coordinate is incremented by 1:

$$
\begin{aligned}
(\ldots, x_{-1}, x_0.x_1, \ldots) & \qquad\qquad (\ldots, x_{-1}, x_0.x_1, \ldots) \\
(j, y_1, \ldots, y_j, y_{j+1}, y_{j+2}, \ldots) & \;\rightarrow\; (j+1, y_1, \ldots, y_j, x_0, y_{j+2}, \ldots)
\end{aligned}
$$

Query nodes. The machine writes 1 to x_0 if the content in $\mathbb{R}^\infty$ in the oracle space belongs to A and 0 otherwise. Once the answer of the query has been written, all the coordinates of the oracle space are reset to 0:

$$
\begin{aligned}
(\ldots, x_{-1}, x_0.x_1, \ldots) \\
(j, y_1, \ldots, y_j, \ldots)
\end{aligned}
\;\rightarrow\;
\begin{cases}
(\ldots, x_{-1}, 1.x_1, \ldots) \\
(0, 0, \ldots), & \text{if } (y_1, \ldots, y_j) \in A \\
(\ldots, x_{-1}, 0.x_1, \ldots) \\
(0, 0, \ldots), & \text{if } (y_1, \ldots, y_j) \notin A
\end{cases}
$$

Definition of polynomial hierarchy for unrestricted machines over field R. Complexity degrees can be generalized for unrestricted machines which can add, multiply, and divide in a field R (e.g., $\mathbb{R}$ or $\mathbb{C}$):

$$\Sigma_0^R := \Pi_0^R := P_R.$$

A predicate A is $\Sigma_n^R(\Pi_n^R)$ (in M) for $n \geq 1$ iff it has an explicit definition $(*)$ with no two adjacent quantifiers of the same kind and the first quantifier existential (universal):

$$(*)A(u) \longleftrightarrow Q_1 x_1 \in \mathbb{R}^{p_1(m)} \ldots Q_n x_n \in \mathbb{R}^{p_n(m)} B(u, x_1, \ldots, x_n)$$

with $u \in \mathbb{R}, B \subseteq (\mathbb{R}^\infty)^{n+1}$, $B \in P_R$ polynomial complex, quantities $Q_i x_i$ either $\exists x_i$ or $\forall x_i$, and polynomial functions $p_1, \ldots, p_n$.

A predicate is $\Delta_n^R(M)$ iff it is both $\Sigma_n^R(M)$ and $\Pi_n^R(M)$.

The levels of the polynomial hierarchy over field or ring R relativized to set M can be characterized by corresponding oracle machines.

$\vdots$	$\vdots$	$\vdots$
$\Sigma_{n+1}^R(M)$	$\Delta_{n+1}^R(M)$	$\Pi_{n+1}^R(M)$
$\Sigma_n^R(M)$	$\Delta_n^R(M)$	$\Pi_n^R(M)$
$\vdots$	$\vdots$	$\vdots$
$\Sigma_2^R(M)$	$\Delta_2^R(M)$	$\Pi_2^R(M)$
$\Sigma_1^R(M)$	$\Delta_1^R(M)$	$\Pi_1^R(M)$

Example of $\Sigma_k^{\mathbb{R}}$-complete problems. Complete problems can be generalized along the levels of the polynomial hierarchy. As an example, we consider $NP_{\mathbb{R}}$-completeness of 4-FEAS:

For every $k \geq 1$, consider the subset $S_k \subset \mathbb{R}^\infty$ with

$$
S_k = \begin{cases}
\{(f, n_1, n_2, \ldots, n_k) | f \text{ is a degree-4} \\
\quad \text{polynomial in } n_1 + n_2 + \cdots + n_k \text{ variables} \\
\quad \text{such that } \exists_{x_1} \in \mathbb{R}^{n_1} \forall_{x_2} \in \mathbb{R}^{n_2}, \ldots, \\
\quad \exists_{x_k} \in \mathbb{R}^{n_k} f(x_1, \ldots, x_k) = 0\}, & \text{if } k \text{ is odd,} \\[1em]
\{(f, n_1, n_2, \ldots, n_k) | f \text{ is a degree-4} \\
\quad \text{polynomial in } n_1 + n_2 + \cdots + n_k \text{ variables} \\
\quad \text{such that } \exists_{x_1} \in \mathbb{R}^{n_1} \forall_{x_2} \in \mathbb{R}^{n_2} \ldots \forall_{x_k} \in \\
\quad \mathbb{R}^{n_k} f(x_1, \ldots, x_k) \neq 0\}, & \text{if } k \text{ is even.}
\end{cases}
$$

Obviously, $S_1 = $ 4-FEAS. The complement of S_2 is a parametrized version of 4-FEAS: There are two groups of variables, x and t, and the problem is to decide whether for all values of the parameter $t \in \mathbb{R}^{n_1}$ there is an $x \in \mathbb{R}^{n_2}$ such that $f(t, x) = 0$. These results can be generalized in the following theorem (Cucker and Koiran, 1995; Cucker and Matamala, 1996):

For all $k \geq 1$, the set S_k is $\Sigma_k^{\mathbb{R}}$-complete.

Complexity classes and formal languages

Complexity classes of solvability and decidability over a ring R are characterized by capacities of machines (e.g., graphic machines, oracle machines) and resources. But complexity classes of solvability and decidability can also be characterized in a machine-independent way by the complexity of their defining and describing languages. Graphic machines were already algebraically described, e.g., $\text{NP}_{\mathbb{R}}$-problems by means of degree-4 polynomial equations: If $S \in \text{NP}_{\mathbb{R}}$, then, for each n, there is a degree-4 polynomial $\Phi_n(y_1, \ldots, y_n, x_1, \ldots, x_m)$ with the property that $a \in S^n$ if and only if the equation $\Phi_n(y_1, \ldots, y_n, x_1, \ldots, x_m) = 0$ is solvable over $\mathbb{R}$ with $a = (a_1, \ldots, a_n)$ and $S^n = S \cap \mathbb{R}^n$. In short:

$$a \in S^n \text{ iff } \exists x_1, \ldots, \exists x_m \Phi_n(a, x_1, \ldots, x_m) = 0$$

is a true statement over $\mathbb{R}$.

In this way, for each n, there is a description of S^n with polynomials Φ_n which are derived from the algebraic characterization of a $NP_{\mathbb{R}}$-decision machine for S. The goal is to find a formal language to describe S without depending on a machine concept.

In the following, we distinguish three logical systems over $\mathbb{R}$-structures:

- First-order logic over $\mathbb{R}$ ($FO_{\mathbb{R}}$)
- Fixed-point first-order logic over $\mathbb{R}$ ($FP_{\mathbb{R}}$)
- Existential second-order logic over $\mathbb{R}$ ($\exists SO_{\mathbb{R}}$)

Sets of $\mathbb{R}$-structures can be associated with subsets of $\mathbb{R}^\infty$, i.e., decision problems of $\mathbb{R}$. It can be proven that a set $S \subset \mathbb{R}^\infty$ is in P_R or $NP_{\mathbb{R}}$ iff the associated sets of $\mathbb{R}$-structures are describable in fixed-point first-order logic or existential second-order logic.

Definition of definability in formal languages.

(a) Formal language:

- A vocabulary L is a set of symbols, each one being relational or functional.
- A structure $\mathcal{A}$ of L is a set A (universe) with interpretations of L.
- If T is a set of sentences over L, $\mathcal{A}$ satisfies T iff all sentences of T are satisfied in $\mathcal{A}(\mathcal{A} \vDash T)$
- Structures $\mathcal{A}$ with $\mathcal{A} \vDash T$ are called models of T.
- The sentences in T are called axioms and T is called a theory.

(b) Definability:

- A relation $R \subseteq A^n$ is (first order) definable in $\mathcal{A}$ over L iff there exists a formula φ over L with n free variables $x_1, \ldots, x_n$ for all $a \in R$ if and only if $\mathcal{A} \vDash \varphi(a)$.
- A partial function $f : A^n \longrightarrow A$ is term definable in $\mathcal{A}$ over L iff there exists a term $t(x_1, \ldots, x_n)$ such that is $t^{\mathcal{A}}(a) = f(a)$ for all $a \in A^n$.
- A set $S \subset \mathbb{R}^\infty$ is $P_{\mathbb{R}}(NP_{\mathbb{R}})$ iff its associated sets of $\mathbb{R}$-structures are definable within fixed-point first-order logic (resp. existential second-order logic).

Example: $\mathbb{R}$-definable structures and semi-algebraic sets.

Consider language $L = \{<, 0, 1, +, -, \times\}$, structure $\mathcal{A} = \langle \mathbb{R}, < \rangle$, and vocabulary $L(\mathbb{R})$. The subsets of $\mathbb{R}^n$ definable in $\mathcal{A}$ over $L(\mathbb{R})$ are exactly the semi-algebraic sets. The term definable functions

Table 2. Complexity in systems science, computer science, and logics.

Dynamical Complexity	Computational Complexity	Logical Complexity
dynamical circuits tractable over $\mathbb{R}$	$\mathrm{NC}^1_\mathbb{R}$ (class of sets decidable by algebraic circuits with size polynomial in n and depth $O(\log^1 n)$)	$\mathrm{FO}_\mathbb{R}$
parallel dynamical systems tractable over $\mathbb{R}$	$\mathrm{PL}^1_\mathbb{R}$ (class of sets decidable by parallel machines with polynomial number of processors and time $O(\log^1 n)$)	
	$\vdots$	$\vdots$
	$\mathrm{NC}^k_\mathbb{R}$ $\mathrm{PL}^k_\mathbb{R}$	$\mathrm{FP}^1_\mathbb{R}$ $(O(log^k n))$
	$\vdots$	$\vdots$
all dynamical systems tractable over $\mathbb{R}$	$\mathrm{P}_\mathbb{R}$ (class of sets decidable by polynomial time machines over $\mathbb{R}$)	$\mathrm{FP}_\mathbb{R}$
Traveling Salesman problems, Knapsack problem, etc.	$\mathrm{NP}_\mathbb{R}$ (class of sets decidable by non-deterministic polynomial time machine over $\mathbb{R}$)	$\exists \mathrm{SO}_\mathbb{R}$
circuit satisfiability, etc.	$\mathrm{NP}_\mathbb{R}$-complete (class of $\mathrm{NP}_\mathbb{R}$-sets polynomial reducible to all $\mathrm{NP}_\mathbb{R}$-sets)	
dynamical systems with polynomial-bounded number of used coordinates in state space	$\mathrm{P\,SPACE}_\mathbb{R}$ (class of sets decidable in polynomial space resp. polynomial time)	$\mathrm{pFP}^2_\mathbb{R}$ (primitive fixed-point second-order logic for parallel ranked $\mathbb{R}$-structures)
dynamical systems with exponential development	$\mathrm{EXPTIME}_\mathbb{R}$ (class of sets decidable by exponential time machine)	$\mathrm{pFP}^2_\mathbb{R}$

$f : \mathbb{R}^n \to \mathbb{R}$ are the polynomial functions. The function $\sqrt{} : [0, \infty) \to \mathbb{R}$ is definable, but not term definable.

Complexity classes of formal languages also connect logic with complexity classes of computability. Obviously, there are deep relations between complex dynamical systems in mathematics, computational complexity in computer science, and modeling formal languages in logic. These complexity classes are not restricted to the digital world, but refer to real-world problems and real algorithms of ordinary mathematics in, e.g., the theory of differential equations, functional analysis, and numerical analysis. Thus, we end this chapter with Table 2 illustrating bridges between logic, computer science, and mathematics in the world of real numbers which is also the "real" world of practical science (according to results in Blum *et al.* (1998, Chapters 18 and 23)). In computer technology, von Neumann architectures with only one processor are extended to complexity classes of real parallel machines with several parallel working processors and real circuit families which are interesting for analog neural nets (cf. Chapter 12).

Chapter 12

Real Computing and Neural Networks

Turing computability was technically realized by von Neumann computers with a processor and a separated store. This computational architecture was built by silicon semiconductor technology. Its computer power could be increased exponentially, but with immense energy costs world-wide. Contrary to technology, in evolution, cognitive abilities were increased by brains and neural networks of high efficiency and low costs of energy. We can prove that the evolutionary strategy of neural networks and the technical strategy of (von Neumann) computers are, at least in principle, logical-mathematically equivalent. That means that, for (nearly) all neural networks of a certain degree of complexity and efficiency, we can find an equivalent automaton or computing machine with the same abilities (e.g., recognition of languages) and vice versa.

Natural neural networks of living brains are mathematically modeled by artificial neural networks. They are complex systems of firing and non-firing neurons with topologies like brains (Fig. 18). There is no central processor ("mother cell"), but a self-organizing information flow in cell assemblies according to rules of synaptic weights ("synaptic plasticity") (Churchland and Sejnowski, 1992; Mainzer, 1994, 1997).

Mathematically, we start with finite-size networks which consist of connections of synchronously evolving processors. Each processor updates its state with a sigmoidal function which is applied to a linear combination of the previous states of all units. The strength

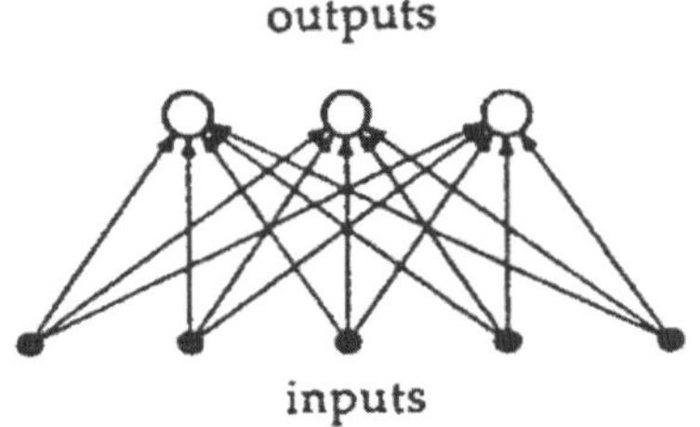

Feedforward with one synaptic layer

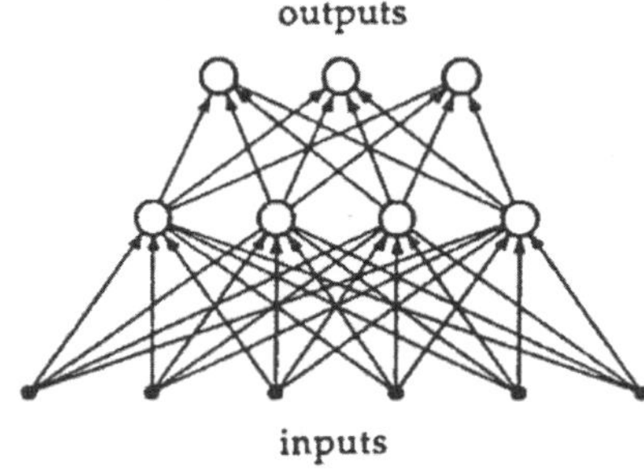

Feedforward with two synaptic layers (hidden units)

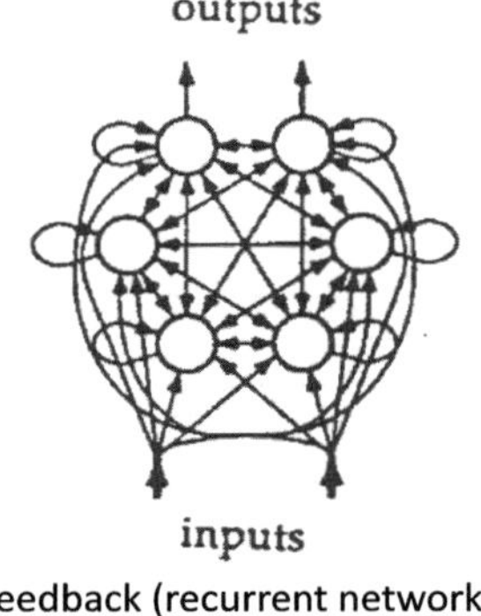

Feedback (recurrent network)

Fig. 18. Topologies of neural networks.

of the synaptic connections of neurons is represented by numerical weights.

At a first glance, neural networks seem to be completely different from machines and automata. Neural networks are understood as models of natural brains. Machines and automata are products of human technology. But, actually, automata and machines can be generated in a hierarchy of more or less complex prototypes corresponding to appropriate models of neural networks which are able to recognize different types of languages. Mathematically, the increasing complexity and abilities of neural networks depend on the numerical weights from integers, and rational numbers up to real numbers. This is a remarkable result, because appropriate "brains" and "machines" are equivalent in a rigorous mathematical sense.

Definition of a (finite-size recurrent) neural network. A (recurrent) neural network $\mathcal{N}$ is presented by a directed graph of nodes called neurons. Each neuron updates its activation value by applying a composition of a one-variable function with a linear combination of the activations of all neurons $x_j (j = 1, \ldots, N)$, the external inputs $u_k (k = 1, \ldots, M)$, and synaptic weights of rational coefficients a_{ij}, b_{ij}, c_i. Formally, each processor's (cellular) state is updated by

$$x_i(t+1) = \sigma \left(\sum_{j=1}^{N} a_{ij} x_j(t) + \sum_{j=1}^{M} b_{ij} u_j(t) + c_i \right)$$

with x_i states of activation, u_j inputs at the previous instants, synaptic weights a_{ij}, b_{ij}, c_i, and sigmoid (e.g., saturated-linear) function σ:

$$\sigma(x) := \begin{cases} 0, & \text{if } x < 0, \\ x, & \text{if } 0 \le x \le 1, \\ 1, & \text{if } x > 1. \end{cases}$$

A neural net can be understood as a dynamical system. Their dynamics is defined by the time-dependent change of their system states. At each instant, the system state is represented by a vector $x(t) \in \mathbb{Q}^N$ of rational numbers with the ith coordinate denoting the activation level of the ith processor. Given an initial state $x(1)$ and an infinite input sequence $u = u(1), u(2), \ldots$, the dynamical state $x(t)$ at time t can be computed for each integer $t \ge 1$ as the value obtained by recursively solving the previously defined equation of updating processor's states.

What does it mean that a function is computed by a neural net (instead of a machine)? We consider a function $\Phi : \{0,1\}^* \to \{0,1\}^*$ with the semi-group $\{0,1\}^*$ of binary strings (Siegelman and Sontag, 1995). An appropriate formal net must transform binary input strings into binary output strings. Such a formal net is equipped with two binary input lines which are called data resp. validation line. The data line carries a binary input signal D. The validation line indicates when the data line is active with $V = 1$ and when it is inactive with $V = 0$. Therefore, for each t, the input $u(t)$ is defined by

$$u(t) = (D(t), V(t)) \in \{0,1\}^2.$$

The initial state $x(1)$ is always taken to be zero and an equilibrium state. The output is represented by two output processors as data and validation lines, denoted by H and G:

$$y(t) = (H(t), G(t)) \in \{0,1\}^2.$$

In general, the (discrete-time) dynamics of system states with two binary inputs is described by the map $\mathcal{F} : \mathbb{Q}^N \times \{0,1\}^2 \to \mathbb{Q}^N$. The state $x(t)$ at time t (for each integer $t \geq 1$) is defined by recursively solving the equations

$$x(1) := x^{\text{init}}$$

$$x(t+1) := \mathcal{F}(x(t), u(t)) \quad (t = 1, 2, \ldots).$$

Definition of σ-processor net $\mathcal{N}$ with binary inputs. A σ-processor net $\mathcal{N}$ with two binary inputs is a dynamical system with map

$$\mathcal{F}(x, u) = \sigma(Ax + b_1 u_1 + b_2 u_2 + c)$$

with matrix $A \in \mathbb{Q}^{N \times N}$, vectors $b_1, b_2, c \in \mathbb{Q}^N$, and sigmoid function $\sigma : \mathbb{Q}^N \to \mathbb{Q}^N$. In 1956, Kleene started the proofs of equivalence of automata and neural networks: Finite automata can be simulated by McCulloch Pitts networks recognizing simple regular languages (Kleene, 1956). McCulloch Pitts networks are based on a binary activation function signal(x) which is equal to 0 if $x \leq 0$ and is equal to 1 otherwise. Another type is the Heaviside function which provides 1 also for the value 0. Obviously, McCulloch Pitts networks represent a simple type of "brain". A finite network of two-state elements is equivalent to the performances of finite automata. Finite automata are also simple examples of automata which are technically realized by, e.g., ticket automata at railway stations (Minsky, 1967):

Definition of a finite automaton. A finite automaton (Q, Σ, f, q_o, F) consists of a finite set Q of sets, a finite set Σ of inputs, initial state q_o, accepting states $F \subseteq Q$, and transition function $f : Q \times \Sigma \to Q$. For any state q and input a, the next state is denoted by $f(q, a)$. For accepting states $q \in F$, it is $f(q, a) = q$.

f^* is an extension of the transition function f, mapping a state and a string of symbols to the resulting state:

$$f^*(q, \omega) := \begin{cases} q, & \text{if } \omega = \epsilon(\text{empty}), \\ f(f^*(q, \omega'), a), & \text{if } \omega = \omega'a. \end{cases}$$

for a symbol $a \in \Sigma$ and strings $\omega, \omega' \in \Sigma^*$.

Definition of a finite automaton accepting a language. A string ω is said to be accepted by a finite automaton $\mathcal{M}$ iff $f^*(q_o, \omega) \in F$. The language $\mathcal{L}(\mathcal{M}) \subseteq \Sigma^*$ accepted by $\mathcal{M}$ is the set of all accepted strings. A language is regular iff it is accepted by some finite automaton.

Networks become more powerful if Boolean neurons with binary states are replaced by neurons with analog values of the interval $[0, 1]$. When the weights are restricted to integers, the analog neurons are reduced to binary activations. In this case, the networks are computationally equivalent to McCulloch Pitts networks (Siegelmann, 1999, 25 f.).

Theorem. *The languages accepted by integer networks are exactly the regular ones over the alphabet* $\{0, 1\}$.

Proof. In order to prove the desired equivalence, an offline finite automaton is introduced which may continue computing after completely reading the input string, and decide on the input after some delay. A decision is reached when the computation arrives at either an accepting or a rejecting state.

Formally, an offline automaton $(Q, \Sigma, f_{\text{off}}, q_0, F_{\text{off}})$ consists of the transition function f_{off} mapping $Q \times (\Sigma \cup \{\$\})$ into Q with a special symbol $\$ \notin \Sigma$ which denotes that the complete input string has been read into the machine. $F_{\text{off}} \subseteq Q$ is the set of accepting states, and the transition function is $f_{\text{off}}(q, a) = q$ for all $a \in \Sigma \cup \{\$\}$ and $q \in F_{\text{off}}$. The transition function f_{off} is extended on a state and a string of the type $\omega\$^*$, where $\omega \in \Sigma^*$ and $\$^*$ is a string of 0 or more appearances of the $\$$ sign.

A function $\tilde{f}_{\text{off}}$ maps $Q \times \Sigma^*$ (without $\$$) into the set 2^Q of all subsets of Q by

$$\tilde{f}_{\text{off}}(q, \omega) = \{f^*_{\text{off}}(q, \omega\$^*)\} \text{ for any } \omega \in \Sigma^*.$$

A string $\omega \in \Sigma^*$ is accepted by an offline finite automaton iff $\tilde{f}_{\text{off}}(q_0, \omega) \cap F_{\text{off}} \neq \emptyset$. $\qquad\square$

Lemma. *The class $\mathcal{L}$ of languages accepted by offline finite automata is exactly the class of regular languages.*

Proof. Any finite automaton can be considered an offline automaton with $f_{\text{off}}(q, \$) = q$ for all $q \in Q$. Therefore, the regular languages are included in $\mathcal{L}$.

The other inclusion demands that, given an offline finite automaton $\mathcal{M} = (Q, \Sigma, f_{\text{off}}, q_0, F_{\text{off}})$ accepting language L, there is a finite automaton $\mathcal{M}'$ accepting the same language. We define $F = \{q \in Q | \tilde{f}_{\text{off}}(q_0, \omega) \cap F_{\text{off}} \neq \emptyset\}$. Further on, f is the map f_{off} restricted to $Q \times \Sigma$. This machine $\mathcal{M} = (Q, \Sigma, f, q_0, F)$ accepts L. $\qquad\square$

The previous theorem states a correspondence between integer networks and offline automata which can now be proved in both directions:

$(\Rightarrow)$

For a given integer network $\mathcal{N}$ consisting of N neurons and accepting language L, an offline automaton $\mathcal{M} = (Q, \Sigma, f_{\text{off}}, q_0, F_{\text{off}})$ can be defined with $L(\mathcal{M}) = L$:

(1) Input $(D(t), V(t))$ to network $\mathcal{N}$ with the values in $\{0, 1, \$\}^2$ is encoded by $\zeta[(0, 1)] = 0$, $\zeta[(1, 1)] = 1$, and $\zeta[(0, 0)] = \$$. The case $(1, 0)$ is invalid.

(2) $Q = \{0, 1\}^2$ and $q_o = 0^N$. As the network starts from an initial state 0^N with only integer weights, its neurons assume binary values only. Each state of offline automaton $\mathcal{M}$ is identified with the activations of all neurons.

(3) $G(t)$ and $H(t)$ are assumed to be the first and second neurons in the state encoding. The ith coordinate of state q is denoted by $q[i]$. It follows that $F_{\text{off}} = \{q \in Q | q[1] = 1 \text{ and } q[2] = 1\}$.

(4) The transition function f_{off} is induced by the update function $q^+ = \sigma(Aq + Bu + c)$.

$(\Leftarrow)$

For a given offline automaton $\mathcal{M} = (Q, \Sigma, f_{\text{off}}, q_o, F_{\text{off}})$ accepting language $L \subseteq \{0, 1\}^*$ (and no transition into the initial state q_o), an integer network $\mathcal{N}$ can be defined for simulating $\mathcal{M}$:

The input letters 0, 1, \$ are translated by the function ζ^{-1}. The number of neurons is $N = 3|Q| + 2$. Among these neurons, $3|Q|$ neurons are indexed by $j = 0, 1, \ldots, |Q| - 1$, and a pair (k, l) which may be (0,0), (0,1), or (1,1). Each x_{jkl} has binary values only:

$x_{jkl} = 1$ iff the current state of automaton M is q_j and its last input was $\zeta[(k, l)]$. At each moment, only one neuron has the value 1 and all the other neurons have the value 0.

In the next step, the updated equations of the neurons x_{jkl} are defined. Therefore, $|Q|$ auxiliary binary variables $p_i (i = 0, \ldots, |Q| - 1)$ are needed with value 1 if the state of automaton $\mathcal{M}$ is to become q_i. These variables are defined by the activation values of the neurons x_{jkl}:

$$p_0 = 1 - \sum_{j,k,l} x_{jkl},$$

$$p_i = \sum_{j,k,l} a^i_{jkl} x_{jkl}$$

with constants $a^i_{jkl} = \begin{cases} 1, & \text{if } f_{\text{off}}(q_j, \zeta[(k, l)]) = q_i, \\ 0, & \text{otherwise.} \end{cases}$

The updated equations of the neurons x_{jkl} are defined by

$$x^+_{i11} = \sigma(p_i + V + D - 2),$$

$$x^+_{i10} = \sigma(p_i + 2V - D - 2),$$

$$x^+_{i00} = \sigma(p_i - V - D).$$

For the validation and data output neurons, two additional updated equations are defined with $F_R = \{q \in Q | \tilde{f}_{\text{off}}(q, \$) \cap F_{\text{off}} \neq \emptyset\}$:

$$x^+_v = \sigma \left(\sum_{j=0}^{|Q|-1} a_j x_{j00} \right) \text{ with } a_j = \begin{cases} 1, & \text{if } q_j \in (F_{\text{off}} \cup F_R), \\ 0, & \text{otherwise.} \end{cases}$$

$$x^+_d = \sigma \left(\sum_{j=0}^{|Q|-1} a_j x_{j00} \right) \text{ with } a_j = \begin{cases} 1, & \text{if } q_j \in F_{\text{off}}, \\ 0, & \text{otherwise.} \end{cases}$$

The simulation implies that automaton $\mathcal{M}$ and net $\mathcal{N}$ (with integer weights only) accept the same language. $\qquad\qquad\square$

In the next step, it can be proved that all Turing machines can be simulated by neural nets with rational numbers as weights and vice versa. In particular, Siegelmann and Sontag (1995) proved that there is a net of 886 processors which can compute a universal partial-recursive function in tractable time. No high order nets with products are needed. Non-deterministic Turing machines can be simulated by non-deterministic nets the synaptic weights of which are also finite rational numbers. Obviously, these simulation results have logical and epistemic consequences with respect to decidability and complexity of problem solving. In the end, we are interested in the impact of digital and real computing to neural nets.

It is obvious that a processor net with a rational initial state can be simulated by a Turing machine. Intuitively, this theoretical result corresponds to the computational practice that artificial nets are simulated by (von Neumann) computers. The converse direction is more ambitious: Any function computable by a Turing machine can be computed by a processor net with rational weights. If we understand networks as models of natural brains, then this result determines the type of brain which is (at least) needed to realize Turing computations. Later on, we will discuss the possibilities of future technology to build neuromorphic architectures of computers which correspond to living brains.

Let us consider partially defined maps $\Phi : \{0,1\}^* \to \{0,1\}^*$ that are recursively computed. For these maps, there is a multi-stack Turing machine M so that, when a word $\omega \in \{0,1\}^*$ is given initially on the input/output stack, machine M halts on the word ω iff the function value $\Phi(\omega)$ is defined and the content of that stack when the machine halts. For each word $\omega = \omega_1 \ldots \omega_k \in \{0,1\}^*$ with $\Phi(\omega) = \beta_1 \ldots \beta_l \in \{0,1\}^*$ or undefined (with l length of the response/output) and each $r \in \mathbb{N}$ (with r response time which is needed to provide the first output bit), input and output can be encoded as follows:

Input is encoded as

$$u_\omega(t) = (V_\omega(t), D_\omega(t))(t = 1, \ldots,)$$

with $V_\omega(t) = \begin{cases} 1, & \text{if } t = 1, \ldots, k, \\ 0, & \text{otherwise.} \end{cases}$

and $D_\omega(t) = \begin{cases} \omega_k, & \text{if } t = 1, \ldots, k, \\ 0, & \text{otherwise.} \end{cases}$

Output is encoded as

$$y_{\omega,r}(t) = (G_{\omega,r}(t), H_{\omega,r}(t)) \quad (t = 1, \ldots)$$

with $G_{\omega,r}(t) = \begin{cases} 1, & \text{if } t = 1, \ldots, (r + l - 1), \\ 0, & \text{otherwise.} \end{cases}$

if the output $\Phi(\omega)$ is defined, and is 0 when $\Phi(\omega)$ is not defined,

and $H_{\omega,r}(t) = \begin{cases} \beta_{t-r+1}, & \text{if } t = r, \ldots, (r + l - 1), \\ 0, & \text{otherwise.} \end{cases}$

if the output $\Phi(\omega)$ is defined, and is 0 when $\Phi(\omega)$ is not defined.

Theorem. *Let* $\Phi : \{0,1\}^* \to \{0,1\}^*$ *be any recursively computable partial function. Then, there exists a processor net $\mathcal{N}$ with the following properties:*

If $\mathcal{N}$ is started from the zero (inactive) initial state and the input sequence u_ω is applied, then $\mathcal{N}$ generates the output $y_{\omega,r}$ (with $H_{\omega,r}$ in the output node and $G_{\omega,r}$ in the validation node) for some r.

If a p-stack Turing machine M (of one input stack and several working stacks) computes $\Phi(\omega)$ in time $T(\omega)$, then one may take response time $r(\omega) = T(\omega) + O(|\omega|)$.

A non-deterministic processor net is a modification of a deterministic one with a guess input line $\mathcal{G}$ in addition to the validation and data lines. The dynamics map of the network is $\mathcal{F} : \mathbb{Q}^N \times \{0,1\}^3 \to \mathbb{Q}^N$.

Definition of non-deterministic σ-processor net $\mathcal{N}$ with binary inputs. A non-deterministic σ-processor net $\mathcal{N}$ with two binary inputs and a guess line is a dynamical system with map

$$\mathcal{F}(x, u, g) = \sigma(Ax + b_1 u_1 + b_2 u_2 + b_3 g + c)$$

with matrix $A \in \mathbb{Q}^{N \times N}$, vectors $b_1, b_2, b_3, c \in \mathbb{Q}^N$, and sigmoid function $\sigma : \mathbb{Q}^N \to \mathbb{Q}^N$.

For each pair of words $\omega, y \in \{0,1\}^*$, the input to the network is encoded as

$$u_{\omega,\gamma}(t) = (V_\omega(t), D_\omega(t), \mathcal{G}_\gamma(t))(t = 1, \ldots,)$$

with input lines V_ω (validation line), D_ω (data line), and guess line

$$\mathcal{G}_\gamma(t) := \begin{cases} \gamma_t, & \text{if } t = 1, \ldots, |\gamma|, \\ 0, & \text{otherwise.} \end{cases}$$

The output is encoded in the same way of a deterministic network. The application is restricted to formal non-deterministic networks computing binary output values $\Phi_\mathcal{G}(\omega, \gamma) \in \{0,1\}^*$ with function value $\Phi_\mathcal{G}(\omega, \gamma)$ computed by a formal non-deterministic network when the input is $\omega \in \{0,1\}^*$ and the guess is $\gamma \in \{0,1\}^*$.

An important ability of brains is to handle languages. The language L computed by a non-deterministic formal network in time B is

$$L = \{\omega \in \{0,1\}^* | \exists \text{ guess } \gamma_\omega (\Phi_\mathcal{G}(\omega, \gamma_\omega)$$
$$= 1 \wedge |\gamma_\omega| \leq T_\mathcal{N}(\omega) \leq B(|\omega|))\}$$

with time $T_\mathcal{N}$ required to compute the response to a given input ω, bound B as computation time, and length $|\gamma_\omega|$ of guess γ_ω which is bounded by function $T_\mathcal{N}$. The simulation theorem can be repeated for non-deterministic Turing machines and non-deterministic rational processor networks.

Artificial networks with rational weights are models of Turing computability. Biological networks with sensors and synapses are characterized by graded scaling. Thus, neural networks with real weights have attracted much attention as models of analog computation. In engineering and natural sciences, real numbers correspond to values of resistances, capacitances, physical constants, and so forth. Real numbers may not be directly measurable and they are not always computable. But they affect the behavior of a dynamical system. On a finite time interval, constants could be replaced by rational numbers generating the same qualitative behavior. But the long-term behavior of the system depends on the real values of the constants. Another well-known illustration is the butterfly effect and sensitive dependence of chaotic dynamics on tiny changes of initial values. What is the impact of real computing on analog networks?

Definition of a real (recurrent) network. A real (recurrent) neural network $\mathcal{N}$ is presented by a directed graph of nodes called neurons like a rational (recurrent) neural network, but with real instead of rational weights. Each processor's (cellular) state is updated by a corresponding equation in vector form. For notional simplicity, "$x(t + 1)$" is replaced by "$x^+(t)$" with dropping arguments t:

$$x^+ = \sigma(Ax + Bu + c)$$

with vector x of size N (number of processors), vector u of size M (number of inputs), N-vector c, real matrices A, B of sizes $N \times N$ and $M \times M$, and sigmoid function σ.

A subset $x_{i_1}, \ldots, x_p$ of the N processors contains the p output processors which are used to communicate with the environment of the network. In short, a net is characterized by the real data (A, B, c) with a subset of its nodes.

Again, these networks can be represented as dynamical systems whose state at each instant is now a real vector $x(t) \in \mathbb{R}^N$. The ith coordinate of this vector indicates the activation of the ith processor at time t. The state $x(t)$ at time t can be obtained, for each value $t \geq 1$, by recursively solving the updating equation for initial state $x(1)$ and an infinite input sequence $u = u(1), u(2), \ldots$. The output produced by the input u for the initial state $x(1)$ refers to the sequence of output values in the output processors.

An essential ability of brains and networks is the recognition and computation of languages. McCulloch Pitts networks are very simple with fixed finite weights. They are equivalent with finite automata and can recognize regular languages only. We could consider a hierarchy of networks and automata with increasing abilities up to Turing machines which are equivalent to networks with rational weights. They can recognize recursive languages of the Chomsky type. Real neural networks can even recognize in polynomial time the same class of languages recognized by Turing machines that consult (sparse) oracle in polynomial time. They can recognize all languages, including non-computable ones, in exponential time. This is an epistemically deep insight and we will come back to its philosophical aspects later on.

How can we define formally that a net recognizes a language? To make it simple, a language is understood as a set $L \subseteq \{0, 1\}$ of words (bit sequences) which can be understood as digital codes of any kind of information. Before defining what is meant by a net recognizing a language L, we must first define a formal network which is able to encode its input and output. Again, we define formal nets with two binary input lines for data and validation for each t:

$$u(t) = (D(t)V(t)) \in \{0, 1\}^2.$$

The initial state $x(1)$ is taken to be zero and an equilibrium state:

$$\sigma(A0 + B0 + c) = \sigma(c) = 0.$$

Two output processors $O_d(t)$, $O_v(t)$ are also taken as data resp. validation lines.

For each word $\omega = \omega_1, \ldots, \omega_k \in \{0, 1\}^*$, the input is encoded by

$$u_\omega(t) = (V_\omega(t), D_\omega(t))(t = 1, \ldots,)$$

with $V_\omega(t) = \begin{cases} 1, & \text{if } t = 1, \ldots, k, \\ 0, & \text{otherwise.} \end{cases}$

and $D_\omega(t) = \begin{cases} \omega_k, & \text{if } t = 1, \ldots, k, \\ 0, & \text{otherwise.} \end{cases}$

Definition of neural networks recognizing languages. In a formal net $\mathcal{N}$ with two inputs, a word ω is classified in time τ iff the output sequence $y(t) = (O_d(t), O_v(t))$ generated by input u_ω when starting from $x(1) = 0$ has the form

$$O_d = \underbrace{0 \ldots 0}_{\tau-1} \eta_\omega \, 000 \ldots, \qquad O_v = \underbrace{0 \ldots 0}_{\tau-1} 1000 \ldots,$$

with $\eta_\omega = 0$ or 1.

A language $L \subseteq \{0, 1\}^*$ is recognized in time T by the net $\mathcal{N}$ provided that each word $\omega \in \{0, 1\}^*$ is classified in time $\tau \leq T(|\omega|)$ whether $\omega \in L$ or not, where $T : \mathbb{N} \to \mathbb{N}$ is a function on natural numbers.

In order to simulate neural nets by machines, we consider circuits which logically model the switching elements of technical computers. Non-uniform families of circuits are sets of circuits which are not necessarily recursively described. Therefore, they can miss classical computability.

Definition of a Boolean circuit. A <u>Boolean circuit</u> is a directed acyclic graph with input nodes of in-degree 0 and gate nodes which are labeled by one of the Boolean functions AND, OR, or NOT. One of the nodes, which has no outgoing edges, is called output node.

<u>Nodes</u> are arranged into <u>levels</u> $0, 1, \ldots, d$, where the input nodes are at level 0, the output level is at level d, and each node only has incoming edges from the previous level.

A <u>gate node</u> computes the corresponding Boolean function of the values from the previous level, and the value obtained is considered as an input to be used by the successive level. In this way, each circuit computes a Boolean function of its inputs.

- The <u>size</u> of the circuit is the total number of gates.
- The <u>depth</u> of the circuit is d.
- The <u>width</u> of the circuit is the maximum size of each level.

Definition of a family of circuits. A family $\mathcal{C} = \{c_n | n \in \mathbb{N}\}$ of circuits is a set of circuits c_n with sizes $S_{\mathcal{C}}(n)$, depth $D_{\mathcal{C}}(n)$, and width $W_{\mathcal{C}}(n)$ for $n = 1, 2, \ldots$ as monotone non-decreasing functions.

A language $L \subseteq \{0,1\}^*$ is said to be computed by a family $\mathcal{C} = \{c_n | n \in \mathbb{N}\}$ of circuits iff the characteristic function of $L \cap \{0,1\}^n$ is computed by c_n for each $n \in \mathbb{N}$.

If language L is recognized by the formal net $\mathcal{N}$ in time T, we write $\Phi_{\mathcal{N}} = L$ and $T_{\mathcal{N}} = T$. If language L is computed by the family $\mathcal{C}$ of circuits, we write $\Phi_{\mathcal{C}} = L$. We want to prove that circuit families can be simulated by networks and vice versa with respect to their ability of recognizing languages in polynomial time. Therefore, in the following, we compare the functions $T_{\mathcal{N}}$ and $\Phi_{\mathcal{C}}$ for formal nets and circuits so that $\Phi_{\mathcal{N}} = \Phi_{\mathcal{C}}$ (Siegelmann and Sontag, 1994).

Theorem. *Let* $\mathrm{NET}(T(n))$ *be the class of languages recognized by formal networks with real weights in time* $T(n)$ *and* $\mathrm{CIRCUIT}(S(n))$ *the class of languages recognized by (non-uniform) families of circuits of size* $S(n)$.

For any function F with $F(n) \geq n$, it follows:

(1) $\mathrm{CIRCUIT}(F(n)) \subseteq \mathrm{NET}(nF^2(n))$
(2) $\mathrm{NET}(F(n)) \subseteq \mathrm{CIRCUIT}(F^3(n))$

In order to prove (1), circuit families are simulated by networks. We construct a fixed universal net with roughly $N = 1000$ processors.

It can simulate an entire circuit family which is encoded by a particular real weight of the net.

Theorem (1). Simulation of circuit families by networks.
There exists a positive integer N such that the following property holds: For each circuit family C of size $S_C(n)$, there exists an N-processor formal network $\mathcal{N} = \mathcal{N}(C)$ such that $\Phi_{\mathcal{N}} = \Phi_C$ and $T_{\mathcal{N}}(n) = O(nS_C^2(n))$.

Proof. Let C be a circuit family of size $S_C(n)$. The required network $\mathcal{N} = \mathcal{N}(C)$ is constructed with three networks which are composed:

- An input network $\mathcal{N}_{\mathrm{I}}$ receives the input

$$u_1 = \omega 00\ldots, \quad u_2 = \underbrace{11\ldots 1}_{|\omega|}00\ldots,$$

  and computes code numbers of u_1 (with word ω) and u_2 for each $\omega \in \{0,1\}^*$.
- A retrieval network $\mathcal{N}_{\mathrm{R}}(c)$ receives the code number of u_2 from input net $\mathcal{N}_{\mathrm{I}}$ and computes the code number of circuit $c_{|\omega|}$.
- A simulation network $\mathcal{N}_{\mathrm{S}}(c)$ receives the code number of circuit $c_{|\omega|}$ and word ω and computes

$$x_0 = \underbrace{00\ldots 0}_{T}\Phi_C(\omega)00\ldots, \quad \underbrace{00\ldots 0}_{T}100\ldots.$$

Encoding of family C of circuits. A circuit c with size s, width w, and w_i gates in the ith level is encoded as a finite sequence over the alphabet $\{0, 2, 4, 6\}$ in the following steps:

(i) The encoding at each level i starts with the letter 6. Levels are encoded successively, starting with the bottom level and ending at the top level.

(ii) At each level, gates are encoded successively. The encoding of gate g consists of a starting symbol, a two-digit code for the gate type, and a code to indicate which gate feeds into it:

- It starts with the letter 0.

- A two-digit sequence $\{42, 44, 22\}$ denotes the type of the gate $\{\text{AND}, \text{OR}, \text{NOT}\}$.
- If gate g is in level i, then the input to g is represented as a sequence in $\{2, 4\}^{w_i - 1}$, such that the jth position in the sequence is 4 iff the jth gate of the $(i-1)$th level feeds into gate g.

Example.

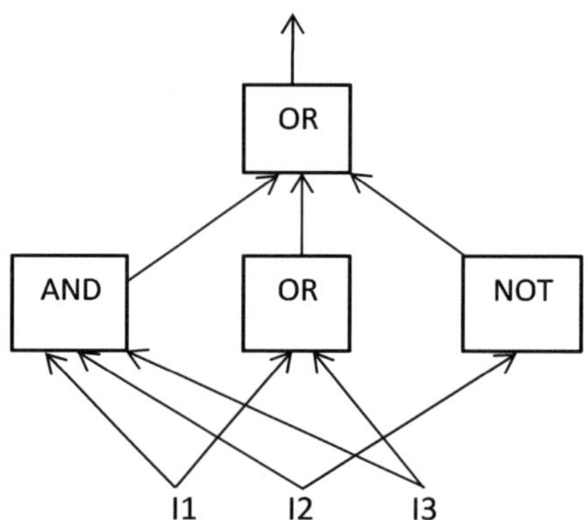

The circuit c_1 in the figure above is encoded as

$$\text{en}[c_1] = \mathbf{6}\,\mathbf{0}\,\underbrace{42}\,\underbrace{444}_{\text{AND I1,I2,I3 feed in}}\,\underbrace{0\,44}\,\underbrace{424}_{\text{OR I1,I3 feed in}}\,\underbrace{0\,22}\,\underbrace{242}_{\text{NOT I2 feeds in}}$$

$$g_1 \qquad\qquad g_2 \qquad\qquad g_3$$

$$\mathbf{6}\,\mathbf{0}\,\underbrace{44}\,\underbrace{444}_{\text{OR AND ,OR,NOT feed in}}$$

$$g_4$$

A non-uniform family $\mathcal{C}$ of circuits of size $S(n)$ is encoded as an infinite sequence

$$e(\mathcal{C}) = 8\,\overline{\text{en}}[c_1]\,8\,\overline{\text{en}}[c_2]\,8\,\overline{\text{en}}[c_3]\ldots$$

with $\overline{\text{en}}[c_i]$ encoding of c_i in reversed order.

Consider a finite or infinite sequence $r = r_1 r_2 \ldots$ of natural numbers smaller than a natural number b. The interpretation of the sequence r in base b is the number

$$r|_b := \sum_{i=1}^{\infty} \frac{r_i}{b^i}.$$

The encoding is one-to-one for the sequences used in the proof. The previous code of the family $\mathcal{C}$ of circuits can be interpreted in base $b = 9$ by

$$\hat{\mathcal{C}} := 8 \; \overline{\mathrm{en}}[c_1] 8 \; \overline{\mathrm{en}}[c_2] 8 \; \overline{\mathrm{en}}[c_3] \ldots |_9.$$

The encoding $\mathrm{en}[c_i]$ of the ith circuit c_i in family $\mathcal{C}$ interpreted in base 9 is denoted by $\widehat{\mathrm{en}}[c_i]$.

A number encoding a family of circuits is a number in the interval [0,1], but not every number of [0,1] can be a code number of this type. If the first digit to the right of the decimal point is 0, then the value of the encoding ranges in $[0,\frac{1}{9}]$, if it is 2, the value is in $[\frac{2}{9},\frac{3}{9}]$, and so forth. In general, the number cannot lie in any of the ranges $[(2i - 1)/9, \; 2i/9]$ for $i = 1, 2, 3, 4$. The set of possible values is a Cantor set which is not continuous, but with holes. The Cantor set is self-similar with bit shifts (in base 9) preserving the holes:

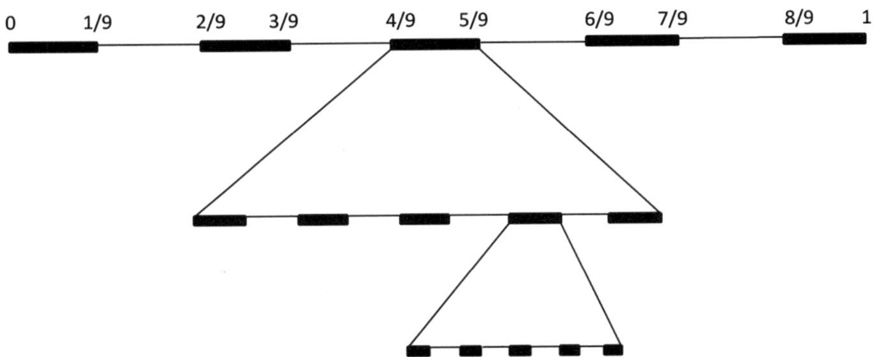

Fig. 19. Circuit encoding as Cantor set.

The structure of the Cantor set clearly distinguishes separated numbers encoding the desired circuit (see Fig. 19). The circuit can be retrieved with operations of finite precision and finite number of neurons.

We can now construct the retrieval network $\mathcal{N}_R(c)$ and the simulation network $\mathcal{N}_S(c)$:

Construction of the retrieval network $\mathcal{N}_R(c)$. For each (non-uniform) family $\mathcal{C}$ of circuits, there exists a 16-processor network $\mathcal{N}_R(c)$with one input line such that, starting from the zero initial state and given the input signal

$$u(1) = \underbrace{11\ldots1}_{n}00\ldots|_2 = 1 - 2^{-n}, \quad u(t) = 0 \text{ for } t > 1,$$

the retrieval network $\mathcal{N}_R(c)$ outputs

$$x_r = \underbrace{000\ldots0}_{2n+2\sum_{i=1}^{n} l(C_i)+4} \widehat{en}[c_n]000\ldots.$$

For the construction of the retrieval network $\mathcal{N}_R(c)$, we define set $\Sigma = \{0, 2, 4, 6, 8\}$ of code numbers. Then, the Cantor set (in base 9) contains all real numbers

$$q = \sum_{i=1}^{\infty} \frac{\alpha_i}{9^i}$$

with $\alpha_i \in \Sigma$. In order to process the real numbers q of the Cantor set (in base 9), we want to define a select left operation with $\Xi[q] = \alpha_1$ and shift left operation with

$$\Lambda[q] = \sum_{i=1}^{\infty} \frac{\alpha_{i+1}}{9^i}$$

such that $\Xi[\Lambda^i[q]] = \alpha_{i+1}$ for each $i \geq 0$.

Let $\Lambda : \mathbb{R} \to [0, 1]$ be the function

$$\Lambda[x] := \begin{cases} 0, & \text{if } x < 0, \\ 9x - \lfloor 9x \rfloor, & \text{if } 0 \leq x \leq 1, \\ 1, & \text{if } x > 1. \end{cases}$$

and $\Xi : \mathbb{R} \to [0, 1]$ the function

$$\Xi[x] := \begin{cases} 0, & \text{if } x < 0, \\ 2\lfloor 9x/2 \rfloor, & \text{if } 0 \leq x \leq 1, \\ 1, & \text{if } x > 1. \end{cases}$$

With select left operation Ξ and shift left operation Λ, Siegelmann *et al.* (1994) suggested a program to perform the required tasks of the retrieval network $\mathcal{N}_{\mathrm{R}}(c)$:

Procedure Retrieval $(\hat{\mathcal{C}}, n)$

Variables *counter, y, z*

Begin

    Counter $\leftarrow 0.\, y \leftarrow 0, z \leftarrow \hat{\mathcal{C}}$,

    While *counter* $< n$

        Parbegin

            $z \leftarrow \Lambda[z]$

            if $\Xi[z] = 8$ then increment *counter*

        Parend,

      While $\Xi[z] < 8$

          Parbegin

             $z \leftarrow \Lambda[z]$

             $y \leftarrow \frac{1}{9}(y + \Xi[z])$

          Parend,

        Return(y)

End

But the select left operation Ξ and shift left operation Λ cannot be programmed in the neural network because of their discontinuity. Therefore, we define functions $\tilde{\Lambda}$ and $\tilde{\Xi}$ which coincide with Λ and Ξ on the Cantor set (in base 9)

$$\tilde{\Lambda}[q] = \sum_{j=0}^{8} (-1)^j \sigma(9q - j),$$

$$\tilde{\Xi}[q] = 2\sum_{j=0}^{3} \sigma(9q - (2j + 1))$$

with sigmoid function σ (Figs. 20 and 21).

With these functions, the retrieval procedure can be realized by the following network:

$$x_i^+ = \sigma(9x_{10} - i)(i = 0, \ldots, 8)$$
$$x_9^+ = \sigma(2u)$$

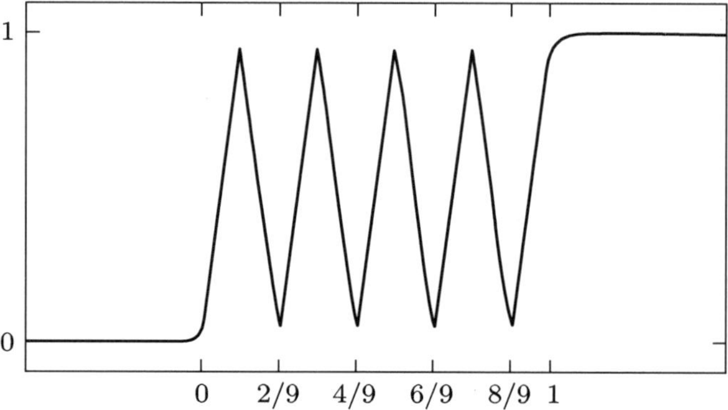

Fig. 20. Function $\tilde{\Lambda}[q]$.

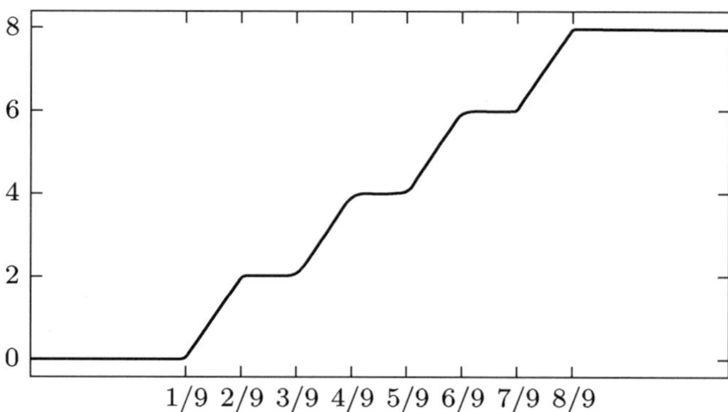

Fig. 21. Function $\tilde{\tilde{\Xi}}[q]$.

$$x_{10}^+ = \sigma(\hat{C}x_9 + x_0 - x_1 + x_2 - x_3 + x_4 - x_5 + x_6 - x_7 + x_8)$$

$$x_{11}^+ = \sigma\left(\frac{1}{9}x_{12} + \frac{2}{9}\left(x_1 + x_3 + x_5 + x_7\right) - 2x_{13}\right)$$

$$x_{12}^+ = \sigma(x_{11})$$

$$x_{13}^+ = \sigma(u + x_{14} + x_{15})$$

$$x_{14}^+ = \sigma(2x_{13} + x_7 - 2)$$

$$x_{15}^+ = \sigma(x_{13} - x_7)$$

$$x_{16}^+ = \sigma(x_{12} + x_7 - 1)$$

If the input u arrives at time 1, then $x_{10}(2k+3) = \tilde{\Lambda}^k[\hat{C}]$ because of the definition of $\tilde{\Lambda}$. Processor nodes x_{13}, x_{14}, x_{15} implement the counter, and processor node x_{16} is the output processor.

Construction of the simulation network $\mathcal{N}_\mathrm{S}(c)$

For a binary sequence $\omega \in \{0,1\}^n$, the encoding $\mathrm{en}[\omega]$ denotes the sequence from $\{2,4\}^n$ that substitutes $(2\omega_i + 2)$ for each ω_i. The interpretation of encoding $\mathrm{en}[\omega]$ in base 9 is denoted by $\widehat{\mathrm{en}}[\omega] = \mathrm{en}[\omega]|_9$.

There exists a network $\mathcal{N}_\mathrm{S}$ such that for each circuit c and binary sequence ω, starting from the initial state and applying the input signal

$$u_1 = \widehat{\mathrm{en}}[c]00\ldots, \quad u_2 = \widehat{\mathrm{en}}[\omega]00\ldots,$$

the simulation network $\mathcal{N}_\mathrm{S}(c)$ outputs

$$x_d = \underbrace{00\ldots0}_{T}y00\ldots, \quad x_v = \underbrace{00\ldots0}_{T}100\ldots,$$

where y is the response of circuit c on the input ω with $T = O(l(c) + |\omega|)$.

The simulation network $\mathcal{N}_\mathrm{S}(c)$ of a family c of circuits is constructed in two steps:

(i) Any circuit can be simulated by a three-tape Turing machine in time $O(l(c) + |\omega|)$: The three tapes are used to store the circuit encoding, the input and output encoding, and the calculation of current level.

(ii) The three-tape Turing machine can be simulated by a net: In general, if M is a k-tape Turing machine with s states which computes in time T a function f on binary input strings, then there exists a rational network $\mathcal{N}$. This network consists of $9^k s + s + 28k + 2$ processors that compute the same function f in time $O(T)$.

With the construction of retrieval network $\mathcal{N}_\mathrm{R}(c)$ and simulation network $\mathcal{N}_S(c)$, we finish the proof of theorem (1). Among the tree networks for input, retrieval and simulation, only the retrieval network $\mathcal{N}_{\mathrm{R}}$ depends on the specific family $\mathcal{C}$ of circuits. All weights can be taken to be rational numbers, except for the weight that encodes the entire circuit family.

The time complexity T to compute the response of $\mathcal{C}$ to the input ω is determined by the description of needed circuits. Formally, the complexity is of order:

$$T = O\left(\sum_{i=1}^{|\omega|} l(c_i)\right).$$

The length of encoding $l(c_i)$ is of order $O(W_{\mathcal{C}}(i)S_{\mathcal{C}}(i)) = O(S_{\mathcal{C}}^2(i))$. Obviously, $S_{\mathcal{C}}(i) \leq S_{\mathcal{C}}(i+1)$ for $i = 1, 2, \ldots$. Therefore, we get the claimed polynomial bound $T = O(|\omega|S_{\mathcal{C}}^2(|\omega|))$.

The reverse simulation of nets by non-uniform families of circuits is stated in Theorem (2) (Siegelmann and Sontag, 1994).

Theorem (2). Simulation of networks by circuit families. *Let $\mathcal{N}$ be a formal network that computes in time $T : \mathrm{N} \to \mathrm{N}$. There exists a non-uniform family of circuits $\mathcal{C}(\mathcal{N})$ of size $O(T^3)$, depth $O(T\log(T))$, and width $O(T^2)$ that accepts the same language as $\mathcal{N}$ does.*

Proof. Theorem (2) is proven in two parts:

2.1 The formal network $\mathcal{N}$ is replaced by a family of formal networks with smaller rational weights.
2.2 The family of formal networks is simulated by a family of circuits.

For these results, we need some definitions.

Definition of an output designated network.

- A processor is called a designated output processor iff its activation value is used as an output of the network and is not fed into any other processor.
- A formal network with two designated output processors is called an output designated network.

- Its processors which are not the designated output processors are called internal processors.

Definition of a q-truncation network. q-truncation is the operation of truncating a bit sequence after q bits. A q-truncation network has an updated equation of the form:

$$x_i^+ = q\text{-truncation}\left[\sigma\left(\sum_{j=1}^{N} a_{ij}x_j(t) + \sum_{j=1}^{M} b_{ij}u_j(t) + c_i\right)\right]$$

2.1. Formal network $\mathcal{N}$ is replaced by a family of formal networks.

Let $\mathcal{N}$ be an output designated network. If $\mathcal{N}$ computes in time T, there exists a family of $T(n)$-truncation output designated networks $\mathcal{N}_o(n)$ such that

- For each n, $\mathcal{N}_o(n)$ has the same number of processors and input and output channels as $\mathcal{N}$ does.
- The weights feeding into the internal processors of $\mathcal{N}_o(n)$ are like those of $\mathcal{N}$, but truncated after $O(T(n))$ bits.
- For each designated output processor in $\mathcal{N}$, if this processor computes $x_i^+ = \sigma(f)$ with linear function f of processors and inputs, then the respective processor in $\mathcal{N}_o(n)$ computes $\sigma(2\tilde{f} - 0.5)$, where $\tilde{f}$ is the same as the linear function f but applied instead to the processors of $\mathcal{N}_o(n)$ and with weights truncated at $O(T(n))$ bits.
- The respective output processors of $\mathcal{N}$ and $\mathcal{N}_o(n)$ have the same activation values at all times $t \leq T(n)$.

At first, the difference between the activations of the corresponding internal processors of $\mathcal{N}_o(n)$ and $\mathcal{N}$ at time $t \leq T(n)$ is measured. The calculation of this error is analogous to that of the chop error of floating points in numerical analysis with the following notations:

- N is the number of processors, M is the number of input lines, $L = N + M + 1$.

- W' is the largest absolute value of the weights of $\mathcal{N}$ with $W = W' + 1$.
- $x_i(t)$ is the value of processor i of network $\mathcal{N}$ at time t.
- $\delta_w \in (0,1)$ and $\delta_p > 0$ are the truncation errors at weights and processors.
- $\varepsilon_t > 0$ is the largest accumulated error at time t in processors of $\mathcal{N}_o(n)$.
- $u \in \{0,1\}^M$ is the input to both $\mathcal{N}$ and $\mathcal{N}_o(n)$ (with $u(t) = 0^M$ for $t > n$).
- a_{ij}, b_{ij}, and c_i are the weights influencing processor i of network $\mathcal{N}$.
- $\tilde{x}_i(t)$, $\tilde{a}_{ij}$, $\tilde{b}_{ij}$, and $\tilde{c}_i$ are the respective activation values of processors and weights of $\mathcal{N}_o(n)$.

Network $\mathcal{N}_o(n)$ computes at each step

$$\tilde{x}_i^+ = q\text{-truncation}\left[\sigma\left(\sum_{j=1}^{N} \tilde{a}_{ij}\tilde{x}_j(t) + \sum_{j=1}^{M} \tilde{b}_{ij}u_j(t) + \tilde{c}_i\right)\right].$$

For all <u>internal processors</u> i, j, we can assume the following errors between the two networks by induction on t:

$$|\tilde{x}_i(t) - x_i(t)| \le \varepsilon_t$$
$$|\tilde{a}_{ij}(t) - a_{ij}(t)| \le \delta_w$$
$$|\tilde{b}_{ij}(t) - b_{ij}(t)| \le \delta_w$$
$$|\tilde{c}_i(t) - c_i(t)| \le \delta_w$$

Because of the Lipschitz property $|\sigma(a) - \sigma(b)| \le |a - b|$, it follows that

$$\varepsilon_t \le N(W' + \delta_w)\varepsilon_{t-1} + (N + M + 1)\delta_w$$
$$+ \delta_p \le LW\varepsilon_{t-1} + L\delta_w + \delta_p.$$

Thus,

$$\varepsilon_t \le \sum_{i=0}^{t-1}(LW)^i(L\delta_w + \delta_p) \le (LW)^t(L\delta_w - \delta_p).$$

For the behavior of the output processors, we must prove

$$\sigma(2\tilde{f} - 0.5) = \begin{cases} 0, & \text{if } \sigma(f) = 0, \\ 1, & \text{if } \sigma(f) = 1, \end{cases}$$

That is,

$$f \leq 0 \Rightarrow \tilde{f} < \frac{1}{4},$$

$$f \geq 1 \Rightarrow \tilde{f} > \frac{3}{4}.$$

This is the case for $|f - \tilde{f}| < \frac{1}{4}$. Obviously, the condition $\varepsilon_t < \frac{1}{4}$ is sufficient. The inequality

$$(L\delta_w - \delta_p) \leq \frac{1}{4}(LW)^{-t}$$

holds if δ_w and δ_p are bounded by $\frac{1}{8}(LW)^{-(t-1)}$. This is the case when the weights and the processors are truncated after $O(t\log(LW))$ bits. With constants L and W, a sufficient truncation for a computation of length T is $O(T)$.

2.2. A family of networks is simulated by a family of circuits.

Let $\mathcal{N}_o$ be a family of $T(n)$-truncation output designated networks, where all networks $\mathcal{N}_o(n)$ consist of N processors and the weights are all rational numbers with $O(T)$ bits. Then, there exists a circuit family $\mathcal{C}$ of size $O(T^3)$, depth $O(T\log(T))$, and width $O(T^2)$ such that circuit c_n accepts the same language as $\mathcal{N}_o(n)$ does on $\{0,1\}^n$.

An appropriate circuit c_n corresponding to a given $T(n)$-truncation output designated network $\mathcal{N}_o(n)$ can be constructed in the following way:

The given network $\mathcal{N}_o(n)$ has two input lines for data resp. validation. The validation line indicates a sequence of n 1's followed by 0's. The n data bits on the data line which appear simultaneously with the 1's in the validation line are considered data input of size n. These n bits are fed simultaneously into c_n via n input nodes.

The sequential input in $\mathcal{N}_o(n)$ is simulated by an input subcircuit which preserves the input as it is to be released one bit at a time in later times of the computation. The input subcircuit is of size $nD_C(n)$:

A processor of network $\mathcal{N}_o(n)$ is denoted by p ($p = 1, \ldots, N$). Each processor p is associated with a subcircuit $\mathrm{sc}(p)$. Each processor $p \in \mathcal{N}_o(n)$ computes a truncated sum of up to $N + 2$ numbers. N of these numbers are multiplications of two T-bit numbers. Each processor computes a sum of $(TN + 2)(2T)$-bit numbers. The summation can be computed by a subcircuit of depth $O(\log(TN))$, width $O(T^2)$, and size $O(T^2N)$ (Savage, 1976).

Only $O(T)$ bits are preserved. Therefore, the activation of each processor can be represented in binary by a unit gate p_u, and the most significant gates p_i($i = 1, \ldots, O(T)$), after the operation $\mathrm{AND}(p_i, \neg(p_u))$($i = 1, \ldots, O(T)$).

A subcircuit of largest depth is denoted by $\mathrm{sc}(p')$. All subcircuits $\mathrm{sc}(p)$ associated with processor p are arranged with equal depth. The output of $\mathrm{sc}(p)$ is called the activation of $\mathrm{sc}(p)$.

The N subcircuits $\mathrm{sc}(p)$ ($p = 1, \ldots, N$) are placed to compute in parallel. This subcircuit is called a layer. A layer corresponds to one step in the computation of $\mathcal{N}_o(n)$. As $\mathcal{N}_o(n)$ computes in time $T(n)$, $T(n)$ layers are connected sequentially. Each layer i receives the ith input bit from the input subcircuit and the N activation values of its preceding layer (except for layer 1 which receives input only). This main subarchitecture is of size $O(T^3)$, depth $O(T\log(T))$, and width $O(T^2)$ with $T = T(n)$.

Network $\mathcal{N}_o(n)$ may compute the response to different strings of size n in different times of order $O(T)$. Therefore, an output subcircuit is constructed in order to force the response to every string of size n to appear at the top of the circuit.

For each layer $i = 1, \ldots, T$, the AND function is applied to the output of the subcircuits $\mathrm{sc}(p_1)$ and $\mathrm{sc}(p_2)$ corresponding to the output processors p_1 and p_2 of network $\mathcal{N}_o(n)$. These values are transferred and processed by the OR functions. The resulting value is the output of the circuit. When OR is applied at each layer, only $D_C(n)$ gates are needed for this subcircuit.

With the proof of part 2.2, we finish the proof of Theorem 2.

$$\square$$

The simulation of networks by Boolean circuits and vice versa can also be proven for threshold circuits. In this case, the Boolean functions AND, OR, or NOT computed by each node are replaced by a linear threshold function. Each gate i computes $f_i : \mathbb{B}^{n_i} \to \mathbb{B}$ with Boolean values $\mathbb{B} = \{0, 1\}$. The activation update of the (linear) threshold function is defined by

$$x_i(t+1) = f_i(x_{i_1}, x_{i_2}, \ldots, x_{i_n}) := \mathcal{H}\left(\sum_{j=1}^{n_i} a_{ij} x_{ij}(t) + c_i\right)$$

with activations x_{ij} of the processors feeding into it, integer constants a_{ij} and c_i associated with the gate, and threshold function

$$\mathcal{H}(z) = \begin{cases} 1, & \text{if } z \geq 0, \\ 0, & \text{if } z < 0. \end{cases}$$

Threshold circuits and Boolean circuits are known to be polynomial equivalent in size.

We remind the reader of the definition of $\text{NET}(T(n))$ as the class of all languages recognized by formal networks with real weights in time $T(n)$. Analogously, we define the following.

Definition: Languages recognized by threshold circuits.

$$T\text{-CIRCUIT}(S(n)) := \text{class of languages recognized}$$
$$\text{by (non-uniform) families of threshold}$$
$$\text{circuits of size } S(n)$$

Simulation of networks by circuits and vice versa can be proven not only for Boolean circuits, but also for threshold circuits.

Theorem: Simulation with threshold circuits. *For any function F with $F(n)$, it follows that*

(1) $T\text{-CIRCUIT}(F(n)) \subseteq \text{NET}(nF^3(n) \log F(n))$
(2) $\text{NET}(F(n)) \subseteq T\text{-CIRCUIT}(F^2(n))$

Implication (1) is proven similarly to the Boolean circuit case: Each threshold gate is encoded in a Cantor-like way, including the description of the weights. The reverse simulation of nets by non-uniform families of threshold circuits was proven in Siegelmann and Sontag (1994, 351f.). Neural networks (brains) and technical circuits can be compared with respect to their cognitive abilities. An important cognitive ability which, in the past, was reserved to human brains can also be realized by automata, machines, and their technical circuits. We can distinguish degrees of more or less linguistic abilities along the hierarchy of computational complexity.

Definition: Acceptance of languages by (deterministic) networks and circuits.

- NET-P := Class of languages accepted by formal networks in polynomial time.
- NET-EXP := Class of languages accepted by formal networks in exponential time.
- CIRCUIT-P := Class of languages accepted by families of circuits in polynomial size.
- CIRCUIT-EXP := Class of languages accepted by families of circuits in exponential size.

CIRCUIT-P (sometimes also called P/poly) coincides with the class of languages recognized in polynomial time by oracle Turing machines. The oracles are sparse sets S in which, for length n, the number of words in S of length at most n is bounded by some polynomial function. For a given sparse set S, class $P(S)$ contains all languages computed by Turing machines using S as oracles in polynomial time.

Theorem: Acceptance of languages by (deterministic) networks, circuits, and oracle machines.

(i) NET-P = CIRCUIT-P
(ii) NET-EXP := CIRCUIT-EXP

(iii) NET-P $= \bigcup_{S \text{ sparse}} \text{P}(S)$

(iv) NET-EXP *includes all possible binary languages. Most Boolean functions require exponential time complexity.*

Definition: Acceptance of languages by (non-deterministic) network and circuits.

- A non-deterministic circuit family is defined as a circuit family with an additional input used as oracle.
- A non-deterministic network is defined as a network with an additional input used as oracle.
- A language L is said to be accepted by a non-deterministic formal network $\mathcal{N}$ in time B iff $L = \{\alpha | \exists \text{ guess } \gamma (\Phi_C = 1 \wedge T_{\mathcal{N}}(\alpha, \gamma) \leq B(|\alpha|)\}$.
- NET-NP $:=$ Class of languages accepted by non-deterministic formal networks in polynomial time.
- CIRCUIT-NP $:=$ Class of languages accepted by families of non-deterministic circuits in polynomial size.

Theorem: Acceptance of languages by (non-deterministic) network and circuits.

$$\text{NET-NP} = \text{CIRCUIT-NP}$$

A theory of computation similar to Turing machine computation is also possible for analog computation of networks with real weights. In classical theory of Turing computations, there are good reasons that NP is strictly more powerful than P. It is also likely that NET-NP is strictly more powerful that NET-P. Consider the following arguments: As a non-deterministic Turing machine can be simulated by a non-deterministic network with rational weights, we get NP $\subseteq$ NET-NP. Now, assume NET-NP $=$ NET-P. Then, we get

$$\text{NP} \subseteq \text{NET-NP} = \text{NET-P} = \text{CIRCUIT-P} = \text{P/poly}$$

$$\Rightarrow \text{NP} \subseteq \text{P/poly}.$$

Thus NET-NP = NET-P would imply the collapse of the computational hierarchy.

A large class of networks and dynamical systems has no more computational power than the considered analog (recurrent) neural networks with real weights. These insights lead to an extended Church's thesis for analog computing.

Church's thesis for time-bounded analog computing. Any reasonable analog computer will have no more power (up to polynomial time) than (first-order) recurrent networks.

Dynamical systems which have the power of neural networks to recognize languages are generalized processor networks.

Definition of generalized processor networks. A generalized processor network is a dynamical system D that consists of N processors $x_1, x_2, \ldots, x_N$, and receives its input $u_1(t), u_2(t), \ldots, u_M(t)$ via M input lines. A subset of the N processors $x_{i_1}, x_{i_2}, \ldots, x_{i_p}$ is the set of output processors of the system used to communicate the output of the system to its environment.

A generalized processor network D updates its processors by the dynamic equation

$$x^+ = f(x, u)$$

with vector x as current state of the network, vector u as external input, and function $f = \psi \circ \pi$ as composition of vector polynomial

$$\pi : \mathbb{R}^{N+M} \to \mathbb{R}^N$$

in $N + M$ variables with real coefficients and vector function

$$\psi : \mathbb{R}^N \to \mathbb{R}^N$$

which has a bounded range and is locally Lipschitz.

Generalized processor networks D can be arranged to output binary information: Therefore, two constants α, β with $\alpha < \beta$ are introduced as decision thresholds, such that each output neuron of D outputs a stream of numbers each of which is either smaller than

α or larger than β. The outputs of each output neuron y can be interpreted as a binary value

$$\text{binary}\,(y) = \begin{cases} 0, & \text{if } y \leq \alpha, \\ 1, & \text{if } y \geq \beta. \end{cases}$$

Even if more general analog values are allowed, no increase in computational power is attained (at least up to polynomial time).

Definition of computational time of generalized processor networks. A generalized processor network D computes in time T : $\mathbb{N} \to \mathbb{N}$ iff for every input of size $n \in \mathbb{N}$, D completes its output in no more than $T(n)$ steps.

A neural network is a special case of a generalized processor network in which all coordinates of function ψ compute the same piecewise linear function σ, and the polynomial π is a first-order polynomial (or affine function).

A vector function $f = \psi \circ \pi$ is said to be approximable in time $A_f(n)$ iff there is a Turing machine M that computes $T(n)$-truncation(f) in time $A_f(n)$ on each input with total bit size n.

Theorem: Simulation of generalized processor networks by neural networks. *Let D be a generalized processor network which computes with a function $f = \psi \circ \pi$. Function ψ is assumed to be non-uniformly $F(n)$-approximable in polynomial time. Then there exists a neural network $\mathcal{N}_D$ which recognizes the same language as D (and which does so with at most polynomial-time slowdown).*

If ψ is $F(n)$-approximable in polynomial time and π involves rational coefficients only, the weights of $\mathcal{N}_D$ are rational numbers as well.

Analog dynamical systems can have computer power beyond Turing computability. Siegelmann (1995) introduced the analog shift map which computes exactly like neural networks and analog machines surpassing a Turing machine. Analog machines with their real constants are also assumed to model physical phenomena of nature.

A shift map is defined over a set of bi-infinite dotted sequences. Let E be a finite alphabet. A dotted sequence over E is denoted by $\dot{E}$. It is a sequences of letters with exactly one dot sign (.) and the rest all in E. The dotted sequences can be finite, one-side infinite, or both-side (bi) infinite over E. For integer $k \in \mathbb{N}$, the shift map

$$S^k : \dot{E} \to \dot{E} \text{ with } (a)_i \mapsto (a)_{i+k}$$

shifts the dot k places, where negative values cause a shift to the left and positive ones a shift to the right.

The analog shift is defined by the following procedure: A dotted substring is replaced with another dotted substring (of equal length) according to a function G. Then, this new sequence is shifted an integer number places left or right according to a function F.

Definition of analog shift. The analog shift is the map

$$\Phi : a \mapsto s^{F(a)}(a) \oplus G(a)$$

with function $G : \dot{E} \to \dot{E}$ describing the modification of the sequence and function $F : \dot{E} \to \mathbb{Z}$ indicating the amount of shifting of the dot.

Both F and G have a finite domain of dependence (DoD). Therefore, F and G depend only on a finite dotted substring of the sequence on which they act. The domain of effect (DoE) of G may be finite, infinite, or bi-infinite. The operation $\oplus$ is defined by

$$(a \oplus b)_i = \begin{cases} g_i, & \text{if } g_i \in E, \\ a_i, & \text{if } g_i = \epsilon \end{cases}$$

with the empty element ϵ not contained in E.

Example of analog shift. Let an analog shift defined by table:

DoD	F	G
0.0	1	π
0.1	1	.10
1.0	0	1.0
1.1	1	.0

$\bar{\pi}$ denotes the left infinite string $\ldots 51413$ of π in base 2 rather than in base 10. The table is a description of the dynamical system in which the next step depends on the first letter to the right of the dot and the one to its left. If these letters (DoD) are 0.0, then, according to the first row of the table, the left side of the dot is replaced by $\bar{\pi}$ and the dot moves one place to the right. If the DoD is 0.1 (as in the second row of the table), the second letter to the right of the dot becomes 0 and there is a right shift, and so forth.

Let us start the dynamics of the analog shift with

$$000001.10110.$$

In this case, the DoD is 1.1. According to fourth row of the table, the letter to the right of the dot becomes 0 and the dot is shifted right:

$$1.1 : (000001.00110) \; 0000010.0110$$

In this case, the DoD is 0.0:

$$
\begin{array}{lll}
0.0: & (\bar{\pi}.0110) & \bar{\pi}0.100 \\
0.1: & (\bar{\pi}0.100) & \bar{\pi}01.00 \\
1.0: & (\bar{\pi}1.00) & \bar{\pi}01.00
\end{array}
$$

In this case, the DoD is 1.0 without any change. Therefore, we get a fixed point. The computation with the analog shift system is the evolution of the initial dotted sequence until a fixed point is reached, from which the system does not evolve anymore. In our example, a fixed point is reached in four steps from 000001.10110 to 00. In general, if the computation ends, the input-output map is defined as the transformation from the initial dotted sequence to the final subsequence to the right of the dot. Even under the constraints to systems that start with finite dotted sequences and that stop with either finite or left infinite dotted sequences, the analog shift computes richer maps than the Turing machines.

Analog computation with analog shift maps is in a mathematical rigorous sense equivalent to computability with appropriate neural networks:

$$\text{AS}(k) := \text{class of functions computed by the analog shift (AS) map in time } k.$$

$$\text{NN}(k) := \text{class of functions computed by the artificial neural net (ARNN) in time } k.$$

$$\text{Poly}(k) := \text{class of polynomials in } k.$$

Theorem: Equivalence of analog shifts and neural nets. *If F is a function with $F(n) \geq n$, then*

(1) $\text{AS}(F(n)) \subseteq \text{NN}(\text{Poly}(F(n)))$
(2) $\text{NN}(F(n)) \subseteq \text{AS}(\text{Poly}(F(n)))$

The essential arguments to prove the theorem can be illustrated for the binary alphabet $E = \{0, 1\}$.

Proof. (1) We start with a dotted sequence

$$a = \ldots a_{-3} a_{-2} a_{-1} a_1 . a_2 a_3 \ldots,$$

which is mapped into the two sequences referring to the left and right side

$$a_l = a_{-1} a_{-2} a_{-3} \ldots$$

$$a_r = a_1 a_2 a_3 \ldots.$$

The analog shift map is redefined for application to a_l and a_r as

$$\tilde{\Phi}(a_l, a_r) = (a_l \oplus G_l(d_l, d_r), a_r \oplus G_r(d_l, d_r))$$

with left part d_l and right part d_r of the DoD.

Now, we can define a neural net ARNN which does the computation in polynomial time: The values of a_l as well as a_r are encoded in a neuron using a Cantor set representation which was already introduced in this chapter. The following operation only depends on the first few bits of both neurons (DoD). If G is finite, the operation involves substitution of the first finitely many bits. If G is infinite, G involves loading of the neuron with a new value. These operations need constant time for an efficient simulation.

(2) In order to simulate a Turing machine by an analog shift, its configuration is encoded in a dotted sequence

$$\underbrace{c_1}_{\substack{\text{code of}\\\text{garbage}}} \quad \underbrace{c_2}_{\substack{\text{code of}\\\text{left end}\\\text{marker}}} \quad \underbrace{c_3}_{\substack{\text{code of}\\\text{tape left}}} \quad \cdot \quad \underbrace{c_4}_{\substack{\text{code of}\\\text{state}}} \quad \underbrace{c_5}_{\substack{\text{code of}\\\text{tape right}}} \quad \underbrace{c_6}_{\substack{\text{code of}\\\text{empty part}}}$$

with

c_1: infinite sequence over E

c_2: left end marker 01

c_3: part of the tape to the left of the read write head with 10 representing 0 and 11 for 1

c_4: the internal state of the Turing machine encoded as a sequence of 0's ending with 1

c_5: the letter under the head and the right part of the tape

c_6: infinite sequence $\bar{0}$ of 0's

The initial dotted string is

$$\bar{0} \cdot q_1 x \bar{0}$$

with finite input string x. The function G substitutes the left side of the dot with the infinite string

$$w = \langle \ldots, w_3, w_2, w_1 \rangle.$$

In polynomially many steps, this string becomes the advice

$$\text{code of garbage } 01w_n.p_1 x\bar{0}.$$

The following steps of the Turing machine are simulated straight-forwardly. $\qquad\square$

The computational power of neural networks depends on the type of numbers used as synaptic weights. The power of networks seems to increase with the amount of information and complexity contained in the weight numbers. In the following, an information measure is introduced to determine the complexity of numbers. In the next step, we prove that the power of networks with real weights coincides with the power of oracle Turing machines with oracles related to those weights. Finally, a proper hierarchy of complexity classes of neural networks can be defined with related oracles of Turing machines.

A measure of information contained in an individual object is Kolmogorov complexity (cf. Chapter 13). The intuitive idea is that objects with very little information would require a brief description only. More complex objects require longer descriptions. Thus, in information theory, complexity depends on the describing language. Kolmogorov complexity relates to a description of an object as the shortest program for a universal Turing machine which constructs the object (Li and Vitányi, 1993).

Let L be a language with an alphabet Σ of symbols. Finite or infinite strings of symbols are denoted by α. A word consisting of the first k symbols of α is denoted by $\bar{\alpha}(k)$. The kth symbol of α is denoted by α_k. In a characteristic sequence $\chi_L \in \{0,1\}^\infty$ of language L, the ith bit of the sequence is 1 iff the ith word of Σ^* is in L with respect to some lexicographic order (Kobayashi, 1981).

Kolmogorov complexity of infinite binary sequences

Let U be a universal Turing machine, f and g two functions $f, g : \mathbb{N} \to \mathbb{N}$. A sequence $\alpha \in \{0,1\}^\infty$ has a Kolmogorov complexity of $K[f(n), g(n)]$ iff there exists $\beta \in \{0,1\}^\infty$ such that the universal Turing machine U outputs $\bar{\alpha}(n)$ in time $g(n)$ for all, but finitely many n, when given $\bar{\beta}(f(n))$ and n as inputs. If the running time is not restricted, then $\alpha \in K[f(n)]$.

$K[\mathcal{F}, \mathcal{G}]$ is the set of all infinite binary sequences taken from $K[f, g]$ with $f \in \mathcal{F}$ and $g \in \mathcal{G}$.

The definition is illustrated in Fig. 22 (Siegelmann, 1999, p. 79).

Kolmogorov complexity can be applied to real numbers as encoded infinite binary strings. Let b be a natural number, and $\alpha = \alpha_1 \alpha_2 \dots$ a finite or infinite sequence of natural numbers smaller than b. The interpretation of the sequence α in base b is the number

$$\alpha|_b := \sum_{i=1}^{|\alpha|} \frac{\alpha_i}{b^i}.$$

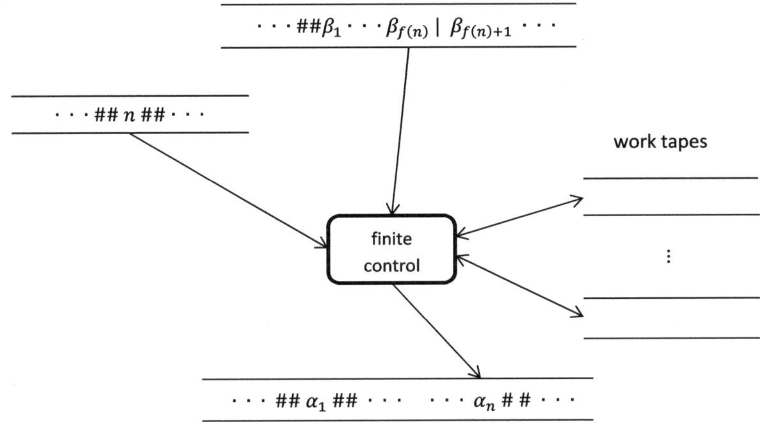

Fig. 22. Universal Turing machine determining Kolmogorov complexity of sequence α.

The encoding functions are defined for two particular cases:

$$\delta_2(\epsilon) = 0 \quad \delta_2(\alpha) = \alpha|_2 = \sum_{i=1}^{|\alpha|} \frac{\alpha_i}{2^i}$$

$$\delta_4(\epsilon) = 0 \quad \delta_4(\alpha) = \alpha|_4 = \sum_{i=1}^{|\alpha|} \frac{2\alpha_i + 1}{4^i}.$$

The image of function δ_4 is the 4-Cantor set which consists of the set $\Delta_4 = \left\{ \sum_{i=1}^{\infty} \frac{\beta_i}{4^i} | \beta_i \in \{1,3\}^{\infty} \right\}$ of δ_4-images for infinite sequences and the set $\bar{\Delta}_4 = \left\{ \sum_{i=1}^{k} \frac{\beta_i}{4^i} | \beta_i \in \{1,3\}, k \geq 0 \right\}$ of δ_4-images for finite sequences.

Kolmogorov complexity of real numbers

The function δ_2 can be used to define the Kolmogorov complexity numbers in interval $[0,1]$.

The functions δ_2 or δ_4 can be used to define the Kolmogorov complexity of numbers in Δ_4:

A number $\omega \in \Delta_4$ has Kolmogorov complexity $K[f(n), g(n)]$ iff $\delta^{-1}(\omega) \in K[f(n), g(n)]$ with δ either δ_2 or δ_4. (The values obtained in Δ_4 by δ_2 or δ_4 differ only in a small constant factor.)

A real number $r \in \mathbb{R}$ has Kolmogorov complexity $K[f(n), g(n)]$ iff the fractional part of r has Kolmogorov complexity $K[f(n), g(n)]$ (disregarding the constant complexity of the integer part of r).

The power of networks with real weights coincides with that of oracle Turing machines with certain oracles related to those weights. First, we define how to combine a fixed number of real weights into a single infinite string. Let S be a subset of finite and infinite binary strings. The set S is said to be closed under mixing if for any finite number k and for any k strings from S with $\alpha^1 = \alpha_1^1 \alpha_2^1 \alpha_3^1 \ldots, \alpha^2 = \alpha_1^2 \alpha_2^2 \alpha_3^2 \ldots, \alpha^k = \alpha_1^k \alpha_2^k \alpha_3^k \ldots$, the mixed string $\alpha_1^1 \alpha_1^2 \alpha_1^3 \ldots \alpha_1^k \alpha_2^1 \alpha_2^2 \alpha_2^3 \ldots \alpha_2^k \alpha_3^1 \alpha_3^2 \alpha_3^3 \ldots$ is also an element of S. A real number can be associated with each tally set by the characteristic function.

Theorem: Equivalence of neural nets with real weights and oracle Turing machines. *Let $S \subseteq \{0,1\}^\infty$ be closed under mixing with $1^\infty \in S$. Let $\mathcal{T} = \{T | \chi_T \in S\}$ be a family of tally sets. The computation time of the following neural nets and oracle Turing machines is polynomially related:*

(1) *Neural nets with weights in the set $\delta_2(S) + \mathbb{Q}$,*
(2) *Oracle Turing machines consulting oracles in $\mathcal{T}$,*
(3) *Neural networks with weights in the set $\delta_4(S) + \mathbb{Q}$.*

The theorem can be proven in three steps (Siegelmann, 1999, 81 f.): A network in (1) can be simulated by a machine from (2). The machine from (2) in turn is simulated by a network from (3). The networks in (3) are a subset of those in (1).

Complexity classes of networks can be ordered in a proper hierarchy. Therefore, an ordering between classes of tally sets is needed. Oracle Turing machines relating to these classes of oracles provide an infinite hierarchy of computational classes. With the previous theorem, the desired hierarchy of neural networks is obtained.

Definition of partial order on function classes.

- An order of magnitude is the set $o(f)$ of functions g such that, for every $c > 0$ and for all but finitely many n, $g(n) < cf(n)$.

- For function classes $\mathcal{F}$ and $\mathcal{G}$, it is $\mathcal{F} \prec \mathcal{G}$ iff

  (a) there is some non-decreasing function $s(n) \in \mathcal{G}$, computable in time polynomial in n, so that $s(n) = o(n)$,

  (b) for every polynomial p and every $r \in \mathcal{F}$, $r \circ p = o(s)$.

The partial order $\prec$ defines an infinite hierarchy. An example is given by the function classes $\theta_i = \{q_1, \ldots, q_i\}$ with an inductive definition of $q_i = \log^{(i)}$ by $q_1 = \log$ and $q_i = \log(q_{i-1})$ for $i > 1$.

Theorem. *Let $\mathcal{T}_{\mathcal{F}}$ be the family of all tally sets T with characteristic function $\chi_T \in K[\mathcal{F}, \mathrm{P}]$ and $\mathrm{P}(\mathcal{T}_{\mathcal{F}})$ the class of all sets decidable in polynomial time by an oracle Turing machine with an oracle in $\mathcal{T}_{\mathcal{F}}$. For (non-empty) function classes $\mathcal{F}$ and $\mathcal{G}$:*

$$\mathcal{F} \prec \mathcal{G} \Rightarrow \mathrm{P}(\mathcal{T}_{\mathcal{F}}) \subset \mathrm{P}(\mathcal{T}_{\mathcal{G}}).$$

Proof. According to the definition of partial order, let $s(n) \in \mathcal{G}$ be a bound on $\mathcal{F}$. Since $\mathcal{F}$ is non-empty, s must be unbounded. Obviously, $\mathrm{P}(\mathcal{T}_{\mathcal{F}}) \subseteq \mathrm{P}(\mathcal{T}_{\{s\}}) \subseteq \mathrm{P}(\mathcal{T}_{\mathcal{G}})$.

In the following, a set L with $L \in \mathrm{P}(\mathcal{T}_{\mathcal{G}})$, but $L \notin \mathrm{P}(\mathcal{T}_{\mathcal{F}})$ is defined: We choose an infinite sequence $\gamma \notin K[\frac{n}{2}]$. For each n, string β_n is defined by

$$\beta_n := \begin{cases} \bar{\gamma}\left(\dfrac{s(n)}{2}\right) 0^{n - \frac{s(n)}{2}}, & \text{if } n \geq \dfrac{s(n)}{2}, \\ 0^n, & \text{otherwise.} \end{cases}$$

Now, L is defined as the tally set with characteristic string $\beta_1 \beta_2 \beta_3 \ldots$. For given $\bar{\gamma}\left(\frac{s(n)}{2}\right)$ and because of $s \in \mathcal{G}$, it can be proved that $\chi_L \in K[\frac{s}{2} + c, \mathrm{P}] \subseteq K[s, \mathrm{P}]$. Thus, $L \in \mathrm{P}(\mathcal{T}_{\mathcal{G}})$.

In the next step, we prove $L \notin \mathrm{P}(\mathcal{T}_{\mathcal{F}})$: If we assume otherwise, then there is some machine that accepts L in time p_1 with an tally set T as an oracle, where $\chi_T \in K[r, p_2]$, p_1 and p_2 are polynomials, and $r \in \mathcal{F}$. In time $p_1(n)$, the machine can query at most the first $p_1(n)$ elements of set T. This machine can print out $\beta_1 \beta_2 \ldots \beta_n$ in time $n p_1(n)$. Hence, $\bar{\gamma}\left(\frac{s(n)}{2}\right)$ can also be printed for given inputs $\frac{s(n)}{2}$ and the first $r(p_1(n)) + O(1) < \frac{s(n)}{4}$ bits of the seed for χ_L. This is a contradiction to the choice of γ. $\qquad\square$

This theorem provides a hierarchy of oracle Turing machine querying oracles that belong to different Kolmogorov complexity classes. In a previous theorem, we proved an equivalence between the set of weights of networks and the oracle Turing machines consulting related tally sets. Both results prove the following hierarchy theorem for function classes $\mathcal{F}$ and $\mathcal{G}$ with $\mathcal{F} \prec \mathcal{G}$ which are closed under $O(\cdot)$.

Definition of $O(\cdot)$. $O(f)$ is the set of all functions g for which there exists a constant c_g with $g(n) \leq c_g f(n)$ for all $n > n_0$ and some n_0.

A class $\mathcal{F}$ of functions is closed under $O(\cdot)$ iff for every f and g:

$$g \in O(f) \text{ and } f \in \mathcal{F} \Rightarrow g \in \mathcal{F}.$$

Hierarchy theorem. *Function classes $\mathcal{F}$ and $\mathcal{G}$ with $\mathcal{F} \prec \mathcal{G}$ are assumed to be closed under $O(\cdot)$. Let $\mathrm{NET}_{K[\mathcal{F},\mathrm{P}]}$ be the class of languages accepted by networks computing in polynomial time. Each of these networks uses weights from $K[\mathcal{F}, \mathrm{P}]$. For $\mathcal{G}$, $\mathrm{NET}_{K[\mathcal{G},\mathrm{P}]}$ is constructed in the same way. Then*

$$\mathrm{NET}_{K[\mathcal{F},\mathrm{P}]} \subset \mathrm{NET}_{K[\mathcal{G},\mathrm{P}]}.$$

From an epistemic point of view, neural nets can be interpreted as prototypes of neurobiological nervous systems or brains which have evolved during evolution. At the first glance, they seem to be completely different from automata and machines which were constructed using technology. But, actually, appropriate classes of machines and automata are equivalent to neural nets and vice versa in a mathematical rigorous sense. Their degrees of intelligent abilities can even be measured in complexity hierarchies. In the end of this chapter, we summarize main results leading beyond Turing computability toward analog models of brains with typical abilities like, e.g., recognition of natural languages. It is remarkable that the complexity degrees of these intelligent abilities depend on the complexity numbers which are allowed as weights of the neural nets:

- McCulloch-Pitts nets with integer weights recognize the same class of languages that finite automata do.

- Recurrent neural nets with rational weights recognize the same class of languages that Turing machines do.
- Analog neural nets with real weights recognize the same class of languages that Turing machines do with not only appropriate oracles in polynomial time (P/poly), but also all possible (e.g., non-computable) ones in exponential time.

These results lead to an extension of Church's Thesis for analog computing: Any analog computer with real values will have no more power (up to polynomial time) than analog neural networks. Analog models are unlikely to solve NP-hard problems, but they are more realistic for analog components of human cognition, brains, and bio-inspired technical networks. According to Church's thesis, every discrete effective procedure can be simulated by a (universal) Turing machine. But, processes in the world are successfully modeled by continuous (analog) dynamical systems (e.g., differential equations, numerical analysis) with different degrees of real computational complexity. Scientific computing uses real and complex models which need foundations of real computing. In the engineering sciences, sensor technology needs analog models with rigorous computational foundations. Last but not least, in philosophy, the question arises whether human brains and human thinking can be restricted to the digital paradigm of Turing computability. With respect to artificial intelligence, degrees of intelligence depend on degrees of computational complexity. Human-like AI-systems will need digital and analog abilities (Mainzer, 2016a). Therefore, the extension of Church's thesis to real (analog) computation is well-motivated with respect to the complexity degrees of the world.

Chapter 13

Complexity of Algorithmic Information

A formal axiomatic system has the great advantage of compressing a lot of theorems into a set of a few axioms. Thus, it delivers a shorter description of mathematical truth. Even a physical theory can be understood as a shorter description of many empirical data. In general, a formal theory can be considered a computer program that calculates true theorems or data. The smaller the program is relative to the output, the better the theory. This strategy is also due to William of Ockham, William of principle of parsimony.

Obviously, besides running time, the size of a computer program is an important measure of computational complexity. As a program is a finite list of symbols, its length can be measured by its number of symbols in binary coding. For example, consider the following sequences of binary digits:

$$s_1 = 111111111111111111$$

$$s_2 = 010101010101010101$$

$$s_3 = 011010001101110100$$

For s_1 and s_2, there are shorter descriptions or printing programs than the actual output: "18 times 1" for s_1 and "9 times 01" for s_2. But for s_3, there seems to be no shorter description than the actual output itself. Chaitin (1969), Kolmogorov (1965), and Solomonoff (1964) came up with the idea that the algorithmic complexity of a symbolic sequence s should be defined by the length of the shortest computer program for generating s (measured in bits).

Algorithmic complexity is sometimes called the algorithmic information content of a symbolic sequence, which is the subject of the algorithmic information theory. As random sequences have no regularities, they cannot be described by shorter programs. They are incompressible with an algorithmic complexity equivalent to their length. But, we can never calculate the program-size complexity because, in general, it is not decidable if a certain program is the shortest one. If we have a program generating a sequence, its size is only an upper bound on the program-size complexity of the sequence. But we can never prove lower bounds. Therefore, we get a kind of incompleteness result in algorithmic information theory (Chaitin, 1998).

Definition of algorithmic information.

- The algorithmic complexity $K(s)$ of a symbolic sequence s is the length of the shortest computer program generating s (measured in Bit) (Chaitin, 1969; Kolmogorov, 1965; Solomonoff, 1964).
- The algorithmic information content of a symbolic sequence is measured by its algorithmic complexity.

The length of the shortest program depends on the machine performing the program. In order to avoid the dependence on a special hardware, we define the algorithmic complexity for a universal machine, which can simulate every other machine. Formally, the algorithmic complexity of a binary sequence s is defined as the shortest program of a universal machine, which can be written as binary sequence s^* and which can reproduce s. For example, the binary codes s_1^* and s_2^* of "14 times 1" and "8 times 01" are shorter than s_1 and s_2 and reproduce s_1 and s_2. Therefore, s_1 and s_2 have low algorithmic complexity.

Algorithmic information is no absolute measure number, but depends on the chosen programming language and coding of sequences. For example, algorithmic complexity is measured with a Turing machine and its coding. Algorithmic complexity or algorithmic information could also be determined for, e.g., a LISP or

Java program. In the following, code numbers are understood as bit sequences. Examples are the sequences 0, 00, 000, representing different codes (but the same natural number).

Algorithmic complexity or algorithmic information can be compared with random sequences. As random sequences (e.g., fair coins with 0 or 1) have no regularities, they cannot be described by shorter programs than their printout. Therefore, they are incompressible with maximal algorithmic complexity which is equivalent to the length of these sequences. A sequence is called algorithmically random if its algorithmic complexity compared with its length is maximal.

But, is there a general procedure to compute the algorithmic complexity or the algorithmic information or the degree of randomness of an arbitrary sequence with 0 and 1 (Mainzer, 2016b)?

Proposition. *There is no general computer program to compute the algorithmic information $K(s)$ of an arbitrary sequence s.*

Proof. The proof is deduced indirectly. We assume the contrary statement:

There is a general computer program P which can compute the algorithmic complexity $K(s)$ of an arbitrary sequence s.

With this program, we could compute for every binary sequence 0, 1, 00, 01, 10, 11, 000, 001,... the algorithmic complexity $K(0)$, $K(1)$, $K(00)$, $K(01)$, $K(10)$, $K(11)$, $K(000)$, $K(001)$,... step by step.

For each arbitrary constant N, the program P will stop after finitely many steps and find a sequence s with $K(s) > N$.

Program P consists (in binary code) of a finite number of Bits. The length $|P|$ of program P will additionally depend of the chosen N. Therefore, for coding N, it suffices $|\log_2 N| + 1$ Bits. It follows that

$$|P| \leq c + |\log_2 N|$$

with constant c depending on the chosen coding and programming language.

Let us choose $N = 2^c$ with

$$|P| \leq c + |\log_2 2^c| = 2c.$$

In this case, because of $|P| \leq 2c < 2^c = N$ (for $c > 2$), N is larger than the program length $|P|$. As before, program P should determine a sequence s with $K(s) > N$. After finitely many steps, P will stop and determine s. But we remarked previously that program P for determining s is with choice $N = 2^c$ shorter than N, i.e., $K(s) < N$.

So, for this choice of N, it follows that $K(s) > N$ as well as $K(s) < N$ which is a contradiction. Therefore, the assumption was false and the proposition is true that there is no general computer program to compute the algorithmic information $K(s)$ of an arbitrary sequence s. $\qquad\square$

The general result does not exclude that the algorithmic information can be computed for examples. But, there is no general procedure for arbitrary cases.

As explained before, we want to code computer programs by binary sequences. All possible binary sequences can graphically be represented by nodes of an infinite bifurcation tree (Fig. 23). One can run along the branches node by node and test where the first node of a halting program occurs. This is a program which stops resp. terminates after finitely many steps. In the case of a binary

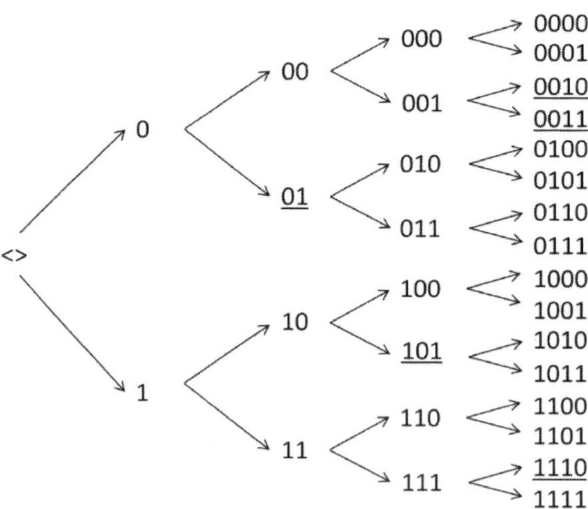

Fig. 23. Binary bifurcation tree with code numbers of terminating programs (underlined).

bifurcation tree, we can also imagine a game of chance with fair coins generating binary sequences of 0 and 1 step by step. After each toss, we test if the sequence represents a computer program and if it is a halting program. We remind the reader of the intuitionistic binary trees of choice sequences in Chapter 6.

Definition of halting probability. Ω_n is called the probability to generate the code number of a halting program with maximally n tosses.

The logician Chaitin suggested the following formulation (Chaitin, 2007; Hoffmann, 2013, p. 351):

The halting probability Ω_n is the probability that a randomly chosen sequence s from all 2^n possible binary sequences of length n stops.

In the finite printout of the bifurcation tree in Fig. 23, we assume five halting programs with code numbers 01, 101, 0010, 0011, and 1110. In the case of a fair toss, code number 01 is one of four possible sequences with two symbols. Therefore, 01 has probability $\frac{1}{4}$ for statistically independent tosses. With the same argument, the probability of generating the program code 101 from three symbols

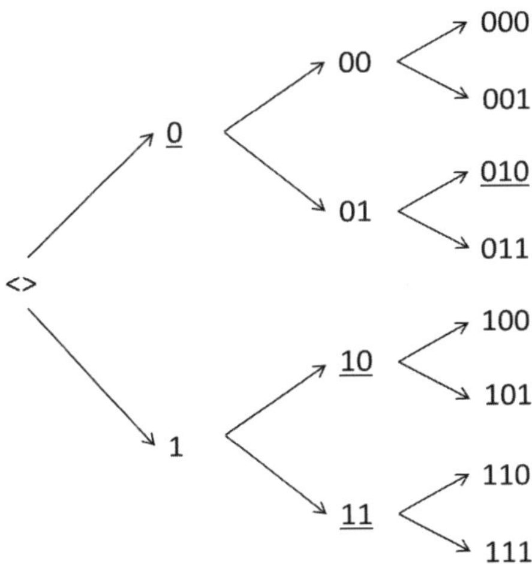

Fig. 24. Binary bifurcation tree with code numbers of terminating programs (underlined).

is $\frac{1}{8}$. For the codes of halting programs 0010, 0011, and 1110 from four symbols, we get the probability $\frac{1}{16}$ in each case.

Thus, the halting probabilitiy Ω_n to get one of the halting programs after maximally n tosses, is determined by the sum of the single probabilities of corresponding code numbers:

$$\Omega_1 = 0,$$

$$\Omega_2 = \frac{1}{4},$$

$$\Omega_3 = \frac{1}{4} + \frac{1}{8} = \frac{3}{8},$$

$$\Omega_4 = \frac{1}{4} + \frac{1}{8} + \frac{1}{16} + \frac{1}{16} + \frac{1}{16} = \frac{9}{16}.$$

This example illustrates Chaitin's general computing procedure for the halting probability Ω_n: Each program P with length $|P| \leq n$ that stops is added up with a single probability $\frac{1}{2^{|P|}}$ in the whole probability, i.e.,

$$\Omega_n = \sum_{\substack{P \text{ halts} \\ |P| \leq n}} \frac{1}{2^{|P|}}.$$

A probability value is expected between 0 and 1. Probability 1 means that an event occurs with certainty. A halting probabilitiy greater than 1 makes no sense. Consider, for example, the binary bifurcation tree in Fig. 23 with different halting programs to Fig. 24. According to the previously explained formula, the halting probability is

$$\Omega_3 = \frac{1}{2^1} + \frac{1}{2^2} + \frac{1}{2^2} + \frac{1}{2^3} = \frac{9}{8} > 1.$$

Obviously, the reason is that, contrary to Fig. 23, the code 0 of a terminating program is also in the beginning of the code 010 of a terminating program in a following branch of 0. Therefore, only codes satisfying the following conditions are allowed.

Definition of prefix-free codes. A prefix-free code is a code without a code word which is a prefix of another code word, i.e., no code word is allowed to be the beginning of another code word.

Example. A code with code words $0, 10, 11$ is prefix-free. A code with the code words $0, 01, 10$ is not prefix-free because the code word 0 is a prefix of the code word 01.

A prefix-free code guarantees that each binary sequence begins with one program at the most resp. no program is the beginning of another program. That means if, in a branch of the binary bifurcation tree (Fig. 23), we meet a terminating program after finitely many steps, then, in the case of prefix-free codes, it is not necessary to test the following infinitely many bifurcations for further terminating programs.

For prefix-free codes, the halting probability Ω_n is uniquely defined. Obviously, the definition implies that the sequence $\Omega_1, \Omega_2, \Omega_3, \ldots$ monotonously increases, i.e., $\Omega_n \leq \Omega_{n+1}$, with upper bound, i.e., $\Omega_n \leq 1$. Therefore, the sequence converges to a limit which allows the following definition.

Definition of Chaitin's constant.

$$\Omega := \lim_{n \to \infty} \Omega_n = \sum_{P \text{ halts}} \frac{1}{2^{|P|}}$$

The examples illustrate that Ω_n is computable for small values n. But the computation of Ω_n can only be possible for finitely many n. Under the assumption that infinitely many Ω_n were computable, Turing's halting problem would be decidable (Hoffmann, 2013, Chapter 6).

Proposition. *There are only finitely many computable halting probabilities Ω_n.*

Proof. Let us assume that there are infinitely many computable Ω_n. Given a program P with length n.

We choose the next larger number n' with computable $\Omega_{n'}$ under precondition.

This halting probability determines how many programs of maximum length n' halt.

We only must decide whether P belongs to these programs:

For this purpose, let all concerning programs run in parallel. As the number of terminating programs with maximum length n' is known, the deciding process is finished after finitely many steps.

But, because of the undecidability of the halting problem, we get a contradiction:

Only finitely many Ω_n can be computable. All Ω_n must become uncomputable after a certain size n. $\square$

Therefore, Chaitin suggested an approximation of Ω with computable halting probabilities.

Definition of halting probability Ω_n^k. The halting probability Ω_n^k is the probability that a randomly chosen binary sequence s of length n begins with the code number of a program halting after maximum k steps.

In order to compute Ω_n^k, we do not need all terminating programs with length n. It is sufficient to consider the programs which do not need more than k steps to stop. We conclude:

$$\Omega_n^n \leq \Omega_n \leq \Omega,$$

$$\lim_{n\to\infty} \Omega_n^n = \lim_{n\to\infty} \Omega_n = \Omega.$$

The values Ω_n^n are always computable: To test this statement, we let run all programs P with length n at the most maximum n steps. With that procedure, the number of the terminating programs is uniquely determined in finitely many steps.

Numbers which can be approximated by a sequence of computable numbers are effectively enumerable. Effective enumerability of sets was already defined in Chapter 2 on Turing computability. Therefore, Chaitin's constant Ω is not only effectively enumerable. but it is also another example of a number which is not effectively computable.

Theorem. Ω *is not computable.*

The reason is that Ω contains the information in order to decide the halting problem for arbitrary programs. Assume that Ω would be computable. In this case, Turing's halting problem would be decidable. The following proposition implies that Ω contains all information to decide the halting problem for arbitrary programs.

Proposition. *Halting probability Ω_n can be determined from the first n bits of Ω.*

Proof. In order to determine the first n bits of Ω, the values Ω_1^1, Ω_2^2, $\Omega_3^3, \ldots, \Omega_m^m, \ldots$ are computed successively, until for $m = m_o$, the first n bits of Ω match $\Omega_{m_o}^{m_o}$.

For $m > m_o$, the first n bits can no longer change: If the first n bits change, then $\Omega_m^m > \Omega$. In the halting probability Ω_m^m, every program P is taken into consideration by the sum $\frac{1}{2^{|P|}}$. As the first n bits cannot be changed, each program P which does not stop before $m > m_o$ steps has a length larger than n.

Therefore, there is no program P with $|P| \le n$ which stops after $m > m_o$ steps.

In order to determine the halting probability Ω_n, we let run all programs P with $|P| \le n$ for m_o steps. Each terminating program is added up with summand $\frac{1}{2^{|P|}}$ in halting probability Ω_m^m. Programs which have not yet stopped after m_o steps will never stop (which was proven before). $\qquad\square$

Chaitin's constant has great importance for algorithmic information theory because it codes deep logical–mathematical information about computability, decidability, and their limits (Schmidhuber, 2002). On this background, algorithmic information theory also allows insights into proof theory which was discovered by Chaitin (2007).

For this purpose, let us compare a formal system S of a theory (e.g., arithmetic) with the smallest possible program P_S which derives the theorems of S step by step with (logical) rules from axioms. Axioms correspond to the inputs of program P_S, rules to the program instructions, theorems to the outputs, and the deductions of a theorem to the program computation of an output for given input. It can be proven that the theorems of a formal system can be effectively enumerated by a computer program along this line (cf. Chapter 4).

We now assume that program P_S contains all axioms and deduction rules of system S. The number of bits the computer program P_S consists of is

$$|P_S| = |S| + c$$

with $|S|$ number of bits of formal systems S and constant c relating to additional bits which are needed for programming the enumeration procedure.

If a computer program of length n exists which effectively enumerates the theorems of a formal system S, then $n = |P_S|$. It is a remarkable result of algorithmic information theory that Gödel's first incompleteness theorem can immediately be derived from properties of algorithmic complexity $K(s)$ of sequences s in a system S with $n = |P_S|$.

Chaitin's incompleteness theorem. *In a correct formal system S with $n = |P_S|$, all propositions of the form $K(s) > n$ are not provable, if n exceeds a certain size.*

Proof. In a formal system S, a formula $\varphi_n(s)$ means that the algorithmic complexity of s is larger than n, i.e., $K(s) > n$.

Is $\varphi_n(s)$ also derivable in S for a given n? The program P_S generates derivable theorems of S step by step. If $\varphi_n(s)$ is enumerated, then it is known that this formula is derivable in S.

Instead of program P_S that generates all derivable formula, we can also assume a program P_S^* which specifically looks for a proof of formula $\varphi_n(s)$, i.e., the program looks for a proof of proposition $K(s) > n$.

As soon as such a proof is found, P_S^* generates the binary sequence s and stops. If no proof is found, P_S^* continues forever.

The number of bits the program P_S^* consists of is

$$|P_S^*| \leq |S| + |\log_n n| + c'$$

with number of bits $|\log_n n|$ of program length $n = |P_S|$ and constant c' relating to additionally needed bits.

For very big n, the program length of P_S^* is certainly bigger than the length of P_S, i.e.,

$$|P_S^*| < n.$$

If we assume that P_S^* finds a formula $\varphi_n(s)$ which is provable in S, then the bit sequence s is provided. In this case, the formula $K(s) > n$ would be provable.

Then, s would be provided with a program P_S^* which is shorter than n, i.e.,

$$K(s) < n,$$

which is obviously a contradiction.

Therefore, no propositions of the form $K(s) > n$ from a certain size of n onwards are provable in a correct formal system. That confirms Chaitin's incompleteness theorem. $\square$

On the other hand, for every value n, the proposition $K(s) > n$ (which means that the complexity of a binary sequence s is bigger than n) is true for matching s. Therefore, in every correct formal system which can express corresponding formulas $\varphi_n(s)$, there are true, but improvable propositions. It can be proven that arithmetic with the fundamental rules of computation satisfies the conditions of Chaitin's incompleteness theorem. It follows that this kind of arithmetic is incomplete.

Chapter 14

Complexity of Information Dynamics

In previous chapters, we considered information systems from a logical–mathematical point of view. Analog systems with parameters of real numbers and continuous scaling are appropriate mathematical models in natural and engineering sciences. But how do information systems develop in nature? What do they have in common with digital information processing in formal proof systems (cf. Chapter 5)? What are the differences? At the first glance, information processing in nature is connected with transmission of energy and entropy satisfying the laws of thermodynamics.

First law of thermodynamics. In a closed system (without exchange of energy and materials with its environment), different forms of energy (e.g., heat, mechanical energy of motion, electrical energy, or chemical energy) can be changed into one another without loss of energy.

Second law of thermodynamics. In a closed system, entropy increases or remains constant.

The second law is found in experience: In a room, a container with a warm fluid cools off spontaneously to the temperature in the room which warms up lightly because of the temperature of the fluid. The reverse procedure of a spontaneous warming up of the fluid with respect to the temperature in the room was never observed. Obviously, heat flows as long as its distribution is homogeneous everywhere. In the state of equilibrium, there is no gradient of temperature. For this value of change of heat, the physicist Rudolf Clausius introduced the term "entropy" (from the Greek words "ergon" for energy and "tropos" for change). Therefore, the change of entropy in a closed system is larger or equal to zero. For closed systems with no change of entropy in the environment, entropy increases or remains constant if thermic equilibrium is reached.

In statistical interpretation, entropy can be understood as measure of uncertainty for the probabilistic distribution of, e.g., molecules in a fluid or a gas. As entropy can spontaneously only increase, but not decrease, this is also true for the uncertainty of a system according to the second law of thermodynamics. Increasing uncertainty corresponds to decreasing information about the single elements of a system.

Therefore, Shannon's measure H of information is interpreted as information entropy of an information source. There seems to be a relation between Shannon's concept of information and entropy as a measure of thermodynamic systems. Thermodynamics considers complex systems (e.g., sets of gas or water consisting of many molecules of gas or water). The total state ("macro state") of a whole system is determined by the states of its single elements at a certain moment of time ("micro states"). Temperature of a gas is explained by the distribution of the total energy to the single molecules.

The macro state of a system is determined by a probabilistic distribution of micro states of system elements. System dynamics (which means the temporal development of the distribution function) is modeled by a stochastic equation (e.g., Master equation). The same macro state (e.g., temperature of a gas) can be generated by different distributions of micro states (e.g., single energies of gas molecules) (Boltzmann, 1896; Maxwell, 1871; Carrington, 1994).

Definition of entropy (Boltzmann). According to Ludwig Boltzmann, entropy is a measure for possible distributions of micro states of elements which generate a macro state.

The number of possible distributions of the micro states determining the total state of the system at a certain moment is called thermodynamic probability W of the system state. Therefore, the entropy S corresponds to the logarithm of W ("logarithmus naturalis"), i.e.,

$$S = k_B \ln W \text{ with Boltzmann constant } k_B.$$

If all W states of the possible distributions happen with equal probability $p_i(1 \leq i \leq W)$, then $p_i = 1/W$, consequently $W = \frac{1}{p_i} = p_i^{-1}$. In general, the states $z_i(1 \leq i \leq W)$ happen with different probabilities p_i. If state z_i happens, the disappearing uncertainty is measured by $\ln p_i^{-1}$. In this case, the entropy S of a system can be computed as expectation value of the probability distribution of its states:

$$S = k_B \sum_{i=1}^{W} p_i \ln p_i^{-1} = -k_B \sum_{i=1}^{W} p_i \ln p_i.$$

The formal correspondence of entropy S of a thermodynamic system with Shannon's information measure H of an information system is obvious. In Shannon's information measure H, the information theoretical constant k is replaced by the physical Boltzmann constant. But we must consider that Boltzmann's concept of entropy is only defined for isolated systems in thermodynamic equilibrium (Nicolis and Prigogine, 1987).

What are the relations between states of molecular systems and symbols of information sources in this case? Actually, these are two examples of random processes which are analyzed in stochastics as a mathematical theory of random processes. In this theory, information entropy is introduced as a measure of the uncertainty of a random variable. A random variable X can relate to the generation of symbols by a sending system as well as the molecular states of a gas which are indicated as dates of a measuring system. In these cases, the (discrete) values x of a set M of random variable X relate to the symbols of a sending system or the state values of a molecular system. The probability of a value x from M is denoted by $p(x)$.

In general, information entropy $H(X)$ of a random variable X can be defined as the expectation value of the probability distribution of their values x of M:

$$H(X) = \sum_{x \in M} p(x) \log p(x)^{-1} = -\sum_{x \in M} p(x) \log p(x).$$

In telecommunications, *logarithmus dualis* (ld) with basis 2 is used because of the measurement of binary bit states. Thermodynamics uses the ln with basis e. The Boltzmann constant $k_B = 1.381 \cdot 10^{-23}$ J/K refers to thermodynamic systems (in equilibrium) with measurement of heat quantity in Joule (J) and temperature in Kelvin (K). In telecommunications, information entropy is understood as mean information content of a sending system, in thermodynamics as entropy of a thermodynamic system (in equilibrium).

Independent of special applications, we can derive general consequences of stochastic systems: If information entropy $H(X)$ of a process X is zero, then event x has probability $p(x) = 1$, which means that x happens with absolute certainty. In this case, the process is deterministic. If $H(X)$ is maximal, then the probabilities $p(x)$ of events x are distributed uniformly, which means that their uncertainty is maximal. Therefore, information entropy can be understood as measure of uncertainty for the probability distribution of possible events.

According to Boltzmann's second law of thermodynamics, the concept of entropy is only defined in the case of equilibrium to be applied to information entropy of material systems. But information systems of nature (e.g., organisms in evolution) exist mainly as open (dissipative) systems far from thermic equilibrium in material and energetic exchange with their environment. Only under these conditions, new structures can emerge in nature. This is the topic of non-equilibrium dynamics of complex dynamic systems (Mainzer, 2007a,b). Therefore, the concept of information systems must be extended for dynamic systems far from thermic equilibrium.

In thermic equilibrium, maximal uncertainty and entropy is realized with uniform distribution of micro states of complex systems. According to the second law of thermodynamics, closed systems develop into a final state where all correlations between system elements decay. Increasing uncertainty, randomness, and entropy intuitively correspond to decreasing information about single system elements. For example, a piece of sugar cube is at first a highly ordered crystal with exactly determined positions of molecules. In coffee, it dissolves, entropy increases, and the positions of the single sugar molecules become more and more undetermined. In thermic equilibrium with uniform distribution, uncertainty is maximal and, therefore, information is minimal.

New information can only be generated in open systems, where entropy can be diminished by energy exchange with their environment. In other words: Information gain is only possible at the cost of energy. Information gain demands open systems far from thermic equilibrium.

Definition of phase transitions in non-equilibrium dynamics. New structures emerge in systems far from thermic equilibrium by changing control parameters (e.g., temperature, energy) until the old system state becomes unstable and descends into a new state. Random fluctuations are the driving forces of phase transitions. At critical values of unstable states, macroscopic order emerges spontaneously from collective interactions of system elements at the microscopic level. In thermodynamics of non-equilibrium, the self-organization of structures in nature is explained by phase transitions of systems in unstable situations with random fluctuations.

Example: Bénard experiment

A system is in thermic equilibrium with its environment if macro-scopic properties (e.g., pressure or temperature) of the system completely are in accordance with its environment. In the Bénard experiment, we consider a fluid layer between two horizontal and parallel plates (Fig. 25). Without influence from outside, the fluid develops into thermal equilibrium resp. a homogeneous state with statistically uncorrelated molecules.

The system is perturbed if the plate below is warmed up. In the case of small differences of temperature, the system returns into the equilibrium state by itself. At a critical value of difference, new macroscopic forms spontaneously emerge in the fluid layer. They

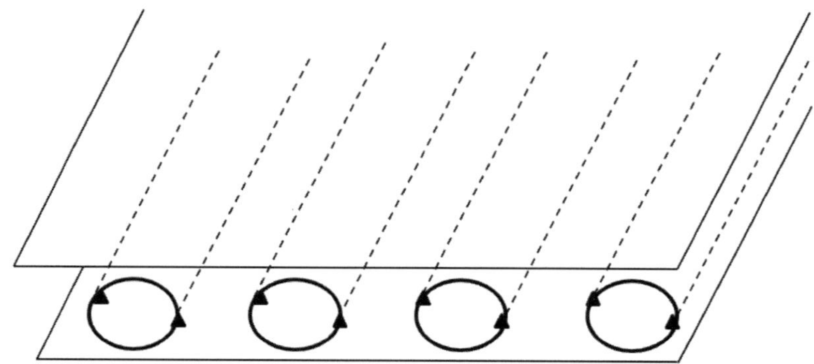

Fig. 25. Bénard cells.

organize in small regular cells of rotating fluid (Bénard convection). The difference of temperature is an example of a control parameter of a dynamic system initiating the self-organization of new macroscopic patterns.

The cause is an ascending and descending flow which is initiated by the different temperature of both plates. The fluid turns in the convection cells alternating to neighboring cells to the right or to the left. By that, an orientation is distinguished and the initial equilibrium is broken. At a critical point of instability, it is decided which order (orientation) succeeds because of random fluctuations.

Further examples of phase transitions in non-equilibrium dynamics

Non-equilibrium states occur very often in nature. An example is the biosphere with an energy flow balancing the radiation between sun and Earth. In meteorology, the spontaneous emergence of cloud patterns can be described by phase transitions which happen at critical values of control parameters (e.g., temperature and flow of certain layers of the atmosphere). Patterns of turbulence in air and water emerge by increasing velocity of flow through energy input. In dependence of energy input, patterns with increasing complexity are generated.

In a stream behind an obstacle, we can observe the emergence of patterns with increasing complexity and depending on the flow rate.

At first, the stream has zero velocity of flow with a homogeneous surface which is in a state of equilibrium. With increasing velocity of flow, vortices emerge on the surface. There are phase transitions from periodic patterns of vortices to quasi-periodic patterns with connected vortices and finally chaotic turbulence.

Definition of emergence und self-organization of new structures.

- Far from thermic equilibrium, complex systems spontaneously generate new macroscopic forms and properties. Emergence of macroscopic structures is explained by interacting elements on the micro level of systems at critical values of control parameters.
- Macroscopic patterns of order are no additive aggregates of elements, but new correlations of order which emerge in phase transitions at critical points of instability of control parameters. This phenomenon is called "self-organization". Emergence and self-organization are mathematically completely explicable in the theory of non-equilibrium dynamics and no mysterious events occur in nature.

Definition of bifurcation trees with critical phase transitions. Phase transitions of complex systems far from critical values of control parameters are geometrically illustrated by bifurcating diagrams of trees. The values of the system states depend on a control parameter of the system (Fig. 26).

In the example of Bénard convection, small values of the control parameter (small difference of temperature between the upper and lower plates) correspond to thermic equilibrium. In this case, the system can damp its internal fluctuations. Beyond a critical value, the branch of the bifurcation tree becomes unstable. The system can no longer damp its own fluctuations. In a new phase transition, the system decides to realize one of the two possible branches with alternative orientations in the Bénard cells.

The points of bifurcation geometrically illustrate spontaneous symmetry breaking of complex dynamical systems. They are

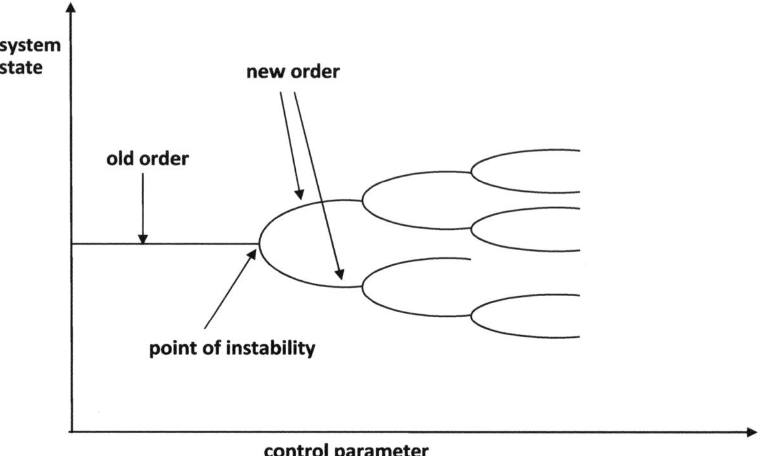

Fig. 26. Bifurcation diagram of complex dynamical systems.

characterized by random fluctuations as driving forces (Beltrami, 1999). In the branches of bifurcation, the patterns of order are locally stable. With increasing control parameter (e.g., velocity of flow), they become unstable again with the emergence of new more complex patterns in the following branches of the bifurcation tree.

Example of information entropy in non-equilibrium dynamics

In Fig. 26, there are two stable states after each bifurcation point which could be labeled by the symbols 0 and 1. In this case, system dynamics generate a sequence of symbols, e.g., 10111001011110..., in sequential phase transitions. In this way, Shannon's information measure can be applied to non-equilibrium dynamics (Nicolis and Nicolis, 2012, p. 117). Let λ be the control parameter and λ_c a critical value at a bifurcation point. Because of Shannon's definition

$$H = -\sum_{i=1}^{N} p_i \ln p_i,$$

the system is at first in the ground state before the first bifurcation with control parameter $\lambda < \lambda_c$, $N = 1$, and $p_1 = 1$. Therefore, information entropy before the first bifurcation is $H = 0$ (no information entropy). The process continues beyond this equilibrium state $(\lambda > \lambda_c)$ with two equivalent states of stability: $N = 2, p_1 = p_2 = 1/2$ with $H = \ln 2$.

Fluctuations (e.g., fluid molecules) in the points of stability can slightly change these results. But it is decisive that each bifurcation resp. phase transition is connected with a finite jump of information entropy H (plus a small negligible correction of fluctuations):

$$\Delta H = H(\lambda > \lambda_c) - H(\lambda < \lambda_c)$$

$$= \ln 2.$$

This result can be transferred to other types of bifurcation (e.g., periodic and chaotic behavior). In this way, information can be understood as an emergent property of dynamical systems leading from a state with no information entropy $(\lambda < \lambda_c)$ to the emergence information entropy $(\lambda > \lambda_c)$. The emergence of information is driven by fluctuations in the bifurcation points.

Phase transitions far from thermic equilibrium lead to emergence and self-organization order with increasing complexity. In general, macroscopic structures cannot be reduced to the micro states of system elements, but to their interactions according to the laws of non-equilibrium dynamics.

At points of instability, only few unstable elements start to oscillate and influence (Hermann Haken: "enslave") the majority of stable elements (Haken, 1988). This oscillating behavior of many elements initiates macroscopic pattern formation. Therefore, it is sufficient to identify the few unstable elements in order to determine the macroscopic trend of the whole system. The parameters which characterize the macroscopic pattern of distributed micro states were called "order parameter" by the Russian physicist Lew D. Landau.

Definition of order parameters. Order parameters are macroscopic quantities which determine the patterns of new order and structures in a complex dynamical system. They determine the type of dynamical complexity of a system after phase transition (Haken, 1990).

Following this line, recommended researchers like the Nobel Prize winners Ilya Prigogine (chemistry) and Erwin Schrödinger (physics) believed that pattern formation only needs non-linear laws and a source of energy. Under these conditions, pattern formation should be initiated by unstable states.

An example of information technology shows that these conditions are not sufficient. An electrical circuit with a battery as source and non-linear switching elements is not sufficient to generate pattern formation (e.g., oscillation). This procedure can be realized by a transistor which can sum up and amplify input signals in order to generate an action potential for pattern formation. In this case, the transistor is said to be locally active. It is surprising that pattern formation can also be initiated in stable states if locally active elements like transistors are involved.

One of the first scientists who thought about pattern formation in nature was Alan Turing (Turing, 1952). After his ingenious contributions to the theory of computability (Chapter 2), we study his last papers on morphogenesis. They are the key to modern mathematical theory of pattern formation.

Pattern formation in dissipative coupling systems

Turing only considers two isolated cells with two molecules, which are assumed to be "stable". Stability means that mathematically there are no positive real parts of the "eigenvalues" (as complex numbers) of the isolated cellular states. Biologically, stability can be interpreted as "dead" state of the isolated cells. If the cells are coupled by a molecular interchange (diffusion) in the simple form of linear equations, then the whole system is destabilized: Mathematically, there is at least one positive real part of the eigenvalues (as complex numbers) of the coupled cellular states.

But if a more complex molecular exchange is allowed, then the non-linear equations have (oscillating) solutions corresponding to complex pattern formation. Oscillation can be interpreted as a phase transition in which isolated and "dead" cells are awakened to "life". Steven Smale (we discussed his contributions to real computing in

Chapters 9 and 10) proved this effect for two cells with non-linear coupling (Smale, 1974).

In the next step, we (together with Leon Chua) considered the general case of arbitrarily many cells (Mainzer and Chua, 2013). In general, we imagine a spatial grid of identical elements ("cells") which interact in different dissipative (e.g., physical, chemical, or biological) ways (Fig. 27).

The change of celluar states and their interactions (diffusion) in a spatial grid can be modeled by (non-linear diffusion and reaction) equations. The solutions of these equations correspond to pattern formation which can be simulated by computers. We imagine a grid in which the cells are ordered in a plane (Fig. 28). The different cellular states are indicated by, e.g., different colors. Pattern formation corresponds to phase transition from simplicity to complexity.

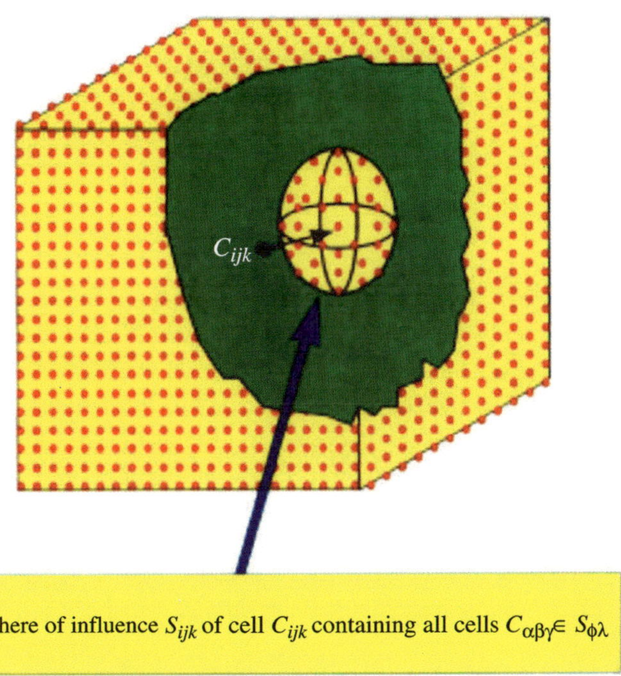

Fig. 27. Complex cellular system with locally active cells and local sphere of influence (Mainzer and Chua, 2013).

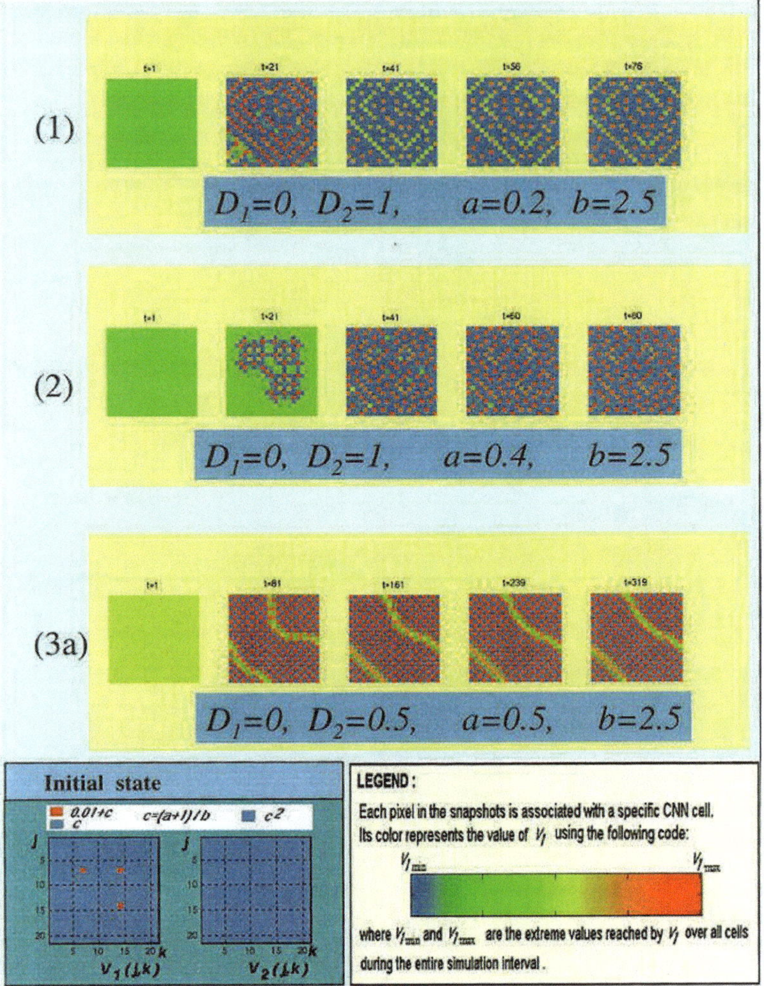

Fig. 28. Complex pattern formation of dynamical systems.

Definition of simple and complex patterns.

- Simplicity means a totally homogeneous (equal) distribution of cellular states.
- Complexity means an inhomogeneous (unequal) distribution of cellular states. The degrees of complexity increase with different distribution of cellular states.

For example, a transistor is called locally active if it can amplify a small signal input to a stronger signal output at the cost of a battery in order to generate a non-homogeneous ("complex") pattern of voltage in a circuit. No radio, TV, or computer could be used without local activity of transistors. Locally active cells amplify small inputs of their neighboring cells and generate a larger output initializing the pattern formation of the whole system.

The principle of local activity has fundamental importance for pattern formation of complex systems which was not recognized in former research. The principle can be formalized mathematically without relating to special applications in physics, chemistry, biology, or technology. In general, we relate to the type of non-linear differential equations which are applied to reaction–diffusion processes (but by no means restricted to fluid media and chemical diffusion). We imagine a spatial grid with identical cells at each grid point and local interaction (Fig. 27). Each cell (e.g., protein in a biological cell, neuron in brain, transistor in a computer) is mathematically a dynamical system with input and output. A cellular state develops locally according to dynamical laws in dependence of neighboring cellular states. The dynamical laws are defined by state equations of isolated cells and their coupling laws. Additionally, initial and secondary conditions must be considered.

Definition of the principle of local activity.

- A cell is called locally active iff, at a cellular equilibrium point, a small local input exists, which can be amplified to a large output at the cost of an energy source.
- The existence of an input initiating local activity can mathematically be tested by certain criteria (Mainzer and Chua, 2013).
- A cell is called locally passive iff there is no equilibrium point with local activity. The principle of local activity demands that systems without locally active elements cannot generate complex structures and patterns.

Pattern formation in nature and technology can be classified by using reaction–diffusion equations according to the principle of local activity. Examples are the differential equations of pattern formation in chemistry (e.g., pattern formation in homogeneous chemical media), in morphogenesis (e.g., pattern formation on shells of snails, furs, and plumage of animals), in brain research (patterns of cell assemblies in brains), and in electronic nets (e.g., patterns of circuits in computers).

Pattern formation corresponds mathematically to non-homogeneous solutions of differential equations which depend on different control parameters (e.g., chemical concentration, ATP-energy in biological cells, neurochemical neurotransmitter of neurons). For these examples of differential equations, we define the parameter spaces with points representing all possible values of control parameters of the system. In these parameter spaces, the areas of local activity and local passivity can be distinguished. In the case of local activity, they allow either pattern formation or, in the case of local passivity, they are mathematically "dead". For any point of a parameter space, the possible pattern formation can be simulated by computers (Fig. 28). Therefore, in this mathematical modeling frame, pattern formation can be completely determined and predicted.

Mathematically, local activity is modeled by an integral adding up of the inputs of neighboring cells and generating an output without any physical, chemical, or biological interpretation. The integral of local activity describes an amplifier generating the ignition spark of pattern formation. It can be considered a new law of thermodynamics that famous scientists have searched for: The First law of thermodynamics only guarantees the conservation of energy in a closed system. The second law of thermodynamics (after Boltzmann) only describes the decay of structure and order in a closed system.

Important insight. The principle of local activity explains the emergence of structure and order in an open system which exchanges energy and materials with its environment.

But, pattern formation is not initiated only by locally active and unstable elements. Even locally active and stable elements can enable pattern formation which was already proved by Turing and Smale. Therefore, instability is not the only driver of pattern formation (which was believed by many scientists in the past). Local activity is also a driver. At the first glance, this insight is puzzling because stable isolated cells are (in a mathematical sense) dead. But after coupling, the stable cells are pushed by diffusion across the separation line of stability into the area of instability where pattern formation is possible. Thus, locally active and stable elements are said to exist "at the edge of chaos resp. instability". Obviously, the popular slogan "at the edge of chaos" has a precise meaning.

Definition of a state at the edge of chaos. A dynamical system is at the edge of chaos iff it is locally active, but (asymptotically) stable.

The origin of life on Earth is a possible example which can perhaps be explained at the edge of chaos: Imagine a totally hostile environment in the dark deep ocean at the edge of a hot volcano. Chemical substances which may at first be separated and stable ("dead") are coupled by dissipative reactions and generate chemical building blocks of primitive organisms.

In general, a complex dynamical system consists of a large number of elements with certain micro states. In a planetary system, the state of motion of a planet is determined by its location and velocity at a certain point of time. Further examples are the state of motion of a molecule in a gas, the state of excitation of a nervous cell in a neural net or the state of a population in an ecological system. The dynamics of a system (which means the temporal change of system states) is modeled by time-dependent equations (e.g., differential equations). In the case of deterministic systems, each future state is determined by the present state.

A simple example is a harmonic oscillator. With small elongation, a mass connected with a spring has small oscillation. A large elongation causes large oscillation.

Definition of linear dynamical systems. In linear systems, causes and effects are proportional. Mathematically, proportionality corresponds to an equation of the form $f(x) = c \cdot x$ with x values, function values $f(x)$ depending on x, and a constant of proportionality c. In a coordinate system, this equation describes a straight line with slope c. Therefore, the equation is called linear.

Definition of non-linear systems. The solution of a dynamical system can be represented by a time series of the system's location depending on time. Linear equations can easily be solved. But, non-linear equations which are geometrically represented by curves cannot always be computed with exact solutions, even with our best computers. Physically, non-linearity means that causes and effects are no longer proportional: A small cause can generate a global effect. An example is weather forecasting which depends on many (non-linear) interacting factors. A local turbulence can change the global weather.

Non-linear dynamics also evolves in a closed system where more than two elements interact. This important insight of classical mechanics came up with Poincaré's Three Body Problem (Barrow-Green, 1997). The state space of a dynamical system is a useful tool to study non-linear dynamics of complex systems (Mainzer, 2004).

Definition of the state space of a dynamical system. The state of a dynamical system is determined by different quantities (e.g., state of motion of a molecule is determined by its location and velocity at a certain point of time). These components of a state are geometrically interpreted as coordinates of the state space of a dynamical system. They define a point in the state space representing the system state. Contrary to three-dimensional spaces with height, depth, and width, a state space can be defined by more than three coordinates (e.g., the state of a disease of a patient depending on several symptoms).

In equilibrium, the state does not change and the corresponding point in the state space is temporarily fixed (fixed point). If the state is changed, a trajectory is generated in the state space. The phase portrait of a state space (Fig. 28) illustrates how the trajectories of a dynamical system reach a final state of a characteristic pattern (attractor).

Definition of an attractor in a dynamical system. An attractor is a final state toward which a dynamical system converges in the long run:

- An equilibrium state corresponds to a fixed-point attractor which does not change in the long run. In the state space, all trajectories converge to this point as final state. Linear systems have no fixed-point attractors.
- Non-linear systems also have limit cycles with periodically repeating states, or, in the case of turbulence, chaos attractors with completely irregular and non-periodic development in a restricted region of the state space (Fig. 29).

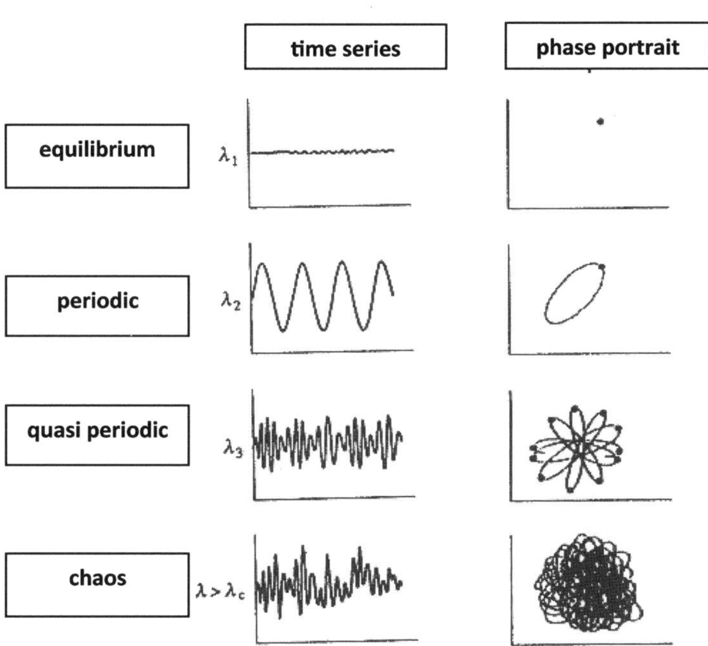

Fig. 29. Degrees of complexity in time series and attractors.

In the case of randomness, all correlations decay into independent events which are distributed irregularly in the whole state space. Dynamical complexity and chaos are situated between regularity (e.g., linear systems) and randomness (Mainzer, 2009a).

In Fig. 29, the time series does not change in the equilibrium state which corresponds to a fixed-point attractor in the state space. In the periodic case, the time series oscillates between two fixed points. In the corresponding state space, the trajectory is closed and returns always to its initial state. In the quasi-periodic case, patterns are repeated periodically in the time series which corresponds to a periodic pattern of the trajectory in the state space. In the chaotic case, the time series develops completely irregularly and non-periodically. In the corresponding state space, a trajectory develops completely irregularly and non-periodically in a restricted area — the chaos attractor. Contrary to the three previous cases, the chaotic development sensitively depends on tiny changes of initial values of a trajectory: After few common initial steps, even tiny differences of two trajectories lead to completely different developments.

Difference of chaos and randomness.

- In deterministic chaos, the dynamics is completely determined by a non-linear law of growth. But, long-term effects are practically not computable. The reason is that the computational capacity increases exponentially because of the sensitive dependence of initial data.
- Contrary to chaos, random developments are not predictable in principle (even short-term predictions are excluded) because all events are independent (example: tossing of a fair coin).

Measurement of sensible dependence of initial data. In order to measure the complexity of a time series and its non-linear dynamics, the degree of non-periodicity or sensitive dependence of dynamics on their initial data can be determined. The so-called Lyapunov exponents measure how the trajectories drift apart in the state space, indicating the degree of sensible dependence.

Trajectories represent the continuous "life line" of a dynamical system in its state space. How can we determine the information processing which is connected with the dynamical process of a system? Therefore, the continuous development must be discretized and mapped on a sequence of symbols. At first, the continuous time must be discretized by assuming temporal distances between measuring points. In Fig. 30, the method of Poincaré cuts is demonstrated (Ebeling *et al.*, 1998, p. 128).

Definition of Poincaré-cuts. A Poincaré cut is intuitively a two-dimensional plane which is cut through a three-dimensional state space. In general, a Poincaré cut reduces an n-dimensional state space to an $(n-1)$-dimensional state space.

The trajectory of a dynamical system penetrates a "Poincaré cut" in a sequence of successive points $x_1, x_2, x_3, \ldots, x_t, \ldots$, representing successive (discrete) system states. If only the previous state x_t causes the immediately following state x_{t+1}, the dynamics is called Markov-like with

$$x_{t+1} = f(x_t, k_t, \ldots, k_m),$$

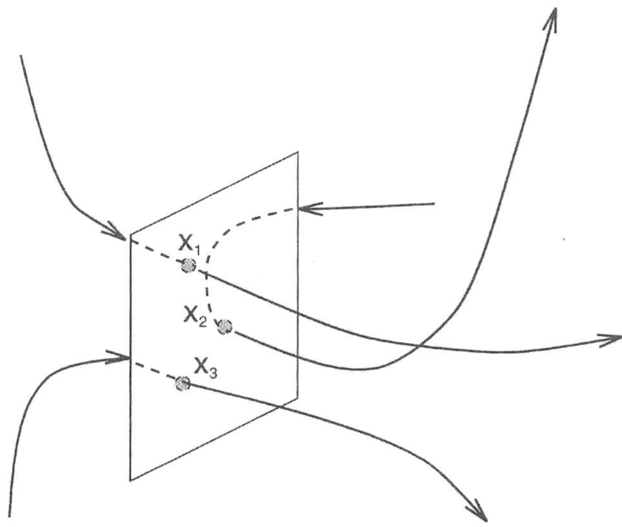

Fig. 30. Poincaré cut.

depending on control parameters $k_t, \ldots, k_m$ and a (non-linear) function f.

In the second step after the continuous time, the continuous state space must be discretized in order to study information processing of a dynamical system with symbolic sequences. We use the method of partitioning (Nicolis and Nicolis, 2012, p. 90).

Definition of partition of a state space. Partition means that a state space is divided into (denumerable many) cells $C_1, C_2, C_3, \ldots$, which do not intersect, but cover the whole state space (Fig. 31).

Definition of symbolic dynamics. After partitioning of the state space, each cell can uniquely be labeled by a symbol. When the trajectory of a dynamical system passes the cells of the state space, a sequence of successive symbols is generated. A Poincaré cut can also be partitioned. In that way, (discrete) symbolic dynamics is connected with the continuous dynamics of a system.

Symbolic dynamics is the foundation of information processing in complex dynamical systems. Spatial structures in a state space are encoded by symbolic dynamics into linear sequences of symbols which can be analyzed by computers. An example is the sequence of

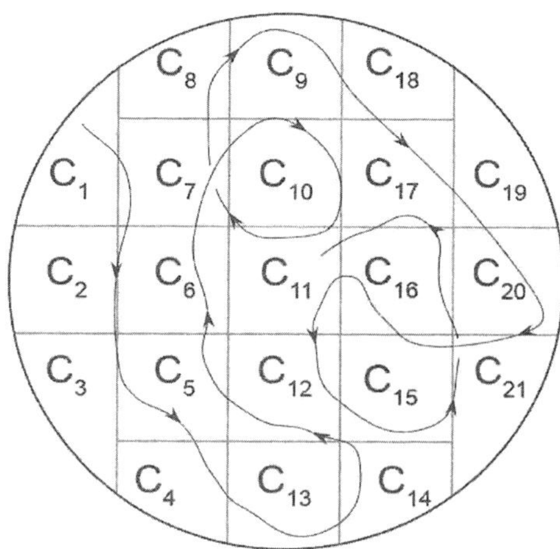

Fig. 31. Partitioning of a state space.

letters in a DNA sequence which encodes the genetic information of an organism.

Example of symbolic dynamics in a chaos attractor. An example of deterministic chaos is the Rössler attractor (Fig. 32) which is defined by three equations of a dynamical system in a three-dimensional state space (Rössler, 1976):

$$\frac{dx}{dt} = -y - z,$$

$$\frac{dy}{dt} = x + ay,$$

$$\frac{dz}{dt} = bx - cz + xz.$$

In order to generate symbolic dynamics by three curves X_1, X_2, and X_3, three levels L_1, L_2, and L_3 are distinguished in Fig. 33. If curve X_i penetrates level L_i ($1 \leq i \leq 3$) with positive slope, a corresponding symbol is quoted. In the case of the Rössler attractor,

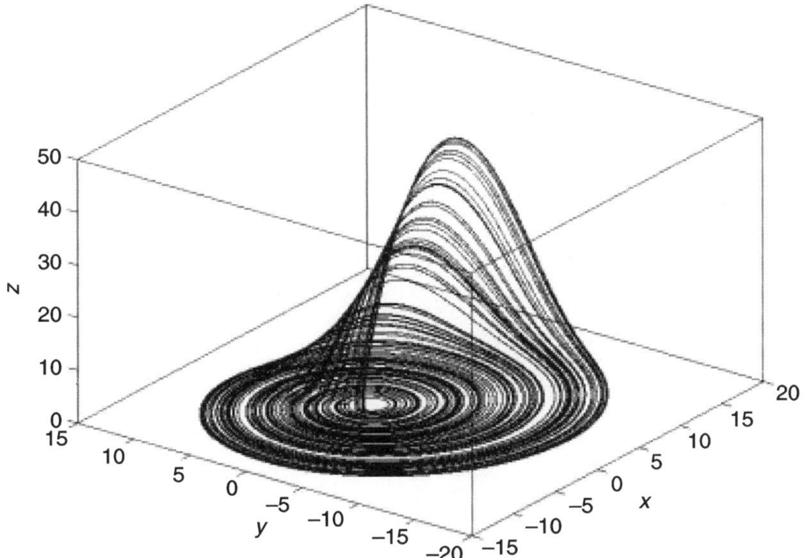

Fig. 32. Rössler attractor.

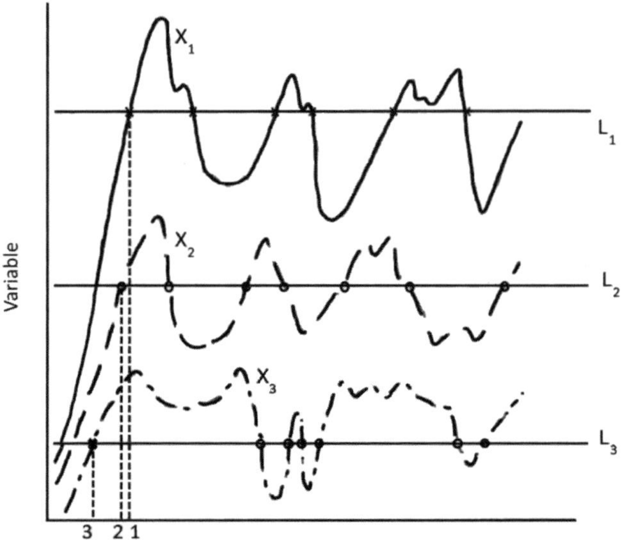

Fig. 33. Generation of symbolic sequences by Rössler attractor in Fig. 32.

the threshold values are defined to be $L_x = L_y = L_z = 3$ with corresponding symbols x, y, z, and constants $a = 0.38$, $b = 0.3$, and $c = 4.5$ (Nicolis and Nicolis, 2012, p. 135).

The curves generate typical sequences

$$zyx \; zxyx \; zxyx \; zyx \; zxyx \; zyx \; zyx \; zx \; zyx \; zyx \; zxyx \; zyx \ldots .$$

If the single sections are encoded by

$$\alpha = zyx, \quad \beta = zxyx, \quad \gamma = zx,$$

then we get the symbolic sequence

$$\alpha\beta\beta\alpha\beta\alpha\alpha\gamma\alpha\alpha\beta\alpha \ldots .$$

Correlations of original sequences can be recognized by statistical analysis of these encodings. In this example, the basic dynamical system is approximately identified as a Markov information source of fifth order which means an information system with a restricted "memory".

Future information development in dynamical systems can also be measured by symbolic dynamics. Dynamical systems can be connected with different information sources. The reason is that entropy h of an information source depends on the chosen partition $\wp$ of the state space of a dynamical system:

$$h = h(\wp).$$

With appropriate partition, the same dynamical system can be related to a periodic source with $h = 0$ as well as to a chaotic source with $h > 0$. The smallest dissolution is realized on the micro level with the single elements of a complex system. Cells mean coarse graining of the dissolution on the micro level which is hidden under the chosen partition.

In the case of a partition $\wp$ with only one cell covering an attractor in the state space, the minimum of entropy $h(\wp)$ of an information source is assumed. For all possible partitions $\wp$, there is a smallest upper bound (supremum) of $h(\wp)$ which only depends on the micro dynamics of the complex system (Ebeling *et al.*, 1998; Mainzer, 2016b).

Definition of Kolmogorov–Sinai entropy. The supremum (smallest upper bound) of entropy $h(\wp)$ of an information source for all partitions $\wp$ is called Kolmogorov-Sinai (KS) entropy:

$$\mathrm{KS} = \sup_{\wp} h(\wp).$$

If the dissolution δ of a partition $\wp$ is defined by the supremum over the diameter of the concerning cells, then the KS entropy can be defined by the limit of entropy $h(\wp)$ for decreasing dissolutions $\delta \to 0$ with

$$\mathrm{KS} = \lim_{\delta \to 0} h(\wp).$$

The KS entropy can be understood as mean information loss of the dynamical system. In the next step, the loss of information resp. the degree of uncertainty of future predictions can be determined. In this case, we consider that predictions of the future relate to data of present and past states. Future states are more or less strongly correlated with the present and past states of a time series. By that, we get degrees of complexity for the range of future predictions.

In the cases of a fixed-point attractor, periodic, and quasi-periodic behavior, the developmental pattern of a time series does not change. It is independent of the initial data. Therefore, the behavior of the system can be predicted without any loss of information arbitrarily long in the future. Geometrically, the repeating development of the curve is known from the past and present data. In these cases, the KS entropy is equal to zero. For chaotic dynamics, the prognosis of future states is bounded after finitely many steps because it depends sensitively on changing initial data. In the chaotic case, there is a finite loss of information. This is the reason why KS entropy has a finite value. Statistically, correlations between future, present, and past data decay after finitely many steps and do not allow further predictions of future. Nevertheless, in the case of chaos, there are correlations with future states for the near future. An example is weather prediction which is possible for the near future, but not in the long run.

If all correlations decay and all future, present, and past events are independent, then random noise happens. In this case, the loss of information is total. No prediction of the future is possible. The situation can be compared with tossing of fair coins. KS entropy becomes infinite. In general, KS entropy is a measure of dynamical complexity from fixed-point attractor of equilibrium via periodic and quasi-periodic behavior to chaos and finally randomness. KS entropy is also a measure of complexity for structure formation from equilibrium structures via periodic and quasi-periodic pattern to highly complex and chaotic structures at the edge of complete decay into random fluctuations.

For chaotic systems, the information loss about system states can also be determined by the degree of (in the mean exponentially) spreading "lifelines" (trajectories) of a dynamical system in the state space. Chaotic systems depend sensitively on smallest changes of initial conditions, which are geometrically illustrated by spreading trajectories. Actually, there is a connection between KS entropy and Lyapunov exponents (Ledrappier and Young, 1985).

Theorem of Pesin. KS = *sum of all positive Lyapunov exponents.*

A dynamical system is chaotic if there is at least one positive Lyapunov exponent. The theorem of Pesin is practically important because the computation of Lyapunov exponents is often easier than KS entropy.

According to its definition, KS entropy measures the information flow in a dynamical system by studying its symbolic dynamics and source entropy h. Source entropy h of an information system depends on the increasing information uncertainty if a symbol is added to a sequence of length n (n-block resp. n-word). After n symbols in the past, we are interested in the information uncertainty which occurs after the first successive symbol $n+1$ in the future. In this sense, KS entropy is a measure of the predictability of a step in the future if n steps of the symbolic dynamics in the past are given. Therefore, KS entropy can be understood as conditional information entropy of the future state resp. symbol z_{n+1} if the states resp. symbols $z_1, \ldots, z_n$ in the past are given.

But how can we predict the long-term future of an information system? In this case, we should not only determine the first successive state z_{n+1} in the symbolic dynamics of an information system, but also the pth successive state z_{n+p} in the future for $p > 1$. We use the transinformation resp. mutual information $I(n; p)$ between the word of the n preceding symbols and the symbol which follows at the pth position. Transinformation measures the statistical correlation between two random variables, in this case between the symbol z_{n+p} and the preceding symbols of the past. How strong are these symbols statistically independent? For chaotic systems, we get

$$I(n; p) > I(n; p + 1).$$

The loss of information increases with each step in the future (Deco and Schürmann, 2001).

Definition of information flow in dynamical systems. The predictability of the pth symbol in the future is determined by its statistical correlation (transinformation) with the whole past of the preceding symbolic dynamics:

$$I_p = \lim_{n \to \infty} I(n, p).$$

KS entropy is a special case of information flow I_p with $p = 1$.

The information flow is small for rigorously ordered sequences. In this case, KS entropy is zero, but it is high in chaotic systems and becomes practically incalculable if all structures decay disorderly. The "edge of chaos" is especially interesting because, before dissipative coupling, dynamical systems are still stable and not chaotic. After dissipative coupling, extensive structures emerge (e.g., living organisms at the edge of hostile environments). Their emergence depends basically on the principle of local activity in open dynamical systems.

Finally, we can clearly distinguish between the complexity of dynamical systems and the algorithmic complexity which was considered in the previous chapter (Bennett, 1995). Algorithmic complexity aims at compression, reduction, and redundancy of symbolic (discrete) sequences which are generated by information systems. Algorithmic complexity does not relate to sensible dependence of initial conditions in dynamical systems developing in (continuous) time. Dynamical systems generate complex structures and patterns in phase transitions far from (thermic) equilibrium. Dissipation even allows pattern formation in initial states of (asymptotic) stability "at the edge of chaos".

Contrary to dynamic complexity, algorithmic complexity is without relation to dynamics and temporal change of system states in physically real systems. Therefore, algorithmic information theory as well as Shannon's information theory must be extended for non-equilibrium dynamics of physically real systems. The bridge between (continuous) dynamical systems and (discrete) information systems is the concept of symbolic dynamics with measurement of information flow in dynamical systems. The concept of information is not only extended for information processing and computer science, but also for physics and natural sciences (Myers and Casiado, 2009; Zurek, 1989). It is a bridge between the digital and the real world.

Chapter 15

Digital and Real Physics

What do we know about the physical origin of information systems? As far as we can measure, the universe came into being 13.81 ± 0.04 billion years ago as a tiny initial quantum system which expanded into cosmic dimensions according to the laws of quantum physics and general relativity. In a hot mixture of matter and radiation, the generation of nuclear particles was finished after 10^{-5} sec. After the separation of matter and radiation in nearly $300\,000$ years, the universe became transparent. Gravitation began to form material structures of galaxies, Black Holes, and first generations of stars. Chemical elements and molecular compounds were generated. Under appropriate planetary conditions, the evolution of life had developed. The evolution of quantum, molecular, and DNA systems is nowadays a model of quantum, molecular, and DNA computing.

In quantum computing, we consider the smallest units of matter and the limits of natural constants (e.g., Planck's constant, velocity of light) as technical constraints of a quantum computer. A technical computer is a physical machine with an efficiency which depends on the technology of circuits. Its increasing miniaturization provided new generations of computers with increasing capacity of storage and decreasing computational time. But increasing diminution leads to the scaling of atoms, elementary particles, and smallest packages of energy (quantum) which obey physical laws different from classical physics (Mainzer, 1988).

Classical machines following the laws of classical physics are replaced by quantum computers which work according to the laws of quantum mechanics. In order to understand a quantum computer, the laws of quantum mechanics must be studied.

Bohr's atomic model

According to Niels Bohr, an atom is illustrated like a planetary model in the Middle Ages. Like planets moving around the Earth, a nucleus of dense matter which consists of neutrons and protons with positive electronic charge is surrounded by shells of increasing diameter with negatively charged electrons. In a carbon atom, the orbits correspond to discrete levels of energy which are determined by integers $1, 2, \ldots$ as multiples of Planck's constant h. The most stable state of an atom is the state of lowest energy. Higher orbits resp. states are called excited states. After Bohr, transitions between different orbits are possible if the energetic difference between the concerning states is either absorbed or radiated as photons of electromagnetic radiation. In this model, Bohr computed the discrete spectrum of carbon with good accordance to the measured spectrum.

Contrary to classical physics, matter of atomic scaling is determined by quantum jumps of electrons with discrete and discontinuous quanta of energy. Therefore, it seems to be obvious that carbon atoms can be used as stores of binary information in a quantum computer. An atom in the ground state represents "zero" (0), in the excited state with electron at a higher level it represents "one" (1). The stored value could be changed into the other one by stimulation with a laser pulse. If the energy of the laser photons is exactly equal to the energetic difference between the excited state and the ground state, the electron jumps from one orbit to the next one.

But quantum objects are no objects of classical physics like footballs, the state of which can be determined by simultaneous measurement of location and momentum. If a ray of neutrons is sent onto a screen with two fissures, then location or momentum of a particle can be measured behind the fissures in dependence of the width of the fissures. Dispersion of measurement can be determined by repetition of measurement. But a smaller dispersion in the measurement of location is related to a larger dispersion in the measurement of momentum and vice versa.

Heisenberg's uncertainty relation

Planck's constant h forbids that both dispersions can become arbitrarily small as in classical physics. Because of this uncertainty relation named after the Nobel Prize winner Werner Heisenberg, the trajectory of a quantum system (e.g., an elementary particle) cannot be computed as, e.g., the trajectory of a billiard ball by measuring its location and momentum with arbitrary exactness at each temporal point.

Wave-particle dualism

Independent of the exactness of measurement, a quantum object (e.g., photon) behaves like a particle as long as the property of a particle (e.g., location and momentum) is measured. If the property of a wave is measured (e.g., frequency of light), a quantum object behaves like a wave. The answer to the question whether the quantum object is a wave or a particle is not determined independently and before measurement like in classical physics. Obviously, the wave-particle dualism is completely different to classical physics. Max Born unified the pictures of wave and particle with his interpretation that the amplitude of a quantum wave is the probability of the location of a particle (e.g., electron) in a very small spatial region.

A central principle of quantum physics is the superposition principle which has immediate consequences for the quantum computer.

Superposition principle

Quantum waves can superimpose themselves like water waves. According to Born's interpretation, a quantum wave describes the location of a particle with a certain uncertainty. In a superposition, a wave consists of two partial waves each of which describes a different location with a certain uncertainty. In that sense, a particle described

by a superimposed wave can simultaneously be at different locations
with a certain uncertainty.

In Bohr's atomic model, superposition intuitively means that an
electron is located in different orbits simultaneously. The uncertainty
ends when the electron emits or absorbs a photon and determines
one of its possible states. It happens in the case of an interaction
with, e.g., a laser pulse. Two waves swinging in accordance as one
wave are called "coherent". The procedure transforming them into
their own states is called "decoherence". Quantum states are discrete
states which can be used to determine "quantum bits" (Schumacher,
1995).

Definition of a quantum bit. If hydrogen atoms are used for
storing information, we must consider a mean state between the
ground state of energy E_0 for 0 and the excited state with energy
E_1 for 1, in which the wave of the ground state and the wave of the
excited state superimpose themselves. Such a quantum bit (qubit) is
half 0 and half 1. A classical bit is always either 1 or 0.

A state of superposition is realized if the appropriate laser light
stimulates the atom only for half of the time in which it would jump
into the excited state.

Signal processing in quantum circuits

Quantum bits can not only be input ("reading in"), but they can
also be output ("reading out"). In that case, laser light of energy
difference $E_2 - E_1$ with energy E_2 larger than E_1 stimulates an
unstable atom. If the atom is in a state of energy E_1, then it can
be transformed into a state of energy E_2 by stimulation with an
appropriate laser pulse. After some time, it can return to a state E_1
with simultaneous emission of a photon. If an atom is in the ground
state, it will not change. If the atom is in the intermediate state, it will
emit a photon in one half of all cases and return to state E_1. In the
other half of cases, it will output no photon and jump into state E_0.

In 1982, Richard P. Feynman already suggested a quantum
computer. He argued that a computer must follow the laws of

quantum physics in order to simulate quantum systems. Machines following the digital paradigm of classical physics would only be poor approximations missing the essential laws of the quantum world (Feynman, 1982). In classical computers, signal processing of bits is realized by logical gates:

NON gate

A flip-flop switch converts an input bit 0 resp. 1 in its opposite value 1 resp. 0. This switch obviously corresponds to logical negation if 0 resp. 1 are understood as truth values.

AND gate

An AND gate has two input channels which can receive two bits and convert them into an output channel with one bit. Only if both input bits are 1, the output bit is 1, otherwise it is 0. The AND gate corresponds to the logical AND connection of two propositions A AND B ($A \wedge B$) which is only true (1) if both partial propositions are true (1).

Copy gate

A copy gate transfers the state of a bit to its copy. The corresponding quantum gates also have to consider intermediate states.

For information processing in a computer, the gates must be connected ("wired up"). In a classical computer, metal wires were used in order to transfer voltage signals from one gate to the next one. An application of hydrogen atoms as switching elements would be rather complicated because they have to be decomposed into their building blocks of nucleus and electrons which must be transferred. Therefore other quantum elements are used for quantum computing.

Quantum spin

Quantum bits cannot only be interpreted as atomic states. An elementary particle can have a quantum mechanical spin. In a magnetic field, a spin only has discrete values like the energy of an electron in the hydrogen atom. Therefore, spins can be used as representation of quantum bits. There are also superposition states of digital spin states which correspond to quantum bits. Spin states can also be used for signal processing in quantum computers.

An example is provided by neighbored ions in a salt crystal (e.g., sodium and chlorine ions in sodium chloride). In a chain of alternating atoms of kinds A and B, all B ions can jump into the other binary state caused by a laser pulse. If the left A neighbor is in state 1, then all A ions to the right of B neighbor are in state 1. In quantum physics, this effect is called double resonance. In that procedure, bit information can be transmitted through the whole atomic chain. All ions which were used as information stores are transferred back to the ground state. In a three-dimensional crystal, signal transmission can be carried out simultaneously.

Polarization states of light quanta

Light quanta (photons) as smallest massless particles can also be used for quantum communication. In this case, single photons can be transferred into two possible states of polarization: Horizontal polarization of a photon corresponds to 0 and vertical polarization to 1. Photons can be sent in a fiber optics network in order to transmit quantum information of quantum bits with light velocity.

Superposition and Schrödinger's cat

Quantum states can superimpose themselves. Erwin Schrödinger illustrated superimposed quantum states in a thought experiment with a cat which is locked in a box with a bottle of hydrocyanic acid (Audretsch and Mainzer, 1996). A mechanism is connected with a random process (e.g., decay of an atomic nucleus). If the atomic nucleus decays, the mechanism is initiated and destroys the bottle of hydrocyanic acid which kills the cat. Nobody can predict whether the atomic nucleus decays which leads to the death of the cat. Therefore, according to Erwin Schrödinger, the cat in the closed box is in a superposition of the two states "dead" and "alive" corresponding to the superimposed quantum states of "decayed" and "non-decayed" nucleus. The superposition collapses if the box is opened. Then the cat is either dead or alive. In general, opening the box means an act of measurement.

There is another (classically) amazing consequence that two separated particles in far distance can nevertheless be in common state of "entanglement" without interacting by any mechanism.

Entanglement and EPR correlation

In Einstein–Podolsky–Rosen (EPR) experiments, pairs of photons are emitted into opposite directions from a central source. From an equal distance, the particles strike polarized filters. Depending on the correlation of the polarization states, a measurement of one particle determines the result of a measurement on the opposite particle instantaneously. This phenomenon cannot be understood in classical physics. In quantum mechanics, it is precisely predicted and confirmed by EPR experiments.

Quantum computers would open new avenues of computational power. Immense advantages are the tiny diminution of atomic gates, the enormous velocity of switching and signal transmission, and minimum energy requirement. The principle of superposition enables simultaneous (parallel) processing of huge data mass as quantum bits. In serial data processing of a classical von Neumann computer, each computational step must be realized one after another.

Quantum parallelism

In the case of parallel data processing, the algorithm of decision can work simultaneously for all data. In a quantum computer, all possible input bits are switched into a superposition of 0 and 1 with equal chances. After processing these inputs in the atomic circuits of a quantum computer, we get a superposition of all possible outputs. The simultaneous processing of all possible inputs is called quantum parallelism. Quantum parallelism can intuitively be compared with the superposition of acoustic waves in different musical instruments which are played simultaneously in an orchestra. You never hear the melody of a single instrument playing the sequence of notes one after another. You hear a melody only in superposition with other sequences of notes.

But there is a fundamental difference between acoustic waves and quantum waves. The intermediate state of superimposed quantum bits jumps into binary bit states 0 or 1 after interaction with the environment (e.g., laser pulse of a readout device). Contrary to acoustic waves, a single "quantum melody" is changed if it is read out from the superposition of all "quantum melodies". In quantum physics, the coherence of superimposed quantum states is lost after interaction with the environment (e.g., after interaction with a measuring instrument) (decoherence). In quantum computers, we still do not know how to stabilize and store quantum bits as coherent quantum bits without perturbations from outside.

Example of quantum parallelism

The laws of quantum mechanics have enormous practical consequences for computation of computers. If we have to solve two partial tasks of a problem, then a classical (von Neumann) computer must solve the two partial problems one after the other. A quantum computer could operate on both tasks simultaneously in a superposition. An example is the task to find a natural number with a certain property. A classical computer enumerates the numbers $1, 2, 3, \ldots$ and tests step by step if a number has the concerning property. If the searched number n is very large, then the criterion must be tested n times with large computational time. A quantum computer could test the criterion for a large set of numbers simultaneously in only one step.

As usual, decimal numbers are represented by binary numbers corresponding to bit sequences. In a quantum computer, a bit is represented by an alternative quantum state of a quantum system. An example is the alternative spin of an elementary particle. In this case, 0 and 1 correspond to the two alternative orientations of spin. A bit sequence represents a sequence of turning elementary particles with different orientations. According to their spins, a combination of binary states of seven particles can represent 2^7 possibilities, e.g., 0000000 (for decimal number 0), 0000001 (for decimal number 1),

0000010 (for decimal number 2), and so forth for all natural numbers between 0 and 127.

In a classical computer, dual numbers 0000000, 0000001, 0000010, ... must be input one after the other, in order to test the property. The spins can be changed by appropriate pulses of energy. For weak pulses, a particle only changes its spin sometimes, sometimes not. In that case of randomness, the spin behavior can only be predicted probabilistically.

In analogy to Schrödinger's cat which is simultaneously dead and alive in the closed box, the particle is in a superposition of alternative spins as long as it is not observed and measured. If all seven particles are stimulated by weak pulses of energy, then all seven particles remain in superimposed states as long as they are not observed and measured. In a superposition, all 128 different states and numbers can be represented and tested simultaneously. We can easily imagine how hundreds of particles can represent giant numbers simultaneously which leads to unbelievable computational times.

Decoherence problem

It is a central technical problem to keep the superposition during a computation undisturbed. Tiny interactions of only one particle with the environment would lead to a collapse of the superposition. As superposition is called coherence, the collapse of coherence is called the decoherence problem of a quantum computer (Deutsch and Eckert, 2000; Alber *et al.*, 2001). The registers of a classical computer consist of storing cells for classical yes (0)/no(0) bits. In a quantum computer, registers are quantum systems the states of which can be superimposed.

Gates of a classical computer

We already mentioned the gates of a classical computer which correspond to elementary logical operations, e.g., the NON gate which converts a bit 0 into 1 and 1 into 0, or the AND gate which converts the bit pair 11 into bit 1, but all other bit pairs 01, 10, and 00 into 0 — corresponding to the logical conjunction.

An OR proposition is only true iff at least one partial proposition is true. Thus, an OR gate converts all bit pairs 01, 10, and 11 into 1, with the exception of 00.

In this way, all logically possible connections and corresponding gates can be defined. For example, the IF–THEN gate corresponds to the IF–THEN proposition which is only false if a false proposition is derived from a true proposition. According to classical logic, the IF–THEN gate can be replaced by a combination of an OR gate and a NON gate (Fig. 34). A computer program is a sequence of gate operations which can be illustrated in a network.

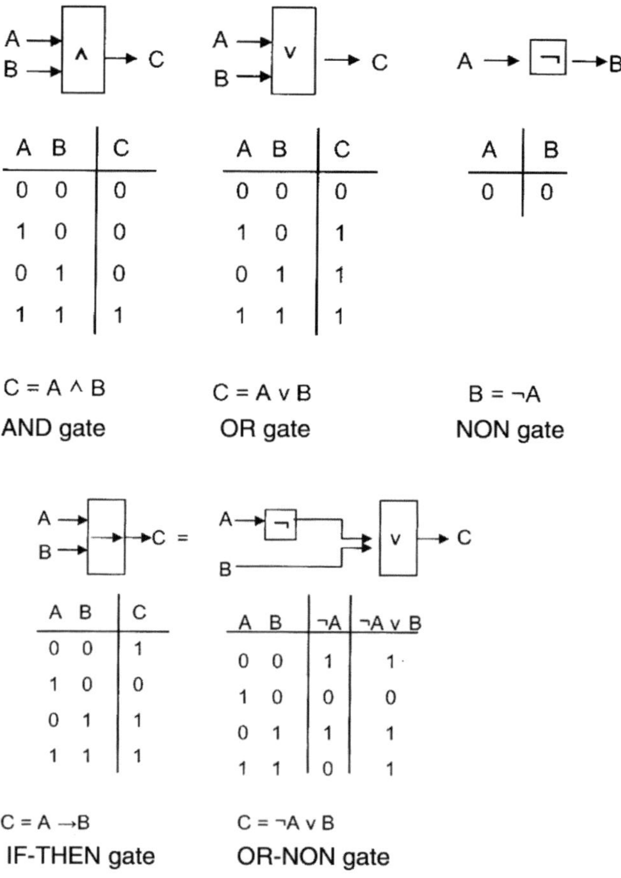

Fig. 34. Elementary and composed gates of a classical computer.

In a classical computer, an input is a sequence with uniquely determined bits of either 0 or 1. The input of a quantum computer needs a preparation of registers. An example is the preparation of registers as superposition of their bit states.

In a classical computer, the output is uniquely determined (Fig. 34). In a quantum computer, a computation can terminate in a superposition of several states. In the next step, the output must be determined by a measurement of appropriate registers. The problem is that a measurement effects a superposition of registers to jump randomly into one of the two possible states 0 or 1 according to the laws of quantum mechanics.

Different transitions of a quantum computer may provide different results which can only be estimated with probabilities. Quantum registers differ from classical registers by the possibility of superposition. In a quantum program, the transformation of bit states 0 and 1 must also be executed for superimposed states according to the principle of superposition.

Reversibility of quantum gates

In quantum mechanics, a superposition of quantum states is transmitted by unitary operators. But, classical gates cannot be used for transmission. The reason is that unitary operators are reversible. Reversibility means that the initial state of a quantum system can be uniquely reconstructed from the output state. Reversibility guarantees that a superposition of inputs is conserved until the output.

Many classical gate operations do not satisfy this condition. In Fig. 34, the OR gate provides 1 in three different cases and the AND gate 0 in three different cases. After application of these gates, it cannot be distinguished which of these three cases was used as input. Reversibility can be realized by an additional instruction which can "remember" the input (Keyl, 2002, Section 4.5.6): The computation of a function f with input x and output $f(x)$ is replaced by a mapping of input x on output $(x, f(x))$. In Fig. 35, input and output are denoted as quantum mechanical state vectors. In

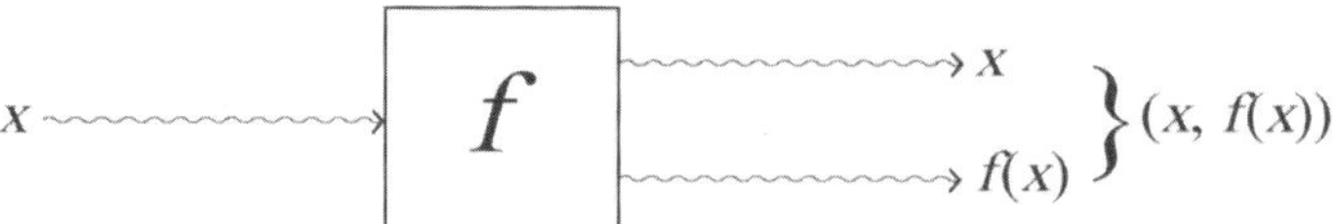

Fig. 35. Reversible computation of a quantum gate.

the following diagrams, quantum states are indicated by wavy lines, classical information by straight arrows.

Classical information is sent between transmitters and receivers which can be realized by different physical, chemical, and biological systems. But they must not be diminished in the scaling of quantum effects. In the quantum world, a transmitter corresponds to the preparation P of a quantum system, a receiver to a measurement M (Fig. 36). Quantum systems (e.g., elementary particles) transmit quantum information during the development from a prepared state to a measurement in a quantum experiment.

Definition of quantum information

Quantum information is defined as information which is transmitted by quantum particles from the preparation of a quantum system to the measurement of the quantum system in a quantum mechanical experiment.

In a quantum transmission, the single result is random because only statistical statements about future events are possible in quantum mechanics. If an experiment is repeated for several times, different results may occur, but with reproducible frequency. According to probability theory, random events are only unpredictable as single occurrence, but their statistical distribution is predictable. A necessary condition is a definition and preparation of the initial state of a system (e.g., dice and roulette in gambling, or photons in an EPR experiment) in order to guarantee reproducibility.

Transmission of quantum information

Measurements and observations of quantum information need a prepared quantum system. Experiments can be repeated under

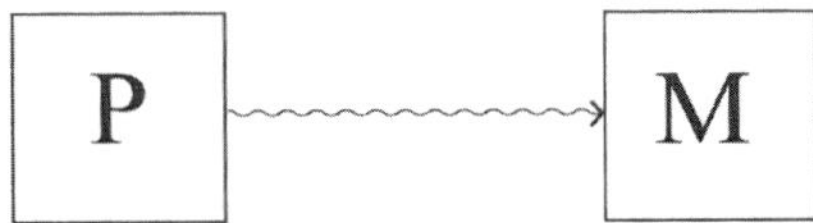

Fig. 36. Statistical transmission of quantum information.

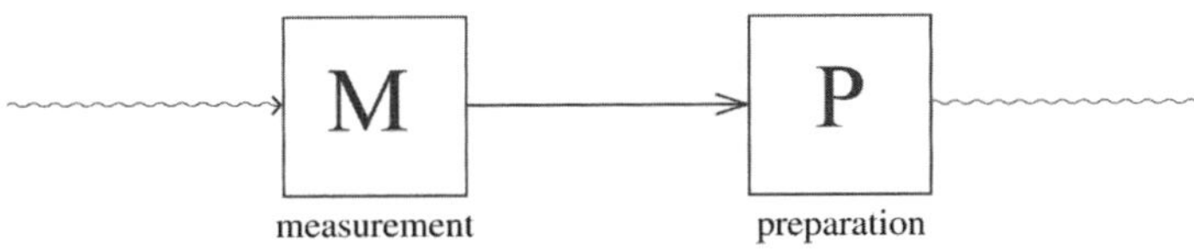

Fig. 37. Classical teleportation of quantum information into classical information and back into quantum information.

identical conditions of preparation P of the initial state of a quantum particle and measurement M of its final state (Fig. 36). In a series of measurements, relative frequency of experimental results is analyzed and used for statistical predictions. The measurement fails if there are coarse deviations to the typical behavior of a random series.

Classical teleportation of quantum information

Quantum Information depends on statistical laws of quantum mechanics. Therefore, it is a new kind of information which cannot be transformed into classical information without loss. Figure 37 illustrates how the transformation resp. teleportation of quantum information into classical information and vice versa would be arranged (Alber *et al.*, 2001, p. 16; Keyl, 2002, p. 435). At first, we would start with a measurement M of a quantum particle as carrier of quantum information (wavy line). The classical data of this measurement (straight arrow) would be emitted to a receiver P which must prepare a quantum particle (wavy line).

Impossibility of a quantum copier

The impossibility of classical teleportation depends on another theorem of impossibility: Quantum information cannot be copied

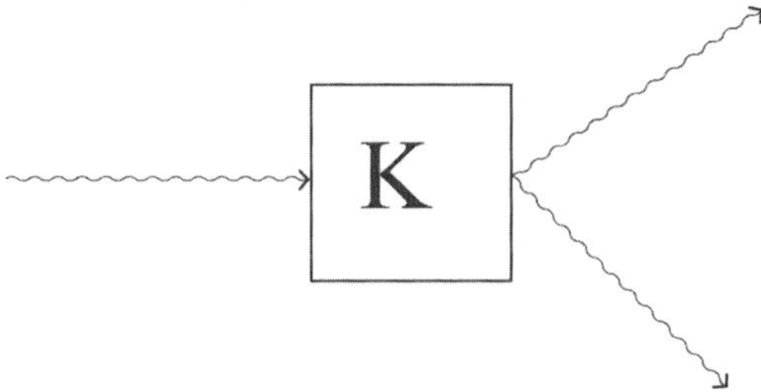

Fig. 38. Scheme of a quantum copier.

(no-cloning theorem) (Wootters and Zurek, 1982). An (impossible) quantum copier is illustrated in Fig. 38. A quantum copier would receive a quantum particle and output two copies (clones). Each copy would be statistically as good as the input. The impossibility of this system has severe consequences:

Data cannot be secured by copies which are usual for classical computers. In a quantum data base, no data can be read out without changing them. But, on the other side, the impossibility of a quantum copier prevents an unnoticed electronic surveillance on quantum data because manipulation changes quantum information. In quantum cryptology, the no-cloning theorem is fundamental.

The impossibility of a quantum copier depends on Heisenberg's uncertainty relation which excludes that certain pairs of quantities (e.g., location and momentum, time and energy) can be measured simultaneously with arbitrary exactness. The no-cloning theorem implies the impossibility of classical teleportation (Keyl, 2002, Section 4.1.1).

No-cloning theorem. *In quantum mechanics, a quantum copier is impossible.*

Proof. Assume the existence of a quantum copier

$\Rightarrow$ There are two copies A_1 and A_2 of a quantum particle A.

$\Rightarrow$ Location of A_1 can be measured with arbitrary exactness. Momentum of A_2 can be measured with arbitrary exactness.

$\Rightarrow$ Contradiction to Heisenberg's uncertainty relation. $\qquad\qquad\square$

Theorem: Impossibility of classical teleportation. *Classical teleportation of quantum information into classical information and back into quantum information is impossible.*

Proof. Assume classical teleportation (Fig. 37)

$\Rightarrow$ A quantum copier can be constructed:

- The transmitted classical information in the classical connection of teleportation can be copied by a classical computer.
- The copies can be transmitted to the receivers P and P′ (Fig. 39) which prepare new quantum particles.

$\Rightarrow$ Contraction to the previous theorem. $\qquad\qquad\square$

But quantum information opens new avenues for quantum teleportation. If we measure one of two entangled quantum bits which are separated with large distance, then the value of the other quantum bit is determined instantaneously.

For teleportation of a physical body from one place to another one, only the information of its physical structure could be transmitted. All physical bodies consist of a more or less huge number of quantum systems (e.g., elementary particles). In principle, the quantum information which describes the structure of a physical body could be transmitted instantaneously to another location in EPR experiments. But in the case of material bodies, matter could not be transmitted. The teleported quantum information would have to meet the same matter at the other location to form the corresponding

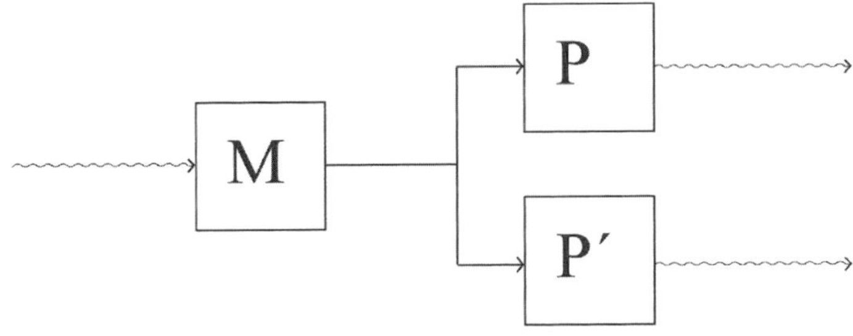

Fig. 39. Quantum copier with classical teleportation.

body. Further on, quantum information cannot be used for direct telecommunication like classical (Shannon) information.

"Beaming" in science fiction

In science fiction movies, teleportation describes the fantastic transmission of objects which can instantaneously be "beamed" in far distances. In this case, information (software) of the atomic structure of an object is separated from the matter of the body in order to transmit the information completely to the distant location. At this location, the corresponding material building blocks must be connected again according to the construction manual of the information.

According to the laws of classical physics and information theory, the procedure seems to be possible. But, in the quantum world, the information of all properties of elementary particles cannot be determined completely because of Heisenberg's uncertainty relation.

Quantum teleportation

Entangled quantum states are used for instantaneous quantum teleportation of quantum information to distant receivers. In Fig. 40, the transmitter of quantum information is called Alice, the receiver is called Bob. The quantum particle which must be teleported is

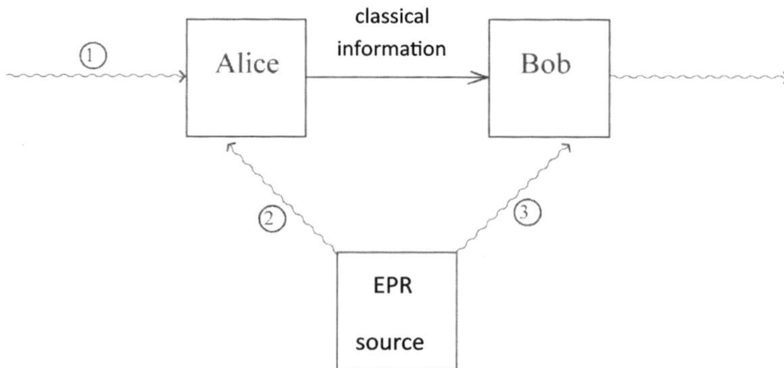

Fig. 40. Quantum teleportation.

denoted with 1 (wavy line). For quantum teleportation, Alice and Bob use an EPR source which generates entangled pairs of quantum particles (Fig. 40).

An entangled pair of particles consists of particle 2 for Alice and particle 3 for Bob. Alice and Bob do not know the single states of the pairs of particles, but they know only their correlations: For example, they are assumed to be in opposite states. Alice also does not know the single state of particle 1 and the single state of particle 2. For teleportation, Alice prepares an entanglement of particle 1 with particle 2. With knowledge of entanglement, we get information of the correlated behavior of both particles without determining their states by measurements (Keyl, 2002, Section 4.1.2).

In analogy to an EPR experiment, a quantum particle (e.g., photon) is assumed to have alternative quantum states (e.g., vertical and horizontal polarizations). In quantum mechanics, a quantum state ψ is denoted by a vector $|\psi\rangle$. In analogy to classical bits 0 and 1, alternative quantum states are understood as quantum bits (qubits) $|0\rangle$ and $|1\rangle$ which can be superimposed. (We use the bra–ket formalism of Dirac to denote quantum states.) In the case of superposition, the quantum system carries quantum bit $|0\rangle$ with a certain probability and simultaneously the quantum bit $|1\rangle$ with the rest probability. A measurement is needed to determine one of these two quantum bits $|0\rangle$ or $|1\rangle$ by random.

In the first step, Alice prepares an entangled pair of particle 1 and particle 2. According to the possible combinations of classical bit pairs 00, 01, 10, and 11, entangled states can be generated for pairs of quantum bits. We get four possible entangled states for pairs of quantum bits $|0\rangle_1|0\rangle_2$, $|0\rangle_1|1\rangle_2$, $|1\rangle_1|0\rangle_2$, and $|1\rangle_1|1\rangle_2$ which are called product states because the whole state of both qubits are given by a product of single states. A product state consists of localized single states. The four local product states correspond to four non-local states which are called Bell's states. They are named after John Bell who proved a crucial theorem in the foundations of quantum mechanics (Bell, 1964): The so-called Bell's inequalities are mathematical consequences of local ("classical") theories which are refuted by measurements and predictions of non-local quantum mechanics

(Audretsch and Mainzer, 1996). Therefore, classical physics can only be an approximation to the quantum world.

Bell's quantum states

$$|\psi^+\rangle = 1/\sqrt{2}(|0\rangle_1|1\rangle_2 + |1\rangle_1|0\rangle_2)$$

$$|\psi^-\rangle = 1/\sqrt{2}(|0\rangle_1|1\rangle_2 - |1\rangle_1|0\rangle_2)$$

$$|\varphi^+\rangle = 1/\sqrt{2}(|0\rangle_1|0\rangle_2 + |1\rangle_1|1\rangle_2)$$

$$|\varphi^-\rangle = 1/\sqrt{2}(|0\rangle_1|0\rangle_2 - |1\rangle_1|1\rangle_2)$$

The minus sign means that both partial states are found in opposite settings. The coefficients indicate that the partial states of superposition happen with equal probability 1:2.

Alice randomly determines one of these four possible Bell's states by measurement, but without learning the states of particle 1 or 2. An example is the Bell's state which correlates both quantum bits in opposite settings. In this case, Alice knows that the quantum bit of particle 1 (which should be teleported) has opposite orientation to the state of the second quantum bit of particle 2.

Under the condition that the entangled pair of particle 2 and 3 has opposite correlation, Bob's particle 3 must be in the same state of Alice's particle which is to be teleported.

Therefore, a measurement of particle 1 (which is to be teleported) instantaneously determines the same quantum bit of Bob's particle 3. In this sense, an instantaneous teleportation of quantum information between transmitter Alice and receiver Bob is realized.

Instantaneous quantum teleportation is also possible if another Bell's state is randomly generated by Alice's measurement. As particles 1, 2, and 3 are correlated one behind the other (Fig. 40), Bob can set his quantum bit into the right state by turning, e.g., the photon in an appropriate way. Bob and Alice know the EPR entanglement of particles 2 and 3. Therefore, it only depends on the entangled state of particles 1 and 2. The message on the correlation of particles 1 and 2 is sent by Alice as classical information (e.g., by telephone) to Bob (Fig. 40). Contrary to instantaneous quantum teleportation, classical information transmission cannot be faster than the velocity of light.

Quantum teleportation and information transmission

But there is a severe weak point of quantum teleportation: The quantum information which is to be sent (particle 1) is unknown. It is decided by random and is not known before a measurement. Therefore, quantum teleportation cannot be used for direct information transmission. Obviously, there is no conflict with Einstein's theory of relativity because no transmission faster than light is assumed.

As long as quantum information is not measured, read out, and observed, it can be transmitted instantaneously and in arbitrary superposition. Quantum parallelism supports a giant increase of computational power and capacity as long as quantum information is transmitted and processed in a quantum computer without observation and measurement. Reading out an output needs a measurement which determines a bit sequence by random. In order to measure its probability, the computational procedure must be repeated very often.

The decisive advantage of a quantum computer is its massive quantum parallelism. Quantum parallelism supports many parallel searches and problem solving procedures in efficient time (Nielsen and Chuang, 2001).

Quantum universe as quantum computer?

Quantum computers are quantum systems the states of which represent quantum information. If any quantum system is considered a kind of quantum computer processing quantum states, then the whole universe might be a huge quantum computer. In the age of mechanism (17th–18th century), Leibniz already explained the universe as mechanical clockwork or a computing machine with causal interactions. The main difference to a quantum universe is the incorporated randomness of quantum events.

There are several suggestions to interpret matter as "condensed" quantum information (cf. chapters by David Finkelstein and von

Weizsäcker *et al.* in von Weizsäcker 1985, 1992). John Wheeler asked if "it is from bit" (Wheeler, 1990). Actually, states of quantum systems can be reduced to alternative yes–no decisions of measurements which determine, e.g., values of location, velocity, or energy of systems. The state space of a quantum system is mathematically a Hilbert space. Space–time, vacuum, mass, energy, and entropy are explained as different forms of quantum information:

In the beginning, there was not matter, but (quantum) information.

Phase transitions of matter during cosmic expansion correspond to information processing with the universe as a giant quantum computer. Konrad Zuse and Jon von Neumann suggested cellular automata to simulate the universe. Cells may be interpreted not only as biological cells, but also as molecules or atoms interacting and generating complex patterns. Cellular automata are mathematically equivalent to Turing machines (Mainzer and Chua, 2011; Wolfram, 1986, 1994; Toffoli, 1977). Therefore, Zuse, von Neumann, and their successors believed in the Turing computability of the universe (Fredkin, 1990). But, they must notice the quantum laws of interacting atoms and elementary particles in the sense of Feynman (1982). Bell's theorem and quantum measurements prove that the ("local") laws of classical physics are not sufficient to simulate a quantum universe. In principle, each piece of matter could be understood as a (quantum) computer processing (quantum) information.

Loss of information by Black Holes?

Contrary to quantum physics, Einstein's relativistic model of the universe is local and deterministic (Weyl, 1923). According to his theory of general relativity, Black Holes are characterized by extreme curvature of space–time which finally concentrates in a singularity of extreme attraction. All matter and even light falling in the singularity is devoured without return.

The question arises whether the structural information of the material systems falling in the Black Holes is also completely destroyed. When a star of sufficient mass balloons out as supernova, explodes, and finally implodes to a Black Hole, all information about this region of the universe must be eliminated in its singularity. Are Black Holes irreversible "holes of memory" in space–time expanding in an aging universe? Will the aging universe suffer from decay and loss of information with increasing number of Black Holes in analogy to Alzheimer's disease of an aging brain (Mainzer, 2000)?

But, according to the laws of quantum physics, quantum information is constantly conserved. Quantum information of a quantum system is completely contained in its wave function. Dynamics of a quantum system is mathematically modeled by a unitary transformation which can be used to reconstruct the initial state of a quantum system from its final state without any loss. Therefore, no quantum information is missed according to the laws of quantum mechanics.

In the deterministic world of Einstein's theory of relativity mass and information are completely eliminated in the singularity of a Black Hole without any chance of reconstruction. But, from the point of view in quantum physics, there cannot be quantum systems with the infinitesimal size of a point because quantum systems depend on the tiny, but finite size of Planck's constant. Therefore, the beginning of the universe cannot be considered a point of singularity in the sense of relativistic cosmic models. Quantum physics demands a tiny initial quantum system of Planck's size expanding into cosmic dimensions. Relativistic models of deterministic physics are only approximations which must be corrected by quantum physics if we approach systems depending on Planck's constant.

Saving quantum information in Black Holes

Stephen Hawking who at first advocated for the information loss in Black holes corrected his original arguments by a quantum physical model (Hawking, 2005, 2009). He considered a quantum vacuum around the Black Hole. In a quantum vacuum, pairs of virtual particles emerge and decay steadily according to Heisenberg's uncertainty principle. A particle of these pairs can fall in the Black Hole and its partner may escape outside. Hawking uses this model to explain the radiation of heat by a Black Hole (Hawking radiation).

The spins of these particles are randomly distributed and do not contain any information about the matter falling in the Black Holes. We register the radiation of escaping particles. But they are independent of their partners falling in the Black Hole. Therefore, in that model, no information can escape from inside to outside of the Black Hole.

A solution is provided by the assumption of entangled quantum states. In the Horowitz–Maldacena model, pairs of particles are entangled (Horowitz and Maldacena, 2004; Ng, 2001). The escaping particle transports not only mass, but also information. The escaping particle is assumed to be entangled with its partner falling in the Black Hole. The information involved in matter can be "beamed" outside of the Black Hole by an EPR correlation. In that way, quantum information can be conserved even in Black Holes according to the laws of quantum physics. But we never could use this information for a direct teleportation.

At the first glance, these results seem to be strange: Stephen Hawking completely deterministic world leads to singularities with complete elimination of information. In the non-deterministic and random quantum world, information is conserved. Laplace's demon of determinism seems to suffer from cosmic Alzheimer's. Only the quantum world never forgets anything, although we cannot directly decode this information.

Encoding of quantum information

For centuries, humans have been searching for sure information transmission. Political and military news, data of banks, insurance companies, and economic markets, intimate and personal messages are decoded in order to save and protect them against undesirable attacks. In the age of globalization, all information is electronically present everywhere. But data security is a sensible part of global information society. There is an armament spiral between those who decode information and those who encode information. When the old decoding procedures are cracked, stronger procedures are developed which can be overcome by even more stronger encoding procedures. So, in the end, there is only practically limited security, but no absolute security of information transmission. But all known decoding procedures are obsolete when a quantum computer is technically realizable.

Classical cryptology

In classical cryptology, the crucial role of randomness for an absolute sure code was already recognized. In 1917, during the First World War, the American mathematician Gilbert S. Vernam (1890–1960) suggested a procedure for decoding military messages. Vernam's principles of decoding are applied even today (Bouwmeester *et al.*, 2000, Section 2.1.2; Wätjen, 2004, p. 30):

The single letters of the alphabet are decoded by sequences of five binary numbers 0 and 1. The reason is that there are $2^5 = 32$ possibilities of 5 tupel which are sufficient to represent the 27 letters of the alphabet. The remaining sequences can be used for decoding other symbols. A message consists of a sequence s of binary numbers with a certain length. According to Vernam, the message should be decoded by a random sequence z of equal length as key. The encoded message $c(s)$ is generated from message s by adding key z to s letter by letter. Addition is defined by the following rules of binary numbers:

$$0 + 0 = 0$$
$$0 + 1 = 1$$
$$1 + 0 = 1$$
$$1 + 1 = 0$$

The receiver of $c(s)$ can encode the decoded message according to the same procedure. We only have to add z to $c(s)$ following the same rules in order to get s back. If, for example, $s = 10010$ and $z = 11011$, we get $c(s) = 01001$. The encoded message is generated from 01001 by adding 11011 in order to get the original message 10010 back. As the key z is random and only the transmitter and receiver are known, an eavesdropper has no chance. Further on, Vernam demands that the random key is only used for one time (one-time pad). Otherwise, a key z would be used for two messages s_1 and s_2. In this case, the decoded messages $c(s_1)$ and $c(s_2)$ could be subtracted from one another in order to eliminate the random sequence z:

$$(s_1 + z) - (s_2 + z) = s_1 - s_2.$$

s_1 could be used as key for s_2 and vice versa. s_1 and s_2 are also meaningful messages and no random sequences which could be found by a sufficiently powerful computer in a list of meaningful combinations.

Vernam's code is an absolutely secure decoding procedure if two problems are solved:

(1) The key must be a guaranteed random sequence. Practically, only pseudo-random sequences are applied. They are generated by computers. They use algorithms which could be discovered by sufficiently powerful computers in order to apply them for encoding.
(2) The key distribution must be absolutely secret. But all procedures of classical information transmission are not secure. Each kind of classical communication can be bugged without being noticed by the transmitter and receiver.

In mathematics, there are decoding procedures which are called "trap doors": It is easy to pass a trap door, but it is difficult to return. In analogy, it is easy to compute the functional value of a trap door function for given functional arguments. But it is very difficult to reconstruct the functional arguments from the computed functional value. For example, it is easy to multiply two arbitrarily large prime numbers p and q and to compute $n = p \cdot q$. Since antiquity, it is well known that each natural number $n \geq 2$ is a prime number or a product of prime numbers. The prime numbers occurring in these products are uniquely determined (e.g., $18 = 2 \cdot 3 \cdot 3$). The task to determine the prime factors of a natural number is known as factorization problem.

But for big numbers the factorization problem has the practical limitation of computation. In algorithmic theory of complexity, the factorization problem is a non-deterministic problem with exponentially increasing computational power. No proof of complexity reduction is known.

Shor's quantum algorithm of the factorization problem

In 1994, Peter Shor found an efficient algorithm of the factorization problem for a quantum computer in polynomial time (Shor, 1997; Ekert and Jozsa, 1996). He applied superimposed states in a giant quantum parallelism which supports many computations simultaneously. In the register of a quantum computer with the size of m quantum bits, the superposition of all 2^m possible m words of quantum bits can be stored. But, in the register of a classical computer with the size of m bits, only one of the 2^m possible m bit words can be stored.

Shor's algorithm used the quantum parallelism to find periods of module functions which lead to the prime factors. For numbers with a length of more than 10^{24} bits, the computation of prime factors is practically excluded for classical computers with serial processing or few parallel procedures. When quantum computers will technically be realized one day, Shor's quantum algorithm will crack any classical decoding procedure.

Quantum cryptology

Quantum cryptology opens new avenues of absolutely secure information exchange because Vernam's criteria for an absolutely secure exchange of an encryption key can be guaranteed in the quantum world:

(1) Guaranteed random sequences must be used as encryption key.
(2) An absolutely secure exchange of the encryption key must be guaranteed.

In 1991, Eckert *et al.* suggested entangled pairs of photons with opposite correlation of polarization (Ekert *et al.*, 2000). In this case, a photon can oscillate vertically (V) or horizontally (H) to the direction of propagation. These opposite states can be considered as quantum

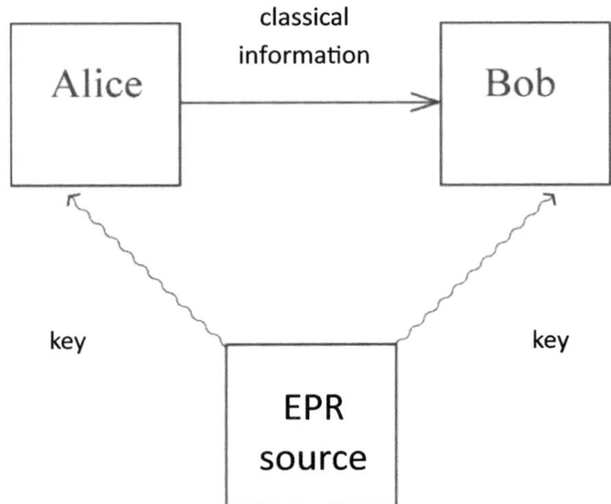

Fig. 41. Random sequence of an EPR source as encryption key.

bits. In an EPR experiment, a sequence of entangled pairs of photons can be sent to a transmitter (Alice) and a receiver (Bob) from a central source (Fig. 41). The pairs of photons are prepared as pairs of correlated quantum bits which are sent to Alice and Bob. How can we get a secret encryption key which can be understood by Alice and Bob only?

Alice and Bob are at distant locations. In an EPR experiment, the local states of Alice's photon and Bob's photon are uncertain as long as no measurement is applied. If Alice measures her photon, it randomly takes one of the two possible states H or V of polarization. At the same moment, Bob's photon has the opposite state of polarization. Therefore, sequential measurements of Alice generate a random sequence of states (e.g., HHVHVVVHHV) corresponding to Bob's sequence with opposite states (e.g., VVHVHHHVVH).

On a public classical information channel, Alice and Bob can communicate that their sequences have opposite polarization. For a common bit sequence, they agree with one another that Alice defines $H = 0$ and $V = 1$ and Bob has the opposite values $H = 1$ and $V = 1$. It follows a common random sequence 0010111001 which is only known by them. According to the laws of quantum physics, this sequence is also random. Vernam's criteria of an absolutely secure exchange of information are perfectly satisfied.

A random sequence of an EPR source provides a secure encryption key to decode and encode messages. Following Vernam, a random sequence is allowed to be used only one time (one-time pad).

With an EPR source, random sequences which are only known to the transmitter and receiver can arbitrarily often be produced. The encryption key is generated randomly by measurements of Alice and Bob without exchange on a classical channel of information. Before the measurement, the encryption key was not determined according to the laws of quantum mechanics.

But how can Alice and Bob be secure that their exchange of quantum information was not bugged? Classically stored information can be read out, copied, and bugged without perturbation. Bugging of quantum transmission needs a measurement which tries to determine the polarization of Alice's photon and to forward it to Bob. If the measuring instrument of the eavesdropper has the same orientation of Alice's instrument, then the correct bit is observed and forwarded to Bob.

But if the measuring instrument has a different orientation, then the measurement effects a change of information about the original state of the photon. If this observed polarization is sent to Bob, he will get a false bit. Contrary to classical information transmission, bugging of quantum information generates failures and traces.

A deep reason is the no-cloning theorem which forbids the existence of a quantum copier (Fig. 38). If the eavesdropper tries to copy transmitted bits, then he changes the original information. Therefore, no random sequence can be reproduced in the quantum world without change.

An eavesdropper could also try to observe the photons of the EPR source before they are received by Alice and Bob. But a previous measurement would cause the collapse of the entangled state which can be determined by measurement (Bell's inequality). When Alice and Bob transmit their encryption key, they can also notice whether they are bugged. For that, it is sufficient to compare few bits of the encryption key in a classical public channel of information.

Quantum complexity and epistemology

Quantum computers would lead to remarkable breakthroughs of information and communication technology. Problems (e.g., the factorization problem) which classically are exponentially complex with practical unsolvability would become polynomially solvable.

In technology, quantum computers would obviously enable an immense increase of problem solving capacity. In complexity theory of computer science, high computational time of certain problems could perhaps be reduced from NP to P. But the question arises if quantum computers can also realize non-algorithmic processes beyond Turing computability.

A quantum computer works according to the laws of quantum physics. In quantum mechanics, the time-dependent dynamics of quantum states is modeled by a deterministic Schrödinger equation. Therefore, the output of quantum states in a quantum computer is uniquely computable for given inputs of quantum states as long as their coherence is not disturbed. A quantum computer can be considered as an algorithm resp. a Turing machine (according to an extended Church's Thesis) working on real and complex numbers in the sense of Chapters 10 and 11.

But are there logical–mathematical breakthroughs that undecidable and unsolvable problems become decidable and computable by quantum computers? Undecidability and unsolvability of problems depend on the logical–mathematical definition of a (real) Turing machine. Therefore, even a quantum computer cannot solve more problems than a (real) Turing machine according to (extended) algorithmic theory of computability (Chapter 10). Algorithmically unsolvable and undecidable problems remain unsolvable even for quantum computers (Deutsch, 1985).

An example is the halting problem for a Turing machine which is not decidable even for a quantum computer. Another problem is the word problem of group theory which demands a test whether two arbitrary words of a symbolic group can be transformed into one another by given transformation rules. These and other problems are not only of academic interest. They are often basic for practical applications (e.g., whether expressions of different languages can be transformed into one another or not).

In the theory of computability, it is proved that there is no general algorithm which can decide everything. Quantum computers will not change anything of these basic results. Even in a civilization with quantum computers, there will be no machine which can solve all problems algorithmically. Gödel's and Turing's logical–mathematical limitations will remain, although there will be a giant increase of computational capacities.

Is the universe digital or real?

The universe is mathematically modeled by equations of quantum and relativistic physics. They are philosophically interpreted as "natural laws" (Feynman, 1967). The emergence of physical particles is mathematically considered as a solution of gauge equations describing the fundamental symmetries of the universe (Bernstein, 1974). If we, for example, try to solve the Schrödinger equation for five or less quarks, there are only two stable possibilities of coordination. They consist of either two up-quarks and one down-quark or two down-quarks and one up-quark. The former cluster is called "proton", the last one "neutron". In the next step we can apply the Schrödinger equation to protons and neutrons as "atomic nuclei". In this case, we get 257 stable possibilities with hydrogen, helium, etc. We can continue the application of Schrödinger equation to molecules, crystal grids, and more complex objects. The names of "protons", "neurons", "atomic nuclei", "hydrogen", "molecules", etc. are, of course, human inventions. The whole information of these objects is involved in the solutions of mathematical equations.

Mathematical equations can be classified in physics (e.g., general/special relativity, electromagnetism, classical mechanics, statistical mechanics, solid state physics, hydrodynamics, thermodynamics, quantum field theory, nuclear physics, elementary particle physics, atomic physics), chemistry, biology, neurobiology, psychology, and sociology with the objects of protons, atoms, cells, organisms, populations, etc. as solutions.

Mathematical structures are systems of objects with (axiomatic) relations and equations. Examples are spaces, manifolds, algebras, rings, groups, fields, etc. Isomorphisms are mappings between structures which conserve their relations. They satisfy the conditions of equivalence relations with reflexivity, symmetry, and transitivity. Therefore, mathematical structures can be classified in equivalence classes. Modern mathematics studies categories of structures with increasing abstraction in the tradition of Dedekind, Hilbert, and Bourbaki. Categories correspond to set-theoretical universes and transfinite cardinal numbers with increasing size. In Chapter 9, we considered the recent univalent foundations project of the Institute for Advanced Study (Princeton) representing and classifying mathematical structures and categories in types of formal systems. Hilbert already tried to extend his axiomatization and classification of mathematical structures to physics (Corry, 2004). Physical structures of space–time, fields, particles, crystals, grids, etc. are only special structures belonging to mathematical equivalence classes and categories.

The question arises whether the universe is finite with only finitely many mathematical structures or at least computable with computable functions and functionals. What about the natural constants which determine fundamental forces of the universe? Mathematically, they are mainly real numbers with infinitely many decimal places, but practically only ca. 16 decimal places are determined by our best measuring instruments and computers.

From a mathematical point of view, physical events are points of a four-dimensional space–time structure which never changes, which was never created, and which will never be destroyed. Space–time is a four dimensional mathematical space with the three usual spatial dimensions of length width, and height and the fourth dimension of time (Audretsch and Mainzer, 1994). Human intuition of space is restricted to three dimensions. In the three-dimensional space of length, width, and height, the orbit of the moon round the Earth

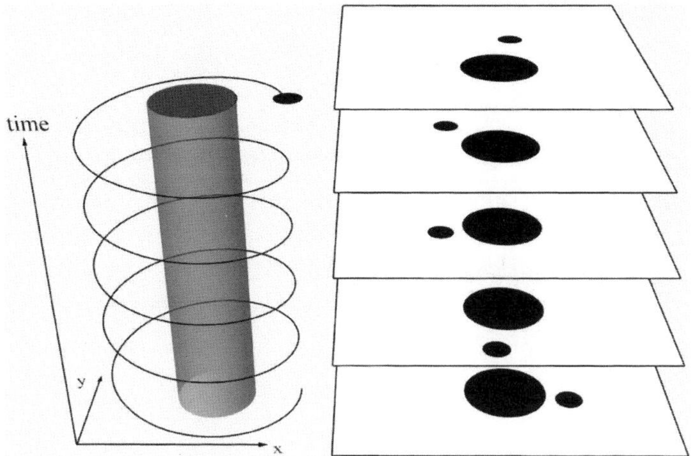

Fig. 42. Orbit of the moon round the Earth as spiral in a classical space–time.

can be illustrated by an (approximated) circle. In this geometrical representation, time dependence can only be considered by the motion of the moon round the Earth. In order to illustrate the dimension of time, the four-dimensional space–time is reduced to the vertical time axis and two spatial coordinates. In this case, the circle is replaced by a spiral along the time axis (Fig. 42). Actually, there is no change and motion of the moon, but only the unchangeable spiral in space–time (Tegmark, 2014, p. 397).

In a three-dimensional illustration, space–time is distinguished in parallel layers of two-dimensional planes each of which represents the three-dimensional space at this moment with all simultaneous events. If an observer belongs to one of these layers, then this layer represents his present time. The layers above his present layer are his future, and below the layer, there is his past. Anyway, the distinction of future, present, and past is relative to an observer or, in the words of Einstein, a "subjective illusion". From a physical point of view, there is no time-dependent change, but only the mathematical structure of space–time with given unchangeable patterns (e.g., spiral of the moving moon).

The mathematical structure of four-dimensional space–time differs in special and general physics (Penrose, 1966). In special relativity, space–time is not Euclidean like in classical physics, but

Minkowskian with respect to relativistic simultaneity. In general relativity, we have to consider a four-dimensional pseudo-Riemannian manifold of curved space–time. A curved (non-Euclidean) Riemannian manifold (e.g., the curved surface of a sphere) consists of infinitesimally small Euclidean planes. In analogy to Euclidean geometry, Minkowskian geometry of special relativity is embedded as local space–time into the global general relativity of curved space–time.

If Einstein's relativistic space–time is combined with quantum physics, the deterministic world is given up. In quantum field physics, we have to consider possible worlds with different pasts, presents, and futures which are weighted by probabilities. The path of a quantum system (e.g., elementary particle) from one point to another cannot be determined by an orbit like a ball in classical mechanics. We have to consider all possible paths with their weights which are summed up in Feynman's path integral. Therefore the initial tiny quantum system of the early universe in the size of Planck's constant could not develop in a classically well-determined way. There are several possible paths of development according to the quantum laws. If we assume that all these possibilities are real in parallel, then we get the concept of a multiverse with quantum parallel universes (Everett, 1957). We cannot get observational data from these parallel worlds because they are separated from our world by different ways of inflation which blow up the universes in different ways. Nevertheless, they can be mathematically modeled by the laws of quantum physics.

In all these cases, there are given mathematical structures (Ludwig, 1978). The development of a quantum system is mathematically given by a pattern in quantum space–time (like the moving moon as spiral in the Euclidean space–time). An organism like our body consists of a bundle of world lines representing the development of elementary particles which are organized as atoms, molecules, cells, organs, etc. They converge during our growth as babies, develop during the period as adults in different ways (e.g., changing by diseases), but diverge again after our death (Fig. 43, cf. Tegmark, 2014, p. 409). Thus, the human body is a complex pattern of converging and diverging word lines in space–time. From a mathematical point of view, the pattern of our body is part of a

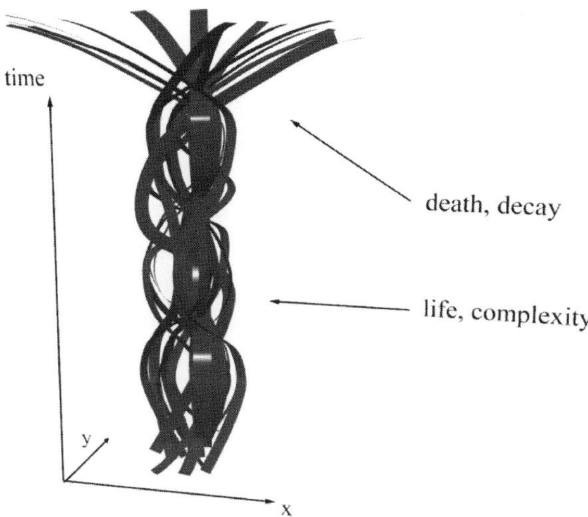

Fig. 43. Human life as a complex mathematical pattern in space–time.

given and unchangeable mathematical structure of space–time. It is a solution of a mathematical equation characterizing this mathematical structure. Only as living bodies we have the "illusion" of change and development with birth and death.

Are these space–time structures continuous or discrete? In classical mechanics, fluids and materials are considered as dynamical systems which are modeled by continuous differential equations. In electrodynamics, each point in magnetic and electric fields is characterized by real numbers of orientation, voltage, etc. Maxwell's equations describe their continuous time-dependent development. In quantum physics, classical electrodynamics is replaced by quantum field theory as foundation of modern particle physics. The quantum wave function provides the probability of a particle in a certain area of space–time. Mathematically, the wave function is an abstract point in the Hilbert space which is an (infinite) mathematical structure. In analogy to fields of photons (light) in quantum electrodynamics, we can also consider quantum field theories of all the other known elementary particles. There are electronic fields, quark fields, etc. But, we still miss a complete "theory of everything" unifying all fundamental forces and their interacting particles in relativistic and quantum physics. The main obstacle stone is the continuous and deterministic structure of relativistic physics contrary to the discrete and probabilistic character of quantum physics. The physical structure of a complete "Theory of everything" would be isomorphic to some mathematical structure which would contain all information of the universe. In that sense, the universe would be completely founded in that mathematical structure. The universe would even be equal to this particular mathematical structure up to equivalence and isomorphism.

But does mathematical infinity have a physical reality? Is being infinitesimally small in the continuum physically real? Fundamental equations of quantum physics (e.g., Schrödinger equation) are continuous differential equations. A quantum system is defined by a Hilbert space which is an infinite mathematical object. Dirac's delta function is a generalized function (distribution) on the real continuum that is zero everywhere except at zero, with an integral of one over the entire continuum. It is illustrated as a graph which is an infinitely high, infinitely thin spike at the origin, with total area of one under the spike. Physically, it is assumed to represent the density of an idealized point mass or point charge. Natural constants of quantum physics are real numbers with infinitely many decimal places which are non-computable, but measurable. According to Emmy Noether, fundamental conservation laws of physics correspond to continuous symmetries of space–time. For example, conservation of energy is equivalent to translation symmetry of time, conservation of momentum to translation symmetry of space, conservation of angular momentum to rotational symmetry, etc. (Mainzer, 1988; Doncel *et al.*, 1987).

Is the continuum with real numbers only a convenient approximation to compute the coarse molecular and atomic world of fluids, materials, and fields with elegant continuous procedures? Actually, the continuum is not only used in classical physics, but also in the quantum world. In quantum field theory, we use quantum fields with discrete computable solutions of crystal grids and emergent particles. Continuous space–time of the universe is assumed to become discrete in the Planck scaling of elementary particles. Information flow in dynamical systems is measured by discretization in bit sequences (e.g., Poincaré cuts in Chapter 14) the complexity of which can be measured by their shortest algorithmic description as a computer program (Chapter 13).

Finite material structures in nature are isomorphic to equivalent finite mathematical structures (e.g., molecular crystals correspond to geometrical grids with group-theoretical symmetries) which can be uniquely encoded by natural numbers. Finite code numbers can be enumerated according to their size. Thus, in principle, all finite structures of the universe can be enumerated. Computable structures can be represented by Turing machines which uniquely correspond to machine numbers as encoding numbers of a Turing program (Chapter 2). Thus, all computable structures of the universe can also be enumerated.

In physics, infinite structures (e.g., space–time, fields, wave functions, Hilbert spaces) are practically handled by software programs (e.g., Wolfram's MATHEMATICA) which can compute arbitrarily large numbers (Wolfram, 2005). Thus, infinite structures are approximated by finite structures which can be processed by computers. But, do we actually need the continuum and infinity beyond Turing complexity in physics? Natural laws are mathematical equations. In reverse mathematics (Chapter 8), we asked for the constructivity of mathematical theorems (e.g., Bolzano–Weierstrass theorem in real analysis). Therefore, we could extend constructive reverse mathematics to physics and try to embed the equations of natural laws into subclasses of arithmetic $\mathcal{Z}_2$ of second order. Are there constructive principles of $\mathcal{Z}_2$ which are equivalent to the equations of natural laws (e.g., Newton's second law of mechanics, Schrödinger's equation)?

Most natural laws which are formulated in real analysis can be reduced to the axiom of arithmetical comprehension CA^0_{ar} in Weyl's and Lorenzen's constructive analysis (Chapter 8). In general, no physical experiment is known which depends on the complete real continuum with non-computable real numbers. In this case, the universe would consist of constructive mathematical structures which are classified in a hierarchy according to their degrees of constructivity. Only an extreme Platonism would believe that all mathematical structures are existent with the subclass of those structures which are isomorphic to physical structures of the universe (Tegmark, 2014). This ontological position cannot be excluded in principle. But in the sense of Ockham's razor, it makes highly abstract assumptions which are not necessary for solving physical problems and which can be avoided by constructive concepts. Modern physical research with its huge amount of data (e.g., high energy and particle physics of accelerators) strongly depends on algorithms and software which must detect correlations and patterns of data. Therefore, constructive proof mining (Chapter 7) should also be extended to physics in order to extract computational procedures and computer programs from proofs in theoretical physics.

If quantum physics is assumed as a general theoretical framework, then nature in Planck's scaling must be discrete. Nevertheless, quantum physicists use infinite and continuous concepts of real and complex numbers in their theories. The standard model of particle physics refers to 32 parameters with real constants. An example is the fine-structure constant $1/\alpha = 137.035999\ldots$.. Only finitely many decimal places are known with respect to contemporary capacities of measurement. From a strictly digital point of view, all infinitely many decimal places had to be input into a computer step by step for computing with real numbers which is impossible. But, physicists compute with these parameters as "whole" quantities with algebraic representations, e.g., α^{-1} with $\alpha = \frac{e^2}{q_P^2}$, e elementary charge, and q_P Planck charge. Thus, real computing, which was introduced in Chapter 10, is an adequate representation of physical research. In real computing, we even can distinguish degrees of computational complexity which were proved in Chapter 11.

Are there "cosmic oracles"?

In real computing (with real numbers), the physical question is still open whether real constants (e.g., the fine-structure constant α^{-1}) or Dirac function correspond to physically "real" entities. Otherwise, they are only approximate terms of a mathematical model. If there are non-computable real numbers (as natural constants of physical theories) which

- definitely cannot be reduced to constructive procedures,
- definitely influence physical experiments,

then the unbelievable happens: Non-computable real numbers would not only be an invention of human mind, but a physical reality. According to Hilbert's classical criterion, the existence of a mathematical object is guaranteed by its consistency in an axiomatic system. But, in constructive mathematics, we need an effective procedure to construct an object and to prove the existential statement. Further on, in physics, the object must be measurable in an experiment or observation (at least approximately exact).

The discovery of non-computable objects would be as surprising as the discovery of incommensurable relations in antiquity. The Pythagoreans were not only horrified by the discovery of incommensurable mathematical relations (Mainzer, 1980). In antiquity, Euclidean geometry was ontologically identified with really existing objects and relations. Therefore, incommensurability was historically understood as loss of harmony and rational proportionality in the real world. But, incommensurable relations in antiquity correspond to irrational numbers which are computable like, e.g., $\sqrt{2}$. Non-computability in nature would be more dramatic. In this case, there would be not only Black Holes in the universe (which are computable in quantum physics), but also "cosmic oracles" all computational models of the universe would depend on.

Since the beginning of modern physics, physicists and philosophers were fascinated by the mathematical simplicity and elegance of natural laws (Mainzer, 2005). For example, the standard model is reduced to the symmetry group $SU(3) \times SU(2) \times U(1)$. Therefore, it is often argued that physics is finally founded in mathematical symmetries which remind us of the old Platonic idea of regular bodies as building blocks of the universe (Doncel *et al.*, 1987; Mainzer, 1988; Weyl, 1931). Natural laws formalize regularities and correlations of physical quantities. In a huge amount of observational data, laws provide a shorter description of the world.

Fig. 44. Complexity of patterns (e.g., binary numbers of $\sqrt{2}$) and their parts.

In Chapter 12, algorithmic information of data was introduced as the bit length of their shortest description. In this sense, natural laws are a shorter description of data patterns. In order to illustrate the relation of simple laws and complex data, we consider a mathematical example in Fig. 44 (Tegmark, 2014, p. 491): The left pattern is very complex and represents the decimal places of $\sqrt{2}$ as binary numbers 0 and 1 which are illustrated by white and black squares. It is assumed (but still not rigorously proved) that the decimal places of $\sqrt{2}$ behave like random numbers with all kinds of patterns. Nevertheless, there is a finite computer program to compute $\sqrt{2}$ which only needs perhaps 100 bits. This computer program corresponds to a short (computable) law which covers an unlimited huge amount of data. The description of the middle partial pattern in Fig. 44 is even more complex than the whole left pattern because it needs 14 additional bits to locate the beginning bits of the partial pattern in the whole pattern. The right pattern only needs nine bits. There would be no advantage to locate this pattern by additional bits in the whole distribution.

These examples illustrate that general laws may be short and simple to describe the whole universe. But special conditions and individual situations may be more complex because we need additional information to locate them in the distribution of the whole. In relativistic and quantum physics, individual entities from atoms and molecules up to organisms are represented by more or less complex patterns of world lines in space–time. Thus, in the whole distribution of patterns, individual parts must be located which needs additional information: Therefore, the description of individual lives and persons may be more complex than the general laws of life and even the whole universe!

In order to predict future events, we need not only laws, but also initial conditions. Sometimes, initial conditions seem to be random contrary to regular laws. But from a universal point of view, initial conditions are not random. They must be located in the whole data pattern of the universe and need more information (bits). In this case, the algorithmic information of initial conditions and individual events is larger than the algorithmic length of the applied laws.

Chapter 16

Digital and Real Computing
in the Social World

Why is it so difficult to mathematize the social world?

In 1960, the Nobel Prize winner of physics Eugene Wigner published a famous paper about "The unreasonable effectiveness of mathematics in the natural sciences" (Wigner, 1960). Wigner was fascinated by the fundamental role of mathematical symmetries in physics. But he was puzzled why the mathematical formalism with its abstract entities (e.g., real and complex numbers) fits physical reality so effectively. For Wigner, "the miracle of the appropriateness of the language of mathematics for the formulation of the laws of physics is a wonderful gift which we neither understand nor deserve". For a Platonist, mathematics is true reality and mathematical structures fitting physics are only a subset. Thus, for a Platonist, the appropriateness of mathematics in the physical world is no wonder.

From a constructive point of view, existence in mathematics can only be guaranteed by constructive procedures (Chapters 4 and 5). Following this line, the mathematical formalism in physics should be reduced to constructive mathematics. Elementary mathematics in arithmetic and geometry fits physical reality because integers and geometrical forms can directly be abstracted from physical objects. In more abstract mathematical theories, we should prefer Ockham's razor and only use abstract assumptions as far as they are necessary to solve problems and do experiments. Their degrees of abstractness and constructivity can be determined by proofs of equivalence with axioms and principles in reverse mathematics (Chapter 8). From this point of view, the appropriateness of mathematics in the physical world is also no wonder. But, it does not need the belief in a Platonic world which cannot be proved constructively in mathematics and which cannot be tested experimentally in physics.

Modern economics and social sciences also apply mathematical models more and more, but, nevertheless, we doubt in the computability of the social world. The reason is obvious: In physical and engineering sciences, mathematics guarantees reliable procedures and exact predictions with respect to measuring conditions. But, even highly mathematized economic theories mainly miss reliable predictions of economic crises. Mathematical theories are formal and can be interpreted in physical as well as economic ways. What are the differences in mathematizing the natural and social worlds (Mainzer, 2014a,b)?

In physics, the fundamental mathematical laws are characterized by natural constants (e.g., Einstein's velocity of light c, Planck's constant h, constant of gravity G). Natural constants are invariant measuring quantities in the mathematical equations of natural laws which do not change in space and time. Their change would change the universe. They are deeply connected with laws of symmetry and invariance. Their mathematical structures explain the fundamental forces of our universe as well as applications in engineering sciences (Feynman, 1967).

Contrary to physical sciences, economics and social sciences do not know universal constants and laws of invariance. Their mathematical models remind us of mathematical astronomy before Galilei and Newton. Ptolemy's astronomy used a mathematically consistent formalism of epicycles and other kinematic assumptions which were adapted to new observations. In the Middle Ages, astronomers had to handle a highly complicated formalism of assumptions and procedures. They could only explain celestial events after their observations by *ad hoc* hypotheses, but they could not predict new events. The situation reminds us of the financial crisis in 2008 when economists tried to explain the event afterwards without being able to predict and to avoid it. In physics, the situation changed with the mathematical laws of Galilei and Newton: Galilei's law of free fall can predict future events, because it depends on the natural constant of gravity which can be explained by Newton's theory of gravity. Is economics still in a stage of development like physics before Galilei and Newton? Must we still wait for a "Newton of economics" in analogy to a "Newton of a blade of grass" who was missed by Kant in the end of the 18th century in order to explain life mathematically and experimentally?

The mathematization of science began with simple models of astronomy and physics and was extended to chemistry and biology. Since its very beginning with only a few parameters, mathematization of science is characterized by increasing complexity — from few planets in early astronomy to billions of molecules and cells interacting in organisms. A huge amount of data is necessary to model interacting people in populations, institutions, or at the markets. Further on, there is an important effect of theories in the social sciences which is not known in the natural sciences. Theories and their predictions can change human behavior: If an economic theory predicts an event with a certain probability, then humans can bet against the main stream speculating on high returns or they can even change their behavior in order to prevent the predicted event. The behavior of atoms and cells is completely independent of their predictions, humans are involved in their predicted behavior.

Besides complexity, we have to consider the factor of time in theories. Physical, chemical, biological, psychological, economic, and sociological theories operate on different scales of time. Scaling of time is also important for natural constants in physics. They are no "eternal" quantities. The natural constant with the longest period is Planck's constant which has been assumed since the beginning of the quantum world. The constant of gravity emerged when, just after the Big Bang, gravity was separated from the unified force in the early stage of quantum world. The fine-structure constant emerged later, when the electromagnetic force was separated from weak force in cosmic expansion.

Even in physics, it is not excluded that the natural constants can change on a long time scaling. Biological laws on Earth began with the evolution of life. Behavioral rules of humans exist since a few million years, our civilizations with their social and legal laws only since some thousands of years. Therefore, mathematical models of these different applications must consider different scales of time. We cannot expect "natural constants" of economic laws like in physics because time scaling in economy is much shorter than in nature. But we need much more data, measurements, and algorithms to construct mathematical models and laws of economy with increasing predictive and explanatory power.

Again, from a constructive point of view, reliability and consistency can only be guaranteed by constructive procedures. Following this line, the mathematical formalism in economics should also be reduced to constructive mathematics. Simple models of human behavior fit social reality because they can intuitively be illustrated and tested by experience. In more abstract mathematical models, we should, again, prefer Ockham's razor and only use abstract assumptions as far as they are necessary to explain and predict economic events. Their degrees of constructivity can be determined by proofs of equivalence with axioms and principles in reverse mathematics. As financial theory is the best mathematized economic theory, we start with a historical and systematic introduction to classical models of financial mathematics. In the next step, constructive procedures are extracted and equivalence with axioms and principles in reverse mathematics is proved.

Beginning of insurance mathematics: Poisson distribution of risks

Mathematical modeling in finance and insurance can be traced back centuries. Insurance of risks against the chances of life is an old topic of mankind (Mainzer, 2007b). Commercial insurance dates back to Renaissance, when great cities of trading introduced bets on safe routes of ships. In the 17th century, the great British insurance company Lloyd arose from this system of bookmakers. The philosopher and mathematician Gottfried Wilhelm Leibniz (1646–1716) already suggested health insurance in which people should pay with respect to their income. In Germany, the ingenious idea of Leibniz was realized not earlier than in the 19th century by Chancellor Bismarck. In the time of Leibniz, life insurances were first applications of probability calculations.

Historical excursion: Huygens and insurances in the 17th century

The Dutch physicist Christiaan Huygens (1629–1695) applied the law of large numbers to calculations of insurance rates. In his approach, insurance is considered as a game between the insurer and clients. The insurer diminishes his risk by adapting the premium payed by a client. Let $c_1, \ldots, c_n$ be the costs of the insurer and $p_1, \ldots, p_n$ the probabilities that the damages happen. The expected damage of the insurer is assumed to be $p_1 c_1 + \cdots + p_n c_n$. The average gain is equal to the premium Q paid by the clients. His risk is zero for a premium $Q = p_1 c_1 + \cdots + p_n c_n$. The risk of clients is also zero, their loss Q and the expected gain $p_1 c_1 + \cdots + p_n c_n$. In this case, Q is called a fair premium to be paid by clients. It is assumed that the probabilities $p_1, \ldots, p_n$ can be estimated according to the law of large numbers. But this assumption was the flaw of Huygens's approach. The law of large numbers cannot be applied in cases of rare damages with extreme costs.

In 1898, the Russian economist and statistician Ladislaus Josephovich Bortkiewicz (1868–1931) published a book about the Poisson distribution, titled *The Law of Small Numbers*. In this book, he first noted that events with low frequency in a large population follow a Poisson distribution even when the probabilities of the events varied. Modern insurance mathematics started with the thesis of the Swedish mathematician Filip Lundberg (1876–1965). He introduced the collective risk model for insurance claim data. Lundberg showed that the homogeneous Poisson process, after a suitable time transformation, is the key model for insurance liability data. Risk theory deals with the modeling of claims that arrive in an insurance business and which gives advice on how much premium has to be charged in order to avoid ruin of the insurance company.

Lundberg started with a simple model describing the basic dynamics of a homogeneous insurance portfolio.

Lundberg's model of a homogeneous insurance portfolio

This means a portfolio of contracts for similar risks (e.g., car or household insurance) under three assumptions (Embrechts *et al.*, 2003, p. 9):

- Claims happen at time T_i satisfying $0 \leq T_1 \leq T_2 \leq T_3 \leq \ldots$ which are called claim arrivals.
- The ith claim arriving at time T_i causes the claim size. The sequence (T_i) is an iid sequence of non-negative random variables.
- The claim size process (X_i) and the claim arrival process (T_i) are mutually independent.

Definition of a risk process of insurance. According to Lundberg's model, the risk process $U(t)$ of an insurance company is determined by the initial capital u, the loaded premium rate c, and the total claim amount $S(t)$ of claims X_i with

$$U(t) = u + ct - S(t) \quad \text{and} \quad S(t) = \sum_{i=1}^{N(t)} X_i (t \geq 0).$$

$N(t)$ is the number of the claims that occur until time t.

Lundberg assumed that $N(t)$ is a homogeneous Poisson process, independent of (X_i). Figure 45 illustrates a realization of the risk process $U(t)$.

Lundberg's model is fine for small claims. But the question arises how the global behaviour of $U(t)$ is influenced by individual extreme events with large claims. Under Lundberg's condition of small claims, Harald Cramér estimated bounds for the ruin probability of an insurance company which are exponential in the initial capital u (Cramér, 1969). Actually, claims are mostly modeled by heavy-tailed distributions like, e.g., Pareto which are much heavier than exponential distribution. Figure 46(a) illustrates a computer simulation

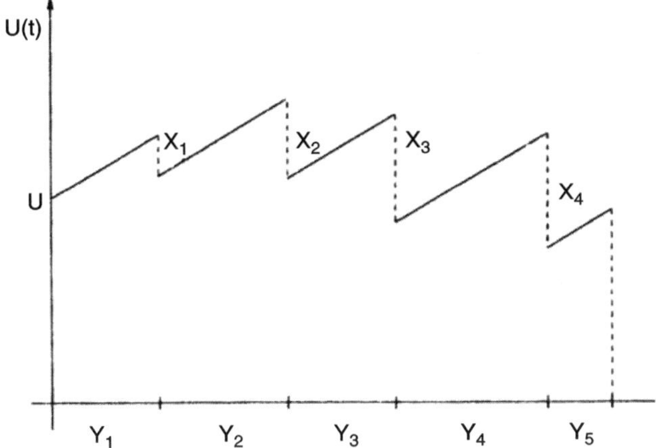

Fig. 45. A realization of Lundberg's risk process.

following Lundberg's classical risk theory for small claims according to the Cramér–Lundberg condition with $U(t)$ for exponentially distributed claims (Embrechts *et al.*, 2003, p. 544). Figure 46(b) does not satisfy the Cramér–Lundberg condition with $U(t)$ for Pareto-distributed claims (Embrechts *et al.*, 2003, p. 546).

Beginning of financial mathematics: Gaussian distribution of risks

With the up-coming stock markets during the period of industrialization, people became more and more interested in their risky dynamics. Asset price dynamics are assumed to be stochastic processes. An early key concept to understand stochastic processes was the random walk. The first theoretical description of a random walk in the natural sciences was performed in 1905 by Einstein's analysis of molecular interactions. But the first mathematization of a random walk was realized not in physics, but in social sciences by the French mathematician Louis Jean Bachelier (1870–1946). In 1900, he published his doctoral thesis with the title "Théorie de la Spéculation" (Bachelier, 1900). During that time, most market analyses looked at stock and bond prices in a causal way: Something happens as a cause and prices react as an effect. In complex

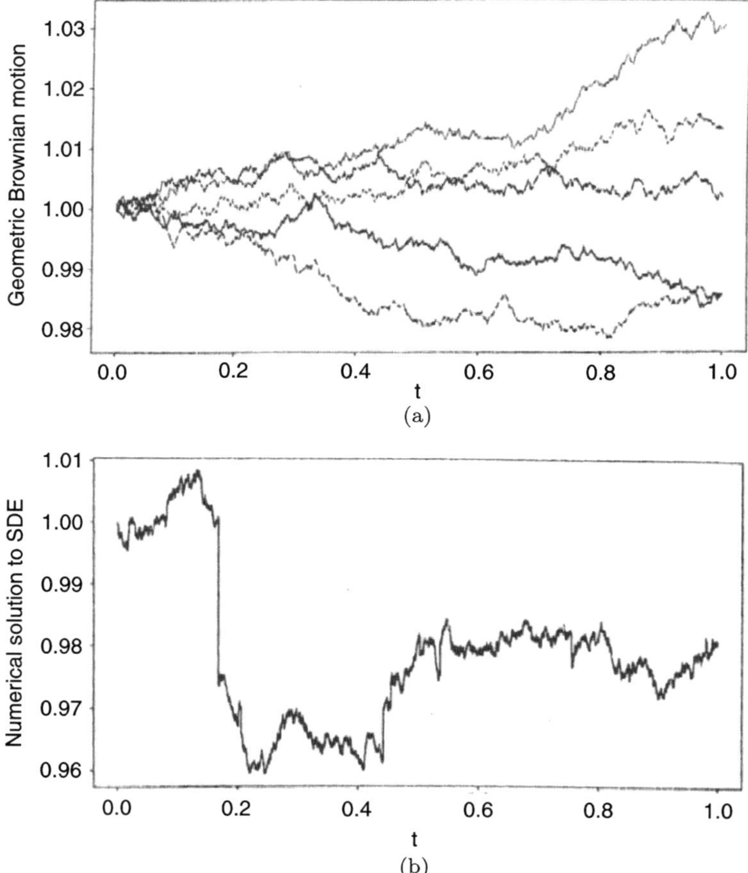

Fig. 46. Risk process $U(t)$ (a) for exponentially distributed claims and (b) for Pareto distributed claims.

markets with thousands of actions and reactions, a causal analysis is even difficult to work out afterwards, but impossible to forecast beforehand. One can never know everything. Instead, Bachelier tried to estimate the odds that prices will move. He was inspired by an analogy between the diffusion of heat through a substance and how a bond price wanders up and down. In his view, both are processes that cannot be forecast precisely. At the level of particles in matter or of individuals in markets, the details are too complicated. One can never analyze exactly how every relevant factor interrelates to spread energy or to energize spreads. But in both fields, the broad pattern of probability describing the whole system can be seen.

Bachelier introduced a stochastic model by looking at the bond market as a fair game. In tossing a coin, each time one tosses the coin, the odds of heads or tails remain 1:2, regardless of what happened on the prior toss. In that sense, tossing coins is said to have no memory. Even during long runs of heads or tails, at each toss, the run is as likely to end as it is to continue. In the thick of the trading, price changes can certainly look that way. Bachelier assumed that the market had already taken account of all relevant information, and that prices were in equilibrium with supply matched to demand, and seller paired with buyer. Unless some new information came along to change that balance, one would have no reason to expect any change in price. The next move would as likely be up as down.

Actually, prices follow a random walk. Imagine a blind drunk staggering across an open field. How far will he have gotten after some time? He could go one step left, two steps right, three backwards, and so on in an aimless path. On average, just as in tossing coins, he gets nowhere. On the average, his random walk will be forever stuck at his starting point. In the same way, the prices on markets can go up or down, by big or small increments. With no new information to push a price in one direction or another, a price on average will fluctuate around its starting point. In that case, the best forecast is the price today. Each variation in price is unrelated to the last. In a stochastic model, the price changes form a sequence of independent and identically distributed random variables. In that case, a chart of changes in price from moment to moment illustrates a more or less uniform distribution over time. The size of most price changes varies within a narrow range. There are also bigger fluctuations. But they barely stand up from the bulk of changes, as some outliers of grass rise above the average height of an unmowed lawn, in that most of the blades of grass fall within a narrow range of heights, while a minority rise above this range (Mainzer, 2007b; Mandelbrot and Hudson, 2004).

In order to illustrate this smooth distribution, Bachelier plotted all of a bond's price changes over a month or year onto a graph. In the case of independent and identically distributed price changes, they spread out in the well-known bell-curve shape of a normal ("Gaussian") distribution: the many small changes clustered in the center of the bell, and the few big changes at the edges. Bachelier assumed that price changes behave like the random walk of molecules in Brownian motion. Long before Bachelier and Einstein, the Scottish botanist Robert Brown had studied the way tiny pollen grains jiggled about in a sample of water. Einstein explained it by molecular interactions and developed equations very similar to Bachelier's equation of bond-price probability, although Einstein never knew that. It is a remarkable interdisciplinary coincidence that the movement of security prices, the motion of molecules, and the diffusion of heat are described by mathematically analogous models.

Bachelier's hypotheses of price changes.

(1) statistical independence (i.e., each change in price appears independently from the last),
(2) statistical stationarity of price changes,
(3) normal distribution (i.e., price changes follow the proportions of the Gaussian bell curve).

Models of efficient markets and computable risks

It took a long time for economists to recognize the practical virtues of describing markets by the laws of chance and Brownian motion (Mainzer, 2007b; Mandelbrot and Hudson, 2004). In 1956, Bachelier's idea of a fair game was used by Paul A. Samuelson and his school to formulate the "Efficient Markets Hypothesis". They argued that in an ideal market, security prices fully reflect all relevant information. A financial market is a fair game in which buyer balances seller. By reading price charts, analyzing public information, and acting on inside information, the market quickly discounts the new information that results. Prices rise or fall to reach a new equilibrium of buyer and seller. The next price change is, once again, as likely to be up as down. So, one can expect to win half the time and loose half the time. If one has special insights into a stock, one could profit from being the first in the market to act on it. But one cannot be sure to be right or first because there are many clever people in a market as intelligent as oneself.

Since Samuelson, Bachelier's theory was elaborated into a mature theory of how prices vary and how markets work. It was more important for the financial world that the theory has been translated into practical tools of finance. In the 1950s, Markowitz (1959) was inspired by Bachelier to introduce "Modern Portfolio Theory" (MPT) as a method for selecting investments. In the early 1960s, Sharpe (1964) devised a method of valuing an asset, called Capital Asset Pricing Method (CAPM). A third tool is the Black–Scholes formula for valuing options contracts and assessing risk. Its inventors were Black and Scholes (1973) in the early 1970s. These three innovations, CAPM, MPT, and Black–Scholes Formula, are still the fundamental tools of classical financial theory until today, resting on Bachelier's hypotheses of financial markets.

Black–Scholes conditions of financial markets

The Black-Scholes formula tries to implement risk-free portfolios. Black and Scholes assumed several conditions of financial markets.

Condition of financial markets (Black–Scholes)

(1) The change of price $Y(t)$ at each step t can be described by the stochastic differential equation of a Brownian motion. This assumption implies that the changes in the (logarithm of) price are Gaussian distributed.
(2) Security trading is continuous.
(3) Selling of securities is possible at any time.
(4) There are no transaction costs.
(5) The market interest rate r is constant.
(6) There are no dividends between $t = 0$ and $t = T$.
(7) There are no arbitrage opportunities.

Arbitrage is a key concept for the understanding of markets. It means the purchase and sale of the same or equivalent security in order to profit from price discrepancies. A stock may be traded in two different stock exchanges in two different countries with different currencies. By buying several shares of the stock in New York and selling them in Frankfurt, the arbitrager makes a profit apart from the transaction costs (Williams, 1991).

Definition of arbitrage. Arbitrage is a strategy of taking riskless gains of a price difference between several markets: It must not cost anything, does not produce any losses, and even a strict gain for at least one possible scenario.

If the market prices do not allow for profitable arbitrage, the prices are said to be in an arbitrage equilibrium (arbitrage-free market).

Traders looking for arbitrage opportunities contribute to a market's ability to evolve the most rational price for a good. The reason is obvious: if someone has discovered an arbitrage opportunity and succeeded in making a profit, he will repeat the same action. After carrying out this action repeatedly and systemically for several opportunities, the prices will be adapted and no longer provide arbitrage opportunities. In short: New arbitrage opportunities continually appear in markets. But as soon as they are discovered, the market moves in a direction to eliminate them gradually (Mandelbrot and Hudson, 2004).

Now, in the absence of arbitrage opportunities, the change in the value of a portfolio must equal the gain obtained by investing the same amount of money in a riskless security providing a return per unit of time. This assumption allows us to derive the Black–Scholes partial differential equation which is valid for both call and put European options. Under some boundary conditions and substitutions the Black–Scholes partial differential equation becomes formally equivalent to the heat-transfer equation of physics which is analytically solvable.

Fundamental theorem of asset pricing. Intuitively, arbitrage-free markets are fair. The assumption of an arbitrage-free market leads to a fundamental theorem of asset pricing (FTAP) (Pascucci, 2011; Delbaen and Schachermayer, 1994): In arbitrage-free ("fair") markets, the "fair" prices are given by expectations under equivalent martingale measures. A martingale can be illustrated by a fair game where knowledge of past events never helps predict the mean of the future winnings. In probability theory, a martingale is a sequence of random variables (stochastic process) for which, at a particular time of the sequence, the expectations of the next value in the sequence is equal to the present observed value with knowledge of all prior observed values.

Definition of a discrete-time martingale. A discrete-time martingale is a discrete-time stochastic process $X_1, X_2, X_3, \ldots$ that satisfies for any time n

$$E(|X_n| < \infty,$$

$$E(X_{n+1}|X_1, \ldots, X_n) = X_n.$$

Due to the linearity of expectations, we get the following equivalences to the second requirement

$$E(X_{n+1}|X_1, \ldots, X_n) = X_n \Leftrightarrow E(X_{n+1} - X_n|X_1, \ldots, X_n) = 0$$

$$\Leftrightarrow E(X_{n+1}|X_1, \ldots, X_n) - X_n = 0$$

which explains that the average winnings from observation n to observation $n + 1$ is 0.

In short, FTAP states that there does not exist an arbitrage strategy (arbitrage-free market) iff there exists an equivalent martingale measure. In order to prove the fundamental theorem, we must formalize the used financial terms and concepts (Föllmer and Schied, 2004):

For asset pricing, we consider a portfolio of m stocks and one bond. The prices of the assets at present time 0 are known. They are denoted by $\pi = (\pi_1, \ldots, \pi_m, \pi_{m+1}) \in \mathbb{R}^{m+1}$. The prices of the stocks at future time 1 are unknown, but there are n possible developments with known prices. The prices of the stocks are positive. The price of the bond is always 1 (also at time 1). The price of the ith asset in case j is denoted by c_{ij} with $C = (c_{ij})$. The first coordinate relates to the bond with $\pi_1 = 1$ and $c_{1j} = 1$ for all $j = 1, \ldots, n$.

We set

$$\mathcal{Y}_n := \left\{ (x_1, \ldots, x_n) \in \mathbb{R}^n \,\middle|\, \sum_{i=1}^{n} x_i > 0 \text{ and } 0 \leq x_i \text{ for all } i \right\},$$

$$\mathcal{X}_n := \left\{ (x_1, \ldots, x_n) \in \mathbb{R}^n \,\middle|\, \sum_{i=1}^{n} x_i = 0 \text{ and } 0 \leq x_i \text{ for all } i \right\},$$

$$\mathcal{P}_n := \left\{ (x_1, \ldots, x_n) \in \mathbb{R}^n \,\middle|\, \sum_{i=1}^{n} x_i = 1 \text{ and } 0 < x_i \text{ for all } i \right\}.$$

Every $\xi = (\xi_1, \ldots, \xi_m, \xi_{m+1}) \in \mathbb{R}^{m+1}$ corresponds to a trading strategy, where $\xi_i (i = 1, \ldots, m, m+1)$ denotes the amount of shares of asset i that the trader buys.

The price of ξ is $\xi \cdot \pi$.

The payoff at time 1 over all possible future scenarios is $\xi \cdot C$.

Arbitrage strategies are trading strategies corresponding to riskless gains, since they do not cost anything ($\xi \cdot \pi \leq 0$), do not produce any losses, and even a strict gain for at least one possible scenario ($\xi \cdot C \in \mathcal{Y}_n$).

A vector $p \in \mathcal{P}_n$ is called equivalent martingale measure if $C \cdot p = \pi$.

A vector $\xi \in \mathbb{R}^{m+1}$ is called arbitrage strategy if $\xi \cdot \pi \leq 0$ and $\xi \cdot C \in \mathcal{Y}_n$.

Lemma. *There exists an arbitrage strategy ξ $\Leftrightarrow$ There exists a vector $\mu \in \mathbb{R}^m$ such that*

$$\mu \cdot (C - \pi) \in \mathcal{Y}_n$$

with

$$C - \pi := (c_{ij} - \pi_i)_{i=2,\ldots,m+1; j=1,\ldots,n}.$$

Proof.

$(\Rightarrow)$ Given an arbitrage strategy $\xi \in \mathbb{R}^{m+1}$,

set $\mu := (\xi_2, \ldots, \xi_{m+1})$

$\Rightarrow \mu \cdot (C - \pi) = \mu \cdot C - \mu \cdot \pi \in \mathcal{Y}_n$

because $\xi \cdot \pi \leq 0$ and $\xi \cdot C \in \mathcal{Y}_n$ for arbitrage strategy ξ.

$(\Leftarrow)$　　Given a vector $\mu \in \mathbb{R}^m$ with $\mu \cdot (C - \pi) \in \mathcal{Y}_n$,

$$\text{set } \xi := \left(-\sum_{i=1}^{m} \mu_i \pi_{i+1}, \mu_1, \ldots, \mu_m \right)$$

$$\Rightarrow \xi \cdot \pi = -\sum_{i=1}^{m} \mu_i \pi_{i+1} + \sum_{i=1}^{m} \mu_i \pi_{i+1} = 0, \text{ since } \pi_1 = 1.$$

$$(\xi \cdot C)_j = \sum_{i=1}^{m} \mu_i (c_{(i+1)j} - \pi_{i+1}) = (\mu \cdot (C - \pi))_j, \text{ since } c_{1j} = 1$$

$$\Rightarrow \xi \cdot C \in \mathcal{Y}_n \qquad \qquad \square$$

With the previous lemma, excluding the existence of arbitrage strategies (arbitrage-free markets) can be formalized by $\{\xi \cdot A | \xi \in \mathbb{R}^m\} \cap \mathcal{Y}_n = \emptyset$ with $A := C - \pi$. If there does not exist an arbitrage strategy, then, according to FTAP, there exists an equivalent martingale measure, or formally, there is a vector $p \in \mathcal{P}_n$ with $C \cdot p = \pi$.

FTAP. $\{\xi \cdot A | \xi \in \mathbb{R}^m\} \cap \mathcal{Y}_n = \emptyset \Leftrightarrow \exists p \in \mathcal{P}_n \, (A \cdot p = 0).$

Proof.

$$(\Leftarrow) \qquad \text{Given } p \in \mathcal{P}_n \text{ with } A \cdot p = 0$$

$$\Rightarrow (\xi \cdot A) \cdot p = \xi \cdot (A \cdot p) = 0 \text{ for all } \xi \in \mathbb{R}^m$$

$$\Rightarrow \neg(\xi \cdot A \in \mathcal{Y}_n).$$

The non-trivial implication of the fundamental theorem will be proved in a recursive constructive way in the following section. $\quad\square$

FTAP in constructive mathematics. In constructive mathematics (in the tradition of Bishop and Brouwer), the logical PEM is not allowed as a proof tool (Chapters 4–6). Thus, Markov's principle and other examples are no longer provable propositions, but additional axioms. Reverse mathematics tries to decide which axioms of this kind are necessary and sufficient to prove certain theorems (Chapter 8). In the following, we prove that the main theorem of asset pricing is equivalent to MP (Berger and Svinland, 2016a,b). Thus, this important theorem of financial mathematics can be guaranteed and confirmed in a computationally constructive way.

The recursive constructive version of FTAP depends on several principles of recursive constructive mathematics. Many classical properties of the real numbers still hold constructively which we used in the previous proofs (Bishop, 1967; Bishop and Bridges, 1985; Bridges and Richman, 1987).

Constructive properties of real numbers.

$$x \geq y \Leftrightarrow \neg(x < y)$$

$$x = y \Leftrightarrow \neg\neg(x = y)$$

$$|x| \cdot |y| > 0 \Leftrightarrow |x| > 0 \wedge |y| > 0$$

$$|x| > 0 \Leftrightarrow x > 0 \vee x < 0$$

But it cannot be proven constructively that every non-empty bounded set of real numbers has an infimum. A constructive version needs the following restrictions.

Definition of total boundedness. Let (X, d) be a metric space with sets $D \subseteq C \subseteq X$ and $\varepsilon > 0$:

D is an ε-approximation of $C \Leftrightarrow \forall x \in C \exists y \in D(d(x, y) < \varepsilon)$.
D is totally bounded $\Leftrightarrow \forall n \exists x_1, \ldots, x_m \in C(\{x_1, \ldots, x_m\}$ is a $1/n$ approximation of C).

Intuitionistically, a set is called inhabited iff there is a constructive proof for the existence of an element. (Non-emptiness of the set is constructively not equivalent to this statement.) Any inhabited totally bounded subset C of X is located which means that $d(x, C) = \inf\{d(x, y)|y \in C\}$ exists for all $x \in X$ (Bridges and Vita, 2006, Section 2.2.9). Under these conditions, the existence of an infimum can constructively be proven: If C is totally bounded and function $f : C \to \mathbb{R}$ is uniformly continuous, then $\inf f = \inf\{f(y)|y \in C\}$ does exist (Bridges and Vita, 2006, Section 2.2.7). In this terminology,

Brouwer's fan theorem (Chapter 6) can be formulated as follows (Julian and Richman, 1984; Berger and Svindland, 2016a).

Brouwer's fan theorem FAN.

$$f : [0,1] \to \mathbb{R}^+ uniformly\ continuous \Rightarrow \inf f > 0.$$

For the constructive proof of FTAP, we need the existence of a (positive) infimum for uniformly continuous and convex functions. Actually, this existence proof can be understood as a constructive version of the fan theorem.

Definition of convex sets and functions. Let C be an inhabited set.

C is convex if, for all $x, y \in C$ and all λ in the interval $[0, 1]$, the point $\lambda x + (1 - \lambda)y$ also belongs to C.

Let C be an inhabited convex set. A function $f : C \to \mathbb{R}$ is convex iff

$$f(\lambda x + (1 - \lambda)y) \leq \lambda f(x) + (1 - \lambda)f(y)$$

for all $\lambda \in [0, 1]$ and $x, y \in C$.

Constructive version of the fan theorem.

$$f : x_m \to \mathbb{R}^+ uniformly\ continuous\ and\ convex \Rightarrow \inf f > 0.$$

The statement is proved by induction on m: Let $A_1, \ldots, A_m$ be elements of a normed space V and a positive valued uniformly continuous convex function f on

$$C(A_1, \ldots, A_m) = \left\{ \sum_{i=1}^{m} \lambda_i A_i | \lambda \in \mathcal{X}_m \right\}$$

$$\Rightarrow \inf f > 0 \quad \text{(Berger and Svindland, 2016a,b, 1163 f.).}$$

Besides the constructive version of the fan theorem, we need a lemma on convex subsets of Hilbert spaces for the proof of the main theorem.

Lemma on convex subsets of Hilbert spaces.

(1) $(H, \langle, \rangle)$ *Hilbert space with inhabited convex subset* $\mathcal{Y} \subseteq H$
(2) *Let* $x \in H$ *and assume that* $d = d(x, Y)$ *exists*

$\Rightarrow$ *There exists a unique* $a \in \bar{\mathcal{Y}}$ *such that* $\|a - x\| = d$,

$\Rightarrow$ $\langle a - x, c - a \rangle \geq 0$,

$\Rightarrow$ $\langle a - x, c - x \rangle \geq d^2$ *for all* $c \in Y$.

The constructive version of the fan theorem and the lemma on convex subsets of Hilbert spaces provides the following lemma on convex closed located subsets of Hilbert spaces. This lemma will be used below to prove the main theorem of asset pricing constructively.

Lemma on convex, closed, and located subsets of Hilbert spaces.

(1) *Let* $(H, \langle, \rangle)$ *be a Hilbert space with* $C \subseteq H$ *convex, closed, and located.*
(2) *Let* $x_1, \ldots, x_n \in H$ *with* $d(x, c) > 0$ *for all* $x \in C(x_1, \ldots, x_n)$ *and* $c \in \mathcal{C}$.

$\Rightarrow$ *There exists* $\varepsilon > 0$ *and* $p \in H$ *such that*
$$\langle p, x - c \rangle \geq \varepsilon \text{ for all } x \in C(x_1, \ldots, x_n) \text{ and } c \in \mathcal{C}.$$

In order to prove the non-trivial implication of the fundamental theorem, we assume that the rows of the matrix A are linearly independent. This assumption economically means that there are no redundant stocks in the market. In short: there are no stocks which can be hedged by others. Further on, we will use an implication of linear independence for convexity. At first, we remind the reader of the definition of linear independence.

Definition of linear independence. Vectors $x_1, \ldots, x_n$ are linearly independent iff for all $\lambda \in \mathbb{R}^n$

$$\sum_{i=1}^{n} |\lambda_i| > 0 \Rightarrow \left\| \sum_{i=1}^{n} \lambda_i x_i \right\| > 0$$

Linearly independent vectors span located subsets (Bridges and Vita, 2006, Section 4.1.5). Therefore, we get

Lemma on linear independence. Vectors $x_1, \ldots, x_n$ are linearly independent $\Rightarrow \{\sum_{i=1}^{m} \xi_i x_i \mid \xi \in \mathbb{R}^m\}$ is convex, closed, and located.

With this collection of results in constructive mathematics, we are able to prove the equivalence of FTAP with MP in the sense of reverse mathematics.

Non-trivial implication of FTAP. For every $\mathbb{R}^{m \times n}$-matrix A with linearly independent rows:

$$\{\xi \cdot A \mid \xi \in \mathbb{R}^m\} \cap \mathcal{Y}_n = \emptyset \Rightarrow \exists p \in \mathcal{P}_n \ (A \cdot p = 0)$$

FTAP will be proved to be equivalent to Markov's principle MP.

MP. $\forall x \in \mathbb{R}(\neg(x = 0) \Rightarrow |x| > 0)$.

Thus, in the sense of reverse mathematics, the proof of the fundamental theorem of asset pricing is recursively constructive. This proof is not constructive in the strict sense of mathematical intuitionism or Bishop's constructive mathematics, but at least in the sense of "Russian" constructivism RUSS (cf. Chapter 8) which accepts MP as an additional axiom to the axioms of intuitionism or Bishop's constructive mathematics.

FTAP is also equivalent to a recursive constructive version of the classical separation theorem which states that any two non-empty closed convex sets may be strictly separated by a hyperplane, provided that their intersection is void and one of the sets is compact.

Recursive constructive separation theorem.

(1) *Let $(H, \langle, \rangle)$ be a Hilbert space with $\mathcal{C} \subseteq H$ convex, closed, and located.*
(2) *Let $x_1, \ldots, x_n \in H$ with $\mathcal{C} \cap \mathcal{C}(x_1, \ldots, x_n) = \emptyset$.*

> $\Rightarrow$ *There exists $\varepsilon > 0$ and $p \in H$ such that*
> $\langle p, x - c \rangle \geq \varepsilon$ *for all $x \in C(x_1, \ldots, x_n)$ and $c \in \mathcal{C}$.*

The equivalence of FTAP with MP in the sense of reverse mathematics is proved with the recursive constructive separation theorem (SEP) (Berger and Svindland, 2016a,b).

Equivalence proof of SEP, FTAP, and MP.
SEP $\Leftrightarrow$ FTAP $\Leftrightarrow$ MP

Proof.
<u>SEP $\Rightarrow$ FTAP</u>:

<table>
<tr><td></td><td>Given $\mathbb{R}^{m \times n}$-matrix A with linearly independent rows</td><td>(assumption of FTAP)</td></tr>
<tr><td>$\Rightarrow$</td><td>$\mathcal{C} = \{\xi \cdot A | \xi \in \mathbb{R}^m\}$ is convex, closed, and located.</td><td>(lemma of linear independence)</td></tr>
<tr><td></td><td>$\mathcal{C} \cap \mathcal{Y}_n = \emptyset$</td><td>(assumption of FTAP)</td></tr>
<tr><td>$\Rightarrow$</td><td>$\mathcal{C} \cap \mathcal{X}_n = \emptyset$</td><td>(definition of $\mathcal{Y}_n$ and $\mathcal{X}_n$)</td></tr>
<tr><td>$\Rightarrow$</td><td>$\exists \varepsilon > 0 \exists q \in \mathbb{R}^n \forall x \in \mathcal{X}_n \forall c \in \mathcal{C}(\langle q, x \rangle \geq \langle q, c \rangle + \varepsilon)$ (*)
Let $e_i \in \mathbb{R}^n$ be the ith unit vector</td><td>(SEP)</td></tr>
<tr><td>$\Rightarrow$</td><td>$\langle q, e_i \rangle \geq \varepsilon$</td><td>(since $0 \in \mathcal{C}$)</td></tr>
<tr><td>$\Rightarrow$</td><td>All components of q are positive.
Assume $c \in \mathcal{C}$ with $|\langle q, c \rangle| > 0$</td><td></td></tr>
<tr><td>$\Rightarrow$</td><td>$\exists d \in \mathcal{C}(\langle q, e_i \rangle < \langle q, d \rangle + \varepsilon)$</td><td>($\mathcal{C}$ linear space)</td></tr>
<tr><td>$\Rightarrow$</td><td>contradiction to (*)</td><td></td></tr>
<tr><td>$\Rightarrow$</td><td>$A \cdot q = 0$</td><td></td></tr>
<tr><td>$\Rightarrow$</td><td>$p = \left(\dfrac{q_1}{\sum_{i=1}^{n} q_i}, \ldots, \dfrac{q_n}{\sum_{i=1}^{n} q_i} \right) \in \mathcal{P}_n$ with $A \cdot p = 0$.</td><td></td></tr>
</table>

$\underline{\text{FTAP} \Rightarrow \text{MP}}$:

Given a real number a with $\neg(a = 0)$ (assumption of MP)

Consider the matrix $A = (|a|, -1)$
$\underline{\text{Statement}}$: $\{\xi \cdot A | \xi \in \mathbb{R}^2\} \cap \mathcal{Y}_2 = \emptyset$

$\underline{\text{Proof (indirect)}}$: Assume that the statement is false

$$\Rightarrow \exists \xi \in \mathbb{R}(\xi \cdot A = \xi \cdot (|a|, -1) = (\xi|a|, -\xi) \in \mathcal{Y}_2)$$

First case: $\qquad \xi|a| > 0$ and $-\xi \geq 0$

$$\Rightarrow \xi > 0 \text{ and } \xi \leq 0.$$

Second case: $\qquad \xi|a| \geq 0$ and $-\xi > 0$

$$\Rightarrow a = 0.$$

$\Rightarrow$ contradiction.

$\{\xi \cdot (|a|, -1) \,|\, \xi \in \mathbb{R}^2\} \cap \mathcal{Y}_2 = \emptyset$ (statement)

$\Rightarrow \exists p \in \mathcal{P}_2 (|a|, -1) \cdot p = 0$ (FTAP)

$\Rightarrow \exists p \in \mathcal{P}_2 p_1 |a| = p_2$

$\Rightarrow |a| > 0.$

$\underline{\text{MP} \Rightarrow \text{SEP}}$:

Assumptions of SEP:

(1) $(H, \langle , \rangle)$ Hilbert space with $\mathcal{C} \subseteq H$ convex, closed, and located,
(2) $x_1, \ldots, x_n \in H$ with $\mathcal{C} \cap \mathcal{C}(x_1, \ldots, x_n) = \emptyset$

$\Rightarrow \quad \neg(d(x, c) = 0)$ for all $x \in \mathcal{C}(x_1, \ldots, x_n)$ and (assumption (2))
$\quad c \in \mathcal{C}$

$\Rightarrow \quad d(x, c) > 0$ for all $x \in \mathcal{C}(x_1, \ldots, x_n)$ and $c \in \mathcal{C}$ (MP)

$\Rightarrow \quad \exists \varepsilon > 0 \exists p \in H \forall x \in \mathcal{C}(x_1, \ldots, x_n)$ (assumption (1))
$\quad \forall c \in \mathcal{C}(\langle p, x - c \rangle \geq \varepsilon)$

The last implication with assumption (1) uses the lemma on convex, closed, and located subsets of Hilbert spaces which was mentioned previously. $\qquad\square$

There is also a constructive version of FTAP (Berger and Svindland, 2016a,b, p. 1169). The equivalence proof with MP provides a recursive constructive version because MP is an additional axiom which is not included in Bishop's constructive mathematics (and not accepted by intuitionistic mathematics) (Chapter 8).

Constructive version of FTAP. *For every* $\mathbb{R}^{m\times n}$*-matrix* A *with linearly independent rows*:

$$\forall \xi \in \mathbb{R}^m \forall x \in \mathcal{X}_n d(\xi \cdot A, x) > 0 \Rightarrow \exists p \in \mathcal{P}_n (A \cdot p = 0)$$

Assumptions of classical economic models

Constructive and computational mathematics provides tools to guarantee the existence of solutions in financial and economic problems. Thus, reverse mathematics opens new possibilities to test provability and constructivity of financial and economic instruments. But, the classical formalism of economics and financial mathematics also rely in axioms which are questionable with respect to real human behavior (Mandelbrot and Hudson, 2004):

(1) <u>Assumption</u>: People are rational in the sense of Adam Smith's homo oeconomicus.

> Consequently, when presented with all the relevant information about a stock or bond, investors will make the obvious rational choice leading to the greatest possible wealth and happiness. Their preferences can be expressed in mathematical formulas of utility functions which can be maximized. By that, rational investors make a rational model of an efficient market. Actually, people do not only think in terms of mathematical utility functions, and are not always rational and self-interested. They are driven by emotions distorting their decisions. Sometimes, they miscalculate probabilities and feel differently about loss than gain.

(2) <u>Assumption</u>: All investors are alike.

> Consequently, people have the same investment goals and react and behave in the same manner. In short: They are like the molecules in an idealized gas of physics. An equation that describes one such molecule or investor can be replaced to describe all of them. Actually, people are not alike. If one drops the assumption of homogeneity, one gets a more complex model of the market. For example, there are at least two different types of investors: A fundamentalist believes that each stock has its own value and will eventually sell for that value. On the other side, a chartist ignores the fundamentals and only watches the price trends in order to jump on or off band wagons. Their interactions can lead to price bubbles and spontaneously arising crashes. The market switches from a well-balanced linear system in which one factor adds predictably to the next to a chaotic non-linear system in which factors interact with the emergence of synergetic and unanticipated effects.

(3) <u>Assumption</u>: Price change is practically continuous.

> Consequently, stock quotes or exchange rates do not jump up or down, but move smoothly from one value to the next. In this way, continuity has been assumed in classical physics, according to the motto of Leibniz *natura non facit saltum* (nature does not make leaps) which was repeated by Alfred Marshall in his text book *Principles of Economics* (1890) for economic systems. From a methodological point of view, the belief in a continuous behavior of nature and economy opens the possibility to apply continuous functions and differential equations in order to solve physical or economic problems analytically. But actually, prices in economy and quantum states in quantum physics do jump, and discontinuity, far from being an anomaly, characterizes the reality. Contrary to Einstein's famous objection against quantum physics: God plays with dice — in nature and society.

(4) <u>Assumption</u>: Price changes follow a Brownian motion.

> The Brownian motion is also a famous model of physics applied to financial markets by Bachelier. In more details, it implies three assumptions: First, each change in price is believed to appear independently from the last (statistical independence). Second, the process generating price changes stays the same over time (statistical stationarity). Third, price changes follow the proportions of the Gaussian bell curve (normal distribution). Financial data clearly contradict a smooth normal distribution of changing prices. The analysis of the real distribution patterns is a challenge of stochastic mathematics and systems theory and opens new avenues to the complexity of modern society.

Brownian motion is mathematically more manageable than any alternative. But, unfortunately, it is an extremely poor approximation to financial reality. Since the end of the 1980s, we can observe financial crashes and turbulences deviating significantly from normal distribution. Investment portfolios collapsed and hedging with options *à la* Black–Scholes failed. From the viewpoint of dynamical systems, patterns of time series analysis illustrate the failures of traditional financial theory. While a record of Brownian motion changes looks like a kind of "grass" with normal length, a record of actual price changes looks like an irregular alternation of quiet periods and bursts of volatility that stand out from the normal length of the grass. This feature demonstrates the apparent non-stationarity of the underlying rules. Further on, discontinuities appear as sharp peaks from the normally distributed Gaussian "grass" (Mandelbrot and Hudson, 2004). These peaks are not isolated but bunched together. Cyclic (but not periodic) behavior can be observed. Instability of the sample variance is expressed by long-tailed distribution price changes. Last but not least, there is a long-term dependence of data (Small, 2005).

Financial markets and fluid turbulence

Financial markets display some common properties with fluid turbulence. As for fluid turbulent fluctuations, financial fluctuations have intermittency at all scales. In fluid turbulence, a cascade of energy flux is known to occur from the large scale of injection to the small scales of dissipation. In the non-linear and fractal approach of the financial system randomness can no longer be restricted to the "normal" Gaussian distribution of price changes (Mandelbrot, 1997). Non-Gaussian distributions with Lévy and Pareto distributions are more appropriate to the wild turbulence of financial markets of today (Embrechts *et al.*, 2003).

Gaussian distribution corresponds to a pattern of time series with a "normal" length of "grass" without extreme peaks. Therefore, it is called mild randomness which can be compared to a solid state of matter aggregation with low energy, stabile structure, and defined volume. Wild randomness resembles the gas phase of matter with high energy, less structure, and no defined volume. Slow randomness means the fluid state between the gas and solid state. From the viewpoint of time series, mild randomness corresponds to short- and long-run evenness. Slow randomness corresponds to short-run concentration and long-run evenness. Wild randomness corresponds to short- and long-run concentration (Mandelbrot, 1999, 2004).

Securitized credit model and increasing networks of risks

Nevertheless, the demand for profit and security has initiated a wave of financial innovation, based on these classical assumptions. They are focused on the origination, packaging, trading, and distribution of securitized credit instruments. Simple forms of securitized credit have existed for almost as long as modern banking. But from the mid-1990s, the system entered explosive growth in both scale and complexity. We observe a huge growth in the value of the total stock of credit securities, an explosion in the complexity of the securities sold, with the growth of structured credit products, and with the related explosion of the volume of credit derivatives, enabling investors and traders to hedge underlying credit exposures, or to create synthetic credit exposures.

This financial innovation sought to satisfy the demand for yield uplift. It was predicated on the belief that by slicing, structuring, and hedging, it was possible to create value, offering investors combinations of risk, return, and liquidity which were more attractive than those available from the direct purchase of the underlying credit exposures. It resulted not only in massive growth in the importance of securitized credit, but also in a profound change in the nature of the securitized credit model. As securitization grew in importance from the 1980s on, its development was praised as a means to reduce banking system risks and to cut the total costs of credit intermediation, with credit risk passed through to end investors, reducing the need for unnecessary and expensive bank capital. Credit losses would be less likely to produce banking system failure (Lord Turner, 2009).

But there is no "free lunch" or financial "perpetuum mobile". When the crisis broke, it became apparent that this diversification of risk holding had not actually been achieved. Instead most of the holdings of the securitized credit, and the vast majority of the losses which arose, were not in the books of end investors intending to hold the assets to maturity, but on the books of highly leveraged banks and bank-like institutions. This reflected an evolution of the securitized credit model away from the initial descriptions. To an increasing extent, credit securitized and taken off one bank's balance sheet was not simply sold through to an end investor, but bought by the propriety trading desk of another bank, sold by the first bank but with part of the risk retained via the use of credit derivatives, resecuritized into increasingly complex instruments (e.g., CDOs and CDO squareds) or used as collateral to raise short-term liquidity (International Monetary Fund, 2009).

The financial innovations of structured credit resulted in the creation of products, e.g., the lower credit tranches of CDOs or even more so of CDO squareds, which had very high and imperfectly understood embedded leverage, creating positions in the trading books of banks which were hugely vulnerable to shifts in confidence and liquidity. This process created a complex chain of multiple relationships between multiple institutions, each performing a different small slice of the credit intermediation and maturity transformation process, and each with a leveraged balance sheet requiring a small slice of capital to support that function (Sinn, 2010). A complex network of dependences has emerged in a hidden and non-transparent world of financial shadows. The new model left most of the risk still somewhere on the balance sheets of banks and bank-like institutions but in a much more complex and less transparent way.

The evolution of the securitized credit model was accompanied by a growth in the relative size of financial services within economy, with activities internal to the banking system growing far more rapidly than end services to the real economy. The growing size of the financial sector was accompanied by an increase in total system leverage. But this process also drove the boom and created vulnerabilities of the whole financial network that have increased the severity of the crisis. According to the Turner Report (Lord Turner, 2009), from about 2003 onwards, there were significant increases in the measured on-balance sheet leverage of many commercial and investment banks, driven in some cases by dramatic increases in gross assets and derivative positions. This was despite the fact that measures of leverage (e.g., value at risk (VaR) relative to equity) showed no such rise. This divergence reflected the fact that VaR measures of the risk involved in taking propriety trading positions in general suggested that risk relative to the gross market value of positions had declined. It is clear in retrospect that the VaR measures of risk were faulty (Stutz, 2009).

The risk of VaR

The increasing complexity of the securitized credit market was obvious to some participants, regulators, and academic observers (Greenspan, 2008). But the predominant assumption was that increased complexity had been matched by the evolution of mathematically sophisticated and effective techniques for measuring and managing the resulting risks (Colander *et al.*, 2008). Central to many of the techniques was the concept of VaR, enabling inferences about forward-looking risk to be drawn from the observation of past patterns of price movement. The risk-forecasting models of VaR are based on the assumption that forecasting credit risk is an activity not unlike that of forecasting weather. It is assumed that one's own action, based on past volatility, does not affect future volatility itself just like forecasting weather does not influence future weather.

This technique, developed in the early 1990s, was not only accepted as standard across the industry, but also adopted by regulators as the basis for calculating trading risk and required capital. Therefore, VaR was incorporated within the European Capital Adequacy Directive (Danielsson *et al.*, 2001). In financial mathematics and financial risk management, VaR is a widely used risk measure of the risk of loss on a specific portfolio of financial assets. For a given portfolio, probability, and time horizon, VaR is defined as a threshold value such that the probability that the mark-to-market loss on the portfolio over the given time horizon exceeds this value in the given probability level. Again, it is assumed that markets are normal. VaR has five main uses in finance: risk management, risk measurement, financial control, financial reporting, and computing regulatory capital (Kleeberg and Schlenger, 2000). VaR is sometimes used in non-financial applications as well. Important related ideas are economic capital, backtesting, stress testing, and expected shortfall.

Mathematically (Föllmer and Schied, 2004, 2008), the uncertainty in the future of a portfolio is usually described by a function $X\colon \Omega \to \mathbb{R}$, where Ω is a fixed set of scenarios. For example, X can be the value of a portfolio. The goal is to determine a number $\rho(X)$ that quantifies the risk and can serve as a capital requirement or the minimal amount of capital which, if added to the position and invested in a risk-free manner, makes the position acceptable.

Definition of VaR. Given some confidence level $\alpha \in (0, 1)$, VaR of the portfolio value X at the confidence level α is given by the smallest number $m \in \mathbb{R}$ such that the probability of a loss is not larger than the confidence level α:

$$\mathrm{VaR}_\alpha(X) = \inf\{m \epsilon \mathbb{R} | P(X + m < 0) \leq \alpha\}.$$

Obviously, VaR only pays attention to see that the boundary of the confidence level is not exceeded. But, it does not consider the degree of loss. Further on, it assumes that the probability distribution of losses is well-known because of historical data. Only in this case, VaR can forecast credit risk like weather, which means that future volatility can be derived from past volatility.

There are, however, fundamental questions about the validity of VaR as a measure of risk. The use of VaR measures based on relatively short periods of historical observation (e.g., 12 months) introduced dangerous procyclicality into the assessment of trading book risk (Lord Turner, 2009). Short-term observation periods and the assumption of normal distribution can lead to large underestimation of probability of extreme loss events. Interconnected market events in complex networks can produce self-reinforcing cycles which models do not capture. Systemic risk may be highest when measured risk is lowest, since low measured risk encourages behavior which creates increased systemic risks.

This kind of mathematics, used to measure and manage risk by VaR, was not very well understood with all its conditions and restrictions by top management and boards to assess and exercise judgement over the risks being taken. Mathematical sophistication ended up not containing risk, but providing false assurance that other indicators of increasing risk (e.g., rapid credit extension and balance sheet growth) could be safely ignored.

The global financial system, combined with macroeconomic imbalances, created an unsustainable credit boom and asset price inflation. Those consequences of the financial crisis transmitted financial system problems into real economy effects. The shock to the banking system has been so great that its impaired ability to extend credit to the real economy has played a major role in enforcing the economic downturn, which in turn undermines banking system strength in a self-reinforcing feedback loop.

From a historical point of view, it is remarkable that the academic professionals were well aware of the methodological weakness of VaR measures. In an "Academic Response to Basel II" (Danielsson *et al.*, 2001), the methodology of VaR was criticized to be insufficient:

(1) VaR risk models treat risk as a fixed exogenous process, but its endogeneity may matter enormously in times of crisis.

(2) VaR is a misleading risk measure when the returns are not normally distributed, as in the case with credit, market, and operational risk. It does not measure the distribution of risk in the tail, but only provides an estimate of a particular region in the distribution. Thus, VaR models generate imprecise and widely fluctuating forecasts.

Crisis of risk modeling

The development of an expanded financial sector and the rapid growth and increased complexity of the securitized model of credit intermediation were accompanied by the development of increasingly sophisticated mathematical techniques for the measurement and management of position taking risks. The techniques entailed numerous variants to cope with, for instance, different categories of option. Their application required significant computing power to capture relationships between different market prices, the complex nature of structured credit instruments, and the effects of diversification across correlated markets. But the underlying methodological assumption was the old idea that analysis of past price movement patterns could deliver statistically robust inferences relating to the probability of price movements in future.

The financial crisis has revealed, however, severe problems with these techniques. They suggest the need for significant changes in the way that VaR-based methodologies have been applied. But, the most fundamental question concerns our ability in principle to infer future risk from past observed patterns. Can financial models still be considered true mappings of an external world in order to derive predictions of future events like in the natural sciences (Lux and Westerhoff, 2009)?

Models frequently assume that the distribution of possible events, from which the observed price movements are assumed to be a random sample, is normal with the shape of a Gaussian bell curve. But there is no clearly robust justification for this assumption. Actually, the financial market movements are inherently characterized by fat-tail distributions. This implies that any use of VaR models needs to be analyzed by the application of stress test techniques which consider the impact of extreme movements beyond those which the model suggests are at all probable.

One explanation of fat-tail distributions may lie in the complex networks of financial dependences. VaR models implicitly assume that the actions of the individual firm, reacting to market price movements, are both sufficiently small in scale so as not to affect the market equilibriums, and independent of the actions of other firms. But this is a deeply misleading assumption if it is possible that developments in markets will induce similar and simultaneous behavior by numerous players. If this is the case, which it certainly was in the financial crisis, VaR measures of risk may not only fail adequately to warn of rising risk, but may also convey the message that risk is low and falling at the precise time when systemic risk is high and rising.

For example, according to VaR measures, risk was low in spring 2007. Actually, the system was overwhelmed with huge systemic risk. This suggests that stress tests are needed to consider the impact of second-order effects, for example, the impact of one bank's likely reaction to the common systemic stress on another bank.

A new paradigm of risk modeling

The most fundamental insight is, however, philosophical: It is important to realize that the assumption that past distribution patterns carry robust inferences for the probability of future patterns is methodologically insecure. It involves applying to the world of social and economic relationships a technique drawn from the world of physics, in which a random sample of a definitively existing universe of possible events is used to determine the probability characteristics which govern future random samples. But it is doubtful when applied to economic and social decisions with inherent uncertainty. Economists sometimes refer to it as "Knightian" uncertainty which is a reference to the classical distinction between risk and uncertainty in Frank Knights's Ph.D. "Risk, Uncertainty, and Profit" from 1921 (Knights, 1921). But it would also suggest that no system of regulation could ever guard against all risks and uncertainties.

Analysis of the causes of the crisis suggests that there is a limit to the extent to which risks can be identified and offset at the level of the individual firm. We explained how the origins of the crisis lay in systemic developments: The crucial shift required in regulatory philosophy is toward one which focuses on macro-analysis, systemic risks, and judgements about business model sustainability, and away from the assumption that all risks can be identified and managed at a firm specific level. As a result most of the changes we propose relate to the redesign of global regulation combined with a major shift in methodology (Colander *et al.*, 2008).

But improvements in the effectiveness of internal risk management and firm governance are also essential. While some of the problems could not be identified at firm specific level, and while some well-run banks were affected by systemic developments over which they had no influence, there were also many cases where internal risk management was ineffective and where boards failed to adequately identify and constrain excessive risk taking. Achieving high standards of risk management and governance in all banks is therefore essential. Detailed proposals are necessary to support a Financial Service Authority (FSA) in all countries.

The origins of the past crisis entailed the development of a complex, highly leveraged, and therefore risky variant of the securitized model of credit intermediation. Large losses on structured credit and credit derivatives, arising in the trading books of banks and investment banks, directly impaired the capital position of individual banks, and because of uncertainty over the scale of the losses, created a crisis of confidence which produced severe liquidity strains across the entire system. As a result, a wide range of banking institutions suffered from an impaired ability to extend credit to the real economy, and have been recapitalized with large injections of taxpayer money.

The mathematical rigor and numerical precision of risk management and asset pricing tools have a tendency to conceal the weakness of models and their assumptions to those who have not developed them and do not know the potential weakness of the assumptions. Models are only approximations to the real world dynamics and partially built upon idealized assumptions. A typical example is the belief in normal distribution of asset price changes completely neglecting the importance of extreme events. Considerable progress has been made by moving to more sensitive models with fat-tailed Lévy processes (Mandelbrot, 1999). Of course, such models better capture the intrinsic volatility of markets. But they might again contribute to enhancing the control illusion of the naïve user.

Therefore, market participants and regulators have to become more sensitive toward the potential weakness of risk management models. Since there is more than one true model, robustness should be a key concern. Model uncertainty should be taken into account by applying more than a single model. For example, one could rely on probabilistic procedures that cover a whole class of specific models. The theory of robust control provides a toolbox of techniques that could be applied for this purpose.

In the field of financial economics there are a number of ways that risk can be defined. To clarify the concept, mathematicians have axiomatically described a number of properties that a risk measure might or might not have (Föllmer and Schied, 2008; York, 2008).

A coherent risk measure (Artzner *et al.*, 1999) is a risk measure ρ that satisfies properties of monotonicity, subadditivity, homogeneity, and translational invariance. Consider a random outcome X viewed as an element of a linear space L of measurable functions, defined on an appropriate probability space.

Definition of coherent risk measure. A functional $\rho : L \to \mathbb{R}$ is said to be a coherent risk measure for L if it satisfies the following properties:

<u>Monotonicity</u>: If $X_1, X_2 \in L$ and $X_1 \leq X_2$, then $\rho(X_1) \leq \rho(X_2)$.

That is, if portfolio X_2 always has better values than portfolio X_1 under all scenarios, then the risk of X_2 should be less than the risk of X_1.

<u>Subadditivity</u>: If $X_1, X_2 \in L$, then $\rho(X_1 + X_2) \leq \rho(X_2) + \rho(X_2)$.

Indeed, the risk of two portfolios together cannot get any worse than adding the two risks separately. This is the diversification principle.

<u>Positive homogeneity</u>: If $\alpha \geq 0$ and $X \in L$, then $\rho(\alpha X) = \alpha \rho(X)$. Loosely speaking, if you double your portfolio, then you double your risk.

<u>Translation invariance</u>: If $m \in R$ and $X \in L$, then $\rho(X + m) = \rho(X) - m$.

The value m is just adding cash to the portfolio X, which acts like an insurance. The risk of $X + m$ is less than the risk of X, and the difference is exactly the added cash m. Therefore, translational invariance is also called cash invariance. In particular, if $m = \rho(X)$, then $\rho(X + \rho(X)) = 0$.

The notion of coherence has been subsequently relaxed. Indeed, the notions of subadditivity and positive homogeneity can be replaced by the notion of convexity.

Definition of convexity. If $X_1, X_2 \in L$ and $0 \leq \lambda \leq 1$, then $\rho(\lambda X_1 + (1 - \lambda)X_2) \leq \lambda \rho(X_1) + (1 - \lambda)\rho(X_2)$.

Consider the collection of possible future outcomes that can be generated with the resources available to an investor. One investment strategy leads to X_1, while a second strategy leads to X_2. If one diversifies, spending only the fraction λ of the resources on the first possibility and using the remaining part for the second alternative, one obtains $\lambda X_1 + (1 - \lambda)X_2$. Thus, the axiom of convexity gives a precise meaning to the idea that diversification should not increase the risk.

It is well known that VaR is positively homogeneous, but it is not in general a coherent risk measure as it does not respect the subadditivity property. Hence, it is not convex. An immediate consequence is that VaR might discourage diversification. VaR is, however, coherent, under the assumption of normally distributed losses when the portfolio value is a linear function of the asset prices. In this case, VaR becomes equivalent to a mean-variance approach where the risk of a portfolio is measured by the variance of the portfolio's return.

Definition of average VaR. $\text{AVaR}_\lambda = \frac{1}{\lambda} \int_0^\lambda \text{VAR}_\alpha(X) d\alpha$ at level $\lambda \in (0, 1]$.

Average VaR is also called conditional VaR, expected shortfall, or tail value at risk, which is a coherent risk measure (Detlefsen and Scandolo, 2005; Riedel, 2006).

We previously underlined that model uncertainty should be taken into account, since we do not know the distinguished true model of financial reality. Therefore, we should consider a whole class of possible probabilistic models with different penalty. In the dual representation theory of convex risk measures, one aims at deriving their representation in a systematic manner. The class M contains possible probabilistic models Q which are taken more or less seriously according to the size of a penalty function $\pi(Q)$. In this way, we take the message of praxis seriously that we should not rely on one single model, but flexibly vary the models with respect to different contextual applications under special attention to the worst case.

A dual representation of a convex risk measure computes the worst case expectation taken over all models Q and penalized by $\pi(Q)$. The class M of possible probabilistic models is a set of probability measures such that the expectation $E_Q(X)$ is well defined for all models Q and portfolios X.

Definition of convex risk measure. According to Föllmer and Schied (2008), the dual representation of a convex risk measure ρ has the form

$$\rho(X) = \sup_{Q \in M}(E_Q(-X)\pi(Q)).$$

These models are no longer considered definitive mappings of reality. But they serve as stress tests. One does not rely on a fixed model, but chooses the sure side for every position and focuses on the corresponding worst case model. Thus, the model ambiguity is explicitly considered during the procedure.

Model ambiguity and rational behavior

Model ambiguity is linked to the economic theory of rational behavior under uncertainty (Cont, 2006; Maccheroni, 2006). Classical economic models are mainly built upon the two assumptions of rational expectations with well-known probabilities of utilities and a representative agent ("homo oeconomicus"). They imply a complete understanding of the economic laws governing the world. These models leave no place for imperfect knowledge discovered in empirical psychological studies of real humans (Frydman and Goldberg, 2007, 2008). Their behavior in financial markets is even strongly influenced by emotional and hormonal reactions. Thus, economic modeling has to take bounded rationality seriously. But, model ambiguity does not mean the collapse of mathematical modeling. Mathematically, a fixed probability measure of expected utilities should be replaced by a convex risk measure which simultaneously considers a whole class of possible stochastic models with different penalties. Financial praxis warned us not to rely on a fixed model, but to vary possible models in a flexible way and to pay attention to the worst case. This is also the mathematical meaning of a convex risk measure.

The differences between the overall system and its parts, macro- and microeconomics, remain incomprehensible from the viewpoint of classical rationality which assumes a representative agent. Since interaction depends on differences in information, motives, knowledge, and capabilities, this implies heterogeneity of agents (Hayek, 1948, 1974). Only a sufficiently rich structure of connections between firms, households and a dispersed banking sector will allow insights in systemic risks and synergetic effects in the financial sector. The reductionism of the representative agent or "homo oeconomicus" has prevented economists from modeling these phenomena.

For natural scientists, the distinction between micro-level phenomena and those originating on a macro originated from the interaction of microscopic units is well-known. In those models, the current crisis would be seen as an emergent phenomenon of the macroeconomic activity (Aoki and Yoshikawa, 2007; Mainzer, 2009b). The reductionist paradigm blocks any understanding of the interplay between micro and macro levels.

Models with interacting heterogeneous agents would also open the door to interdisciplinary research from different sciences. Complex networks of different agents or statistical physics of interacting agents can model dynamic economic systems (Mantegna and Stanley, 2000; McCauley, 2004). Self-organized criticality is another area that seems to explain boom-and-bust cycles of the economic non-equilibrium dynamics (Scheinkman, 1990).

Challenges of future risk modeling

In macroeconomics, data mining is often driven by the pre-analytic belief in the validity of certain models which should justify political or ideological opinions. The political belief in deregulation of the 1990 is a typical example. Rather than misusing statistics as a means to illustrate these beliefs, the goal should be to put theoretical models to scientific tests like in the natural sciences.

A chain of specification tests and estimated statistical models for simultaneous systems would provide a benchmark for the tests of models based on economic behavior. Significant and robust relations within a simultaneous system would provide empirical regularities that one would attempt to explain, while the quality of fit of the statistical benchmark would offer a confidence for more ambitious models. Models that do not reproduce (even) approximately the quality of the fit of statistical models would have to be rejected. This methodological criterion also has an aspect of ethical responsibility of researchers: Economic policy models should be theoretically and empirically sound. Economists should avoid giving policy recommendations on the basis of models with a weak empirical grounding and should, to the extent possible, make clear to the public how strong the support of the data is for their models and the conclusions drawn from them.

A neglected area of methodology is the degree of connectivity and its interplay with the stability of the complex system. It will be necessary for supervision to analyze the network aspects of the financial system, collect appropriate data, define measures of connectivity, and perform macro stress testing at the system level. In this way, new measures of financial fragility would be obtained. This would also require a new area of accompanying academic research that looks at agent-based models of the financial system, performs scenario analyses, and develops aggregate risk measures. Network theory and the theory of self-organized criticality of highly connected systems would be appropriate starting points (Scheinkman and Woodford, 2001; Mainzer, 2007a).

Such scientific analysis must be supported by more practical consequences. The hedge fund market is still widely unregulated. The interplay between connectivity, leverage, and system risks needs to be investigated at the whole level. It is highly likely that extreme leverage levels of interconnected institutions impose dangerous social risks on the public.

On the macroeconomic level, it would be desirable to develop early warning schemes that indicate the formation of bubbles. Combinations of indicators with time series techniques could be helpful in detecting deviations of financial or other prices from their long-run averages. Indication of structural change would be a sign of changes of the behavior of market participants of a bubble-type nature (McCauley, 2004).

Obviously, there is no single causal model as definitive mapping of reality. In this sense, David Hume and his followers were right in their skepticism against classical axiomatization of rationality in the world. But that does not mean a complete denial of mathematical tools and models. We have to consider whole classes of possible stochastic models with different weights. They must be combined with a data-driven methodology and insights in the factual human behavior and its diversity. Therefore, psychological and sociological case studies of human behavior under risk conditions (e.g., stakeholders at stock markets) are necessary. In experimental economics, decision behavior is already simulated under laboratorial conditions. Even philosophical ethics can no longer only argue with arm-chaired considerations and *a priori* principles, but must relate to empirical observations of factual decision behavior. That is done in the new approaches of experimental ethics. We argue for this kind of interdisciplinary methodology which opens new avenues for mathematical modeling in science. In this case, robust stochastic tools are useful because they are used under restricted conditions and with sensibility for the permanent model ambiguity.

In a globalized world, risks are mainly systemic and cannot be reduced to single causes. They emerge from complex interactions in natural, technical, economic, and social systems. Examples are complex information and communication networks, power ("smart") grids as well as cellular interactions in organisms or transactions in financial markets. We must consider their explanatory power as well as their limitations. Then, they can supplement themselves mutually.

Risk measurement and risk awareness

Formal models are not sufficient. Risk awareness even of experts is often subjective and depends on individual experience, societal, and cultural contexts. Remember the extremely different reactions of the public to the Fukushima disaster in Japan, Germany, and other countries. Therefore, formal risk models must be complemented by sociological and cultural studies. Psychic behavior in decision situations must also be taken into account. Therefore, experimental economics and ethics relate to observations of factual behavior of people, e.g., at stock markets. Behavioral studies under experimental lab conditions are even useful for social philosophy and ethics.

The past crises might be characterized as examples of final stages of well-known boom-and-bust patterns that have been repeated so many times in the course of economic history. But, there are several new aspects leading to a shift of methodological paradigm: The preceding boom had its origin in the development of new financial products with increasing complexity which seemed to promise diminishing risks. The financial market detaches itself from the real market. Profit seems to be possible by clever financial innovations loosing their connection to real economy. But, like in nature, there is no "free lunch" or "perpetuum mobile" of profit in finance. Further on, the past crises were due to the increasing complexity of interconnected financial networks. These aspects have been largely ignored by traditional economic models.

Therefore, we cannot trust in a single risk model, but must consider a class of more or less appropriate models, supplemented by experimental behavioral case studies. The lack of methodological understanding of models and the lack of ethical responsibility to warn the public against the limitations of models were the main reasons of the past economic crises. It is the task of philosophy of science to evaluate scientific modeling and the ethical responsibility of scientists. During booming periods, we should better prepare the next crisis in a countercyclical manner.

Digital and neural computing in global information systems

Financial markets are nowadays realized in global information systems. Technical neural networks are complex systems of firing and non-firing neurons with topologies like living brains. There is no central processor (mother cell), but a self-organizing information flow in cell assemblies according to rules of synaptic interaction. According to the local activity principle (cf. Chapter 14), the cells are amplifiers of low input signals generating complex patterns of cellular behavior. There are different architectures with one synaptic layer or more synaptic layers with hidden units for feedforward flow, or feedback loops with backpropagation flow. Learning algorithms change the synaptic weights in order to realize the synaptic plasticity of living brains. At least in mathematical models, we can consider digital as well as real weights. The dynamics of neural nets can be modeled in phase spaces of synaptic weights with trajectories converging to attractors which represent prototypes of patterns.

Neural networks are already applied in cognitive robots. A simple robot with diverse analog sensors (e.g., proximity, light, collision) and motor equipment can generate complex behavior by a self-organizing neural network. In the case of collision, the connections between the analog sensors of proximity and collision layer are reinforced by learning algorithms. A behavioral pattern emerges from simultaneously firing neurons in the technical brain of the robot.

A robot society as complex dynamical system

A robot society is a group of robots which has the ability to communicate, interact, and to perform tasks jointly. A society is defined by its information and control structure which make possible common task planning and execution. In this case, according to the local activity principle, a robot is a locally active agent driven by a battery and low input signals which are amplified and transformed into complex patterns of behavior.

Most of the autonomous mobile robots are operating in neither stabile nor structured environments. Therefore, a major trend in robotics is going toward multi-robot systems. In many cases, decomposing a complex task into parallel subtasks is a possibility to speed up the performance. Sometimes, several robots work with the same subtask, increasing the redundancy of the system. Furthermore, there can be tasks where a successful completion of a task requires close cooperation among the robots. Such a case is, for example, the carrying of a large object together. It requires some sort of interaction between robots, whether it is a direct communication or some sort of indirect communication through sensing the forces in the object to be transported.

This kind of task as well as many other tasks normally related to multi-robot systems has clear analogy to biological systems (Wilson, 1971, 2000). A group of ants solve the problem through sensing the forces and torque in the object. Based on this information, they change the direction of forces accordingly or need some ants to change the position of their hold. Numerous similar examples can be found from nature. Tests by evolution during millions of years are proven to be feasible in dynamic and hostile environments and can thus provide valuable information and inspiration for similar type of engineering tasks.

Digital and analog computing in robot societies

Artificial life represents an attempt to understand all life phenomena through reproduction in artificial systems typically through their simulation on a computer (Mainzer, 2010). To attain this ambitious goal, artificial life relies on the theory of complex dynamical systems with differential equations and real computing. A complex dynamical system is a system that can be described at different levels, in which global properties at one level emerge from the interaction of a number of simple elements at lower levels. Global properties are emergent in the sense that they cannot be predicted from a knowledge of the elements or the rules by which the elements locally interact, given the high non-linearity of these interactions. In engineering science, evolutionary robotics not only shares most of these characteristics with artificial life, but it also stresses the importance of using digital devices in robots instead of simulated agents. They all realize the principle of local activity, i.e., locally active agents with low energy and information input signals generating different patterns of behavior in complex networks and dynamic environments.

Cooperative robots are constantly interacting not only with the dynamic environment, but also naturally with each other and with the persons who are using them (Balch and Parker, 2002). This large variety of interactions will produce behaviors that are going to be superior compared to those performed by current robots. The robots will be equipped with sensors to survive as a part of a complex system, where the cooperation is essential for their survival. Collective intelligence emerging from these interactions gives a reason to call these systems at their highest level "robot societies". Multi-agent systems, whether software or hardware agent based, have to solve some fundamental problems. They must formulate, describe, decompose, and allocate tasks among a group of intelligent agents (Mataric *et al.*, 2003). Agents must be enabled to communicate and to interact by sensors and learning algorithms. They must be ensured to act coherently in their actions. Further on, they must recognize and reconcile conflicts. Therefore, distributed robotic systems are defined and designed in models of real computing, but technically constructed as digital systems. All functions of these systems are obviously realized through their members. The members' behaviors are results from their own needs and from the constraints set of the system, environment, or operator.

The number of distributed autonomous robotic system applications will increase as the technology and knowledge improve. Various robot societies will move from research laboratories into real life situations. Normal applications concern everyday tasks like cleaning, monitoring, caring people in households or hospitals, etc. (COTESYS, 2012). In the future, robot societies will be taken to distant planets, deep sea module collection missions, and mining operations. Robots on the nanolevel will be sent into veins for search of tumors, which they will attack at close range. Just as in natural systems the intelligence of the collective system will emerge from the multiple interactions among the members and with the environment.

In first steps of development, natural populations, for example ant societies, were studied to find the key issues. The next step as to design a physical society. It consists of two types of autonomous mobile robots, and the task for the society is classical, to gather stones from an unknown environment along with the mapping of the environment while operating. The society has been implemented both as a physical society with real computing and as a simulated one in digital models. A possible application is the use of a robot society inside industrial processes.

Societies of simple robots solve complex tasks

A profound challenge of interdisciplinary importance is the question how culture can emerge and evolve as a novel property in groups of social animals. The question can be solved by focusing the attention on the very early stages of the emergence and evolution of simple cultural artefacts. Therefore, one should start by building an artificial society of embodied intelligent agents as physical robots, creating an environment or artificial ecosystem and appropriate primitive behaviors for those robots, then freely running the artificial society. Embodied robots mean agents with analog sensors like living organisms interacting with their physical environment. Even with small populations of simple robots, a large number of interactions between robots can be generated (Brooks, 1999; Braitenberg and Radermacher, 2007; Pfeifer and Scheier, 2001). The inherent heterogeneities of physical robots, and the continuous fields of noise and uncertainty of the real world, increase the space of possibilities and the scope for unexpected emergence in the interactions between robots.

The goal is to create the conditions in which protoculture can emerge in a robot society. Robots can copy each other's behaviors and select which behaviors to copy. Behaviors will mutate because of the continuous noise and uncertainty in the analog robots' sensors and actuators. Successful types of behavior will undergo multiple cycles of copying (heredity), selection, and variation (mutation). With evolutionary time, a genetic algorithm process should be developed to grow and evolve the robots' controllers so that the emerging patterns of behavior become hard-wired into the robots' controllers.

The behavioral artefacts emerging and evolving will not be human but robotic. They will be meaningful only within the closed context of this artificial society. A significant challenge for these projects will be to identify and interpret these patterns of behavior as evidence for an emerging protoculture.

Based on the economic ideas about cost and utility, robot ecosystems were developed. From an ethologist point of view, robots are interesting for investigating biologically realistic issues. The emergence of cooperation can be studied by experiments with robots in real-valued settings. For example, in a robot ecosystem, there are a few robots and a few boxes with infrared lamps. The lamps are called competitors because they (like robots) consume energy from the ecosystem. As the overall amount of electricity is limited, the robots have to compete with the lamps for the electricity. A constant but restricted influx of energy into the system limits the amount of electricity available. The robots have to push against the competitors to reduce the competitors' energy consumption (Nolfi and Floreano, 2001; Mainzer, 2008).

Reducing the infrared lamps' energy consumption dims their lights. The darker the boxes, the more current there is in the charging station. The robots are equipped with the sensors whose activation levels can be coupled to motivations. If the environment is made more taxing by increasing the number of energy-consuming boxes, the robots have to exploit these analog sensors to produce beneficial sensor-to-motor couplings. Beneficial in this case means leading to higher energy levels.

For example, the robots have an analog sensor for internal energy level. If the sensor is coupled directly with the motivation for forward movement, the robot moves more slowly when the energy is low, which makes the robot take longer to move through the charge station. The robots also have a sensor for detecting the charging station. If the sensor is inversely coupled to the motivation for forward movement, this keeps the robot in the charging station even longer. In addition, the robots have light sensors, and the charging station is marked with a light, so they can potentially use phototaxis to approach the charging station. If the robots exploit the various sensory modalities in appropriate ways, they get an advantage in terms of energy management. This implies increasing the complexity of their behavioral repertoire, for example, better exploitation of the motor system.

Digital and analog computing in evolutionary robotics

When an evolutionary perspective is applied, one needs a value system. In the robot's case, it is related to energy supply: Increase in energy level is of value. The task of energy management is difficult, because the beneficial effects of certain actions manifest themselves only much later. This is a fundamental issue in reinforcement learning. The entire experiment can also be put into an evolutionary context. It would be interesting to study what sorts of value systems would evolve.

Cooperation can also be studied in this experiment. The energy influx into the charging station increases if another robot pushes against the energy consuming boxes. In a sense, the robot that pushes against the boxes works for the one in the charging station. If the robots cooperate in this way they can potentially draw more current from the charging station. Thus, they may be able to support more robots because they manage to divert electricity from the competitors. This entire procedure can also be embedded into an evolutionary cycle. In this case, one can see which analog sensors will be exploited given a particular environmental pressure. The fitness criterion in this case would be how long the agents can survive.

According to the local activity principle, it depends on their ability to amplify energy and generate efficient strategies of behavior.

The behavioral economics approach to autonomous agents brings in important ideas. Thinking in terms of utilities rather than specific quantities leads to plausible models of rational behavior. This framework is highly useful in understanding animal behavior and behavior of autonomous agents in general. The idea that rational behavior can occur without rational thinking or that the problems of thresholds can be avoided by using utilities is highly appealing. The fact that behavior is an emergent property of the interaction between robots and their environment has the consequence that simple robots can produce complex behavior. However it also has the consequence that, as in all dynamical systems, the properties of the emergent behavior cannot easily be predicted or inferred from the knowledge of the rules governing the interactions. The reverse is also true: it is difficult to predict which rules will produce a given behavior, since the behavior is the emergent result of the dynamical interaction between the robots and the environment.

However, the behavioral economics approach has some problems. The main drawback of behavioral economics is that the approach is more analytical than synthetic. It does not provide heuristics on how to derive the mechanisms that lead to the desired behaviors. Another difficulty is that the approach is top-down and suggests that careful prior analysis of the problem can produce optimal designs. Experience with software engineering suggests that a top-down approach may not work well in an unstructured domain. Moreover, from a cognitive science perspective, a bottom-up approach seems more promising.

Reminder of the Internet as complex dynamical system

Social networks of more or less autonomous robots are only one possible development in a general trend of future technology. In a technical co-evolution, global information and communication networks are emerging with surprising similarity to self-organizing neural networks of the human brain. The increasing complexity of the World Wide Web (www) needs intelligent strategies of information retrieval and learning algorithms simulating the synaptic plasticity of a brain (Berners-Lee, 1999). The Internet links computers and other telecommunication devices. At the router level, the nodes are the routers, and the edges are their physical connections. At the interdomain level, each domain of hundreds of routers is represented by a single node with at least one route as connection with other nodes. At both levels, the degree distribution follows a power law of scale-free network which can be compared with the networks in systems biology. Measurements of the clustering coefficient deliver values differing from random networks and significant clusters. The average paths at the domain level and the router level indicate the small-world property.

Smart grids as complex dynamical systems

In the future, global information networks will grow together with societal infrastructure. Current examples are complex smart grids of energy. Many energy providers of central generators and decentralized renewable energy resources lead to power delivery networks with increasing complexity. Energy is obviously a real-valued physical quantity. Smart grids mean the integration of the analog power delivery infrastructure with a digital communication and control network in order to provide the right information to the right entity at the right time to take the right action. It is a complex information, supply and delivery system, minimizing losses, self-healing, and self-organizing.

Smart grids are complex organizations of networks regulating, distributing, storing, and generating electrical power. Their structure and dynamics have surprising similarity with complex protein networks in systems biology regulating the energy supply of a cell. The intelligence of smart grids increases with their ability of self-organizing information processing for optimal energy supply. In communication networks, appropriate prices of optimal energy supply could be automatically negotiated by virtual agents. In smart grids, the analog energy system grows together with digital information and communication technology in a kind of symbiosis.

Automatic negotiations of virtual agents

A well-known problem with wind mills and solar cells is the unpredictability of production depending on changing weather conditions. In intelligent networks, the need can be locally satisfied by virtual negotiations. A model assumes the following rules and conditions of negotiating virtual agents (Wedde *et al.*, 2008):

1. The need for renewable energy can be satisfied either in a local regional subnet or between subnets. Reserve capacity is used only in exceptional cases.
2. Energy must be adjusted between different voltage levels or different groups of balance on the same level.
3. Producers are also consumers and vice versa.
4. Negotiations on local energy supply are automatically performed by agents of producers and agents of consumers. They are coordinated by balance group managers working parallelly and synchronized in time on each level.
5. In the model, the negotiations start in periods of 0.5 s. The negotiations as well as the distribution of negotiated energy are expected to be finished before the end of each period. Bids and offers arriving in the meantime are negotiated in the next period.
6. At the beginning of each period, each client decides whether he/she takes part as producer or consumer or not. He/she decides with respect to the current difference between the states of demand and production.
7. Bids and offers occur in frameworks of prices with respect to amortization and maintenance. In the model, there are no long-range contracts or discounts for big and future acquisitions which can occur in reality.

The algorithm of negotiation assumes a framework of prices for each level of negotiation. Each balance group manager on each level accomplishes a cycle of coordination of 10 turns. Each turn takes

1 ms. After each turn, the balance managers test in parallel whether bids and offers are sufficiently similar. If they are sufficiently similar, a contract between the partners is concluded. A fixed amount is added until the stock or demand is spent. The negotiation strategies of a client are given by an opening bid, an opening offer, and parameters of priority and strategy. After n turns, the unsatisfied agents adapt their bids and offers with respect to an exponential law of behavior which is useful to realize a fast convergence between bids and offers. The negotiated price is the arithmetic mean between similar values. Unsatisfied clients are passed on to the next level of negotiation. On this level, the framework of prices is reduced to a constant relation. The needs and interests of finally unsatisfied clients are satisfied by a central reserve capacity (but with very bad prices).

Digital and analog computing in smart grids

Short term fluctuations of consumption in the ms to min interval, which are effected by sudden and unpredicted local or regional causes, are not only observed as perturbations in households, but they can also endanger the stability of large transport networks. In our model, these critical situations are avoided by the activation of agents after each cycle of negotiation. It is assumed that many electrical appliances (e.g., refrigerator, boiler) can temporarily work without power or with a battery. In these cases, reserve energy can be used for other purposes. The reserve energy is more competitive than the traditional one because of low costs of transport and storage in the network. Additionally, the balance managers act on each level in parallel in shortest time. Thus, the agents act according to the principle of local activity: with low cost and power input, they generate complex patterns of efficient behavior.

In this way, smart grids with integrated digital communication systems accomplish an analog dynamical regulation of energy supply. They are examples of large and complex real-time systems according to the principles of cyberphysical systems (Lee, 2008). Traditionally, reserve energy which is used to balance peaks of consumption or voltage drops is stored by large power plants. The problem can be solved by dynamically reacting strategies of negotiation in proper time. The main problem of changing to renewable energies is the great number of constraints depending on questions of functionality as well a security, reliability, temporary availability, tolerance of failures, and adaptability. Cyberphysical systems with local and bottom-up structures are the best answer to the increasing complexity of supply and communication systems (Cyber-Physical Systems, 2008). In a technical co-evolution, mankind is growing together with these technical infrastructures in a kind of superorganism.

Digital and analog computing in IoT

Increasing computational power and acceleration of communication need improved consumption of analog quantities such as energy, better batteries, miniaturization of appliances, and refinement of display and sensor technology (Weiser, 1991; Hansmann, 2001). Under these conditions, intelligent digital functions can be distributed in a complex network with many multimedia terminals. Together with satellite technology and global positioning systems (GPS), digitally connected societies are transformed into cyberphysical systems. They are a kind of symbiosis of man, society, and machine. Digital communication is not only realized between human partners with natural languages, but with the things of this world. Cyberphysical systems also mean a transformation into IoT. Things in the Internet become locally active agents with analog sensors and digital communication.

Security in the digital and analog worlds

The increasing automation and complexity of the digital and analog worlds is a great challenge of human control. Human mind seems to be overwhelmed by the velocity and size of technical information systems which more and more determine human civilization. In Chapter 7, we explained proof mining in mathematics as the basis for program extraction and automated theorem proving. Proof verification and algorithms are a crucial topic in the digital world

(Mayr *et al.*, 1998). In order to prevent costly and life threatening flaws, theorem provers must be applied to verify and guarantee the security of hardware and software design. The verification of digital and analog hardware is important. The commercial costs of errors in microprocessors may be immense. High order logic interactive proof assistants are useful tools to support design verification (Kaufmann and Moore, 2004). In software verification, terrible failures can be caused by simple mistakes. An example was an error in the floating point unit of the Pentium II processor. The Ariane 5 rocket was destructed by a simple integer overflow problem in the software program which could have been detected by a verifying theorem prover of the program (Baier and Katoen, 2008).

A key problem is, of course, network security in the Internet. IoT is obviously the nervous system of human civilization. Banks, companies, and governments must exchange data globally and depend sensitively on network security. Secure data exchange does not only require sure encryption procedures which we discussed in Chapter 13. The whole design of the net must avoid hidden loopholes, dangerous signal transmission, and bad agents. Automated theorem provers could help to prove the safety of security protocols and predict failures (Cohen, 2000). Machine learning and artificial intelligence (AI) open new avenues of extended applications of automated proof mining.

What are intelligent systems?

Intelligent systems are a subclass of complex information systems which can be found in nature, technology, and society. There are different examples of intelligent systems — animals, primates and humans, populations and societies, computers and robots, information and communication networks. They are mainly integrated analog and digital systems with analog sensors and digital functions. Different kinds and degrees of intelligence can be distinguished. Technical systems generate intelligent functions sometimes only in interaction and in dependence of humans. But with increasing autonomy of agents and robots populations and with self-organization of information and communication networks, we observe the technical development of intelligent artificial systems surpassing natural evolution of organisms and populations.

Therefore, we define intelligence as an ability to solve problems. The degree of intelligence of a system depends on the complexity of problems to be solved, the efficiency of the applied algorithms, and the autonomy of the system. In this case, intelligence does not depend on humans. A tick is intelligent with respect to a particular class of problem solving (e.g., finding the blood of a host organism) as well as a computer program of pattern recognition. Of course, there are hybrid systems with several intelligent abilities like, e.g., humans and primates. But, these are only differences of complexity degrees, not in principle. The distinction of natural and artificial systems is only justified by the fact that "artificial" systems are once initiated by human technology. But, in the future, originally "artificial" systems may reproduce and organize themselves in an automated evolution (Mainzer, 2010, 2016a).

In the past, intelligent systems have only been developed during biological evolution by more or less random conditions. During the factual evolution on Earth, many species occurred according to the laws of evolution, but many of them disappeared again because of less adaptability to changing living conditions. The laws of evolution, e.g., self-replication, selection, and mutation, are independent of particular molecular substances. In the past, they were only applied to particular biomolecules on Earth. But, in cellular automata, they are virtually simulated, and in synthetic biology, they are already applied to new constructed chemical building blocks which were unknown in the chemistry of life on Earth. Thus, the criteria of life do not depend on a particular species on Earth. On this line, we also argue that the definition of intelligence does not depend on the abilities of a particular species on Earth. Thus, the traditional Turing test of artificial intelligence is less helpful because it relates machine intelligence to human intelligence (Turing, 1950). Furthermore, in our huge universe, new forms of life and intelligence are highly probable, even though still unknown.

Unified theory of complex networks

Intelligent systems are only subclasses of complex dynamical systems. Different disciplines are growing together in a unified theory of complex networks: systems and evolutionary biology, brain and cognition research, software and hardware engineering, robotics, information and communication networking, construction of cyber-physical systems and living infrastructures. Neural networks are mathematically equivalent to appropriate automata and machines with respect to language and pattern recognition. The equivalence of networks and machines is even true beyond Turing computability for analog networks and machines (cf. Chapter 12).

The common laws of this unified theory of complex networks are part of the theory of complex dynamical systems. Applications are self-organizing gene and protein networks, cellular organisms, agents and robots population, cyberphysical systems and communication networks. They all are typical examples of networks with locally active agents amplifying and transforming low energy and input signals into new patterns, structures, and behavior. The unified theory of networks in these fields is not yet completely accomplished, and, sometimes, we only know certain aspects and main features. But, at least in the natural sciences, it is not unusual to work successfully with incomplete theories: in elementary particle physics, the standard theory of unified forces has still many open questions, but it is nevertheless successfully applied to measure and solve problems.

Complex networks without recognizable patterns are described as random graphs (Albert and Barabási, 2002). They start with N nodes and connect pairs of nodes with a probability p. With this procedure, we get a graph with $pN(N-1)2$ randomly distributed edges. In many networks, there are relatively short connecting paths between two edges ("small worlds"). In complex molecular, cellular, social or technical networks, there are often patterns of, e.g., molecular clusters, cellular assemblies, social groups, circuit diagrams or wireless connections. During biological, social or technical evolution, they were mainly selected as robust structures which are assigned with biological, social or technical functions of regulation, control, logistics or communication. These clusters are also typical examples of locally active centers emerging from random networks and generating highly structured patterns. Protein networks with their integrated control, logistics, and information systems are examples of cyberphysical systems which are currently developing in social and technical networks of society.

Mathematically, the structures of networks are described by real cluster coefficients and real distribution degrees of connections between nodes. Gene and protein networks, cyberphysical systems, and the Internet are characterized by power laws of distribution degrees. The oriented edges are determined by two distribution degrees of incoming and outgoing connections with power law, i.e., $P_{\text{out}}(k) \sim k^{-\gamma_{\text{out}}}$ and $P_{\text{in}}(k) \sim k^{-\gamma_{\text{in}}}$ (Takayasu *et al.*, 1996, 1999). Power laws are real-valued and scale-invariant. Therefore, structures and patterns of networks are invariant if their dimensions are changed. During evolution, they were robust and keep their functions when small perturbations occurred, e.g., by random mutations.

Information systems with $1/f$–spectra

Information systems of gene and protein networks, cellular organs and organisms, social groups (e.g., facebook), and cyberphysical systems of human societies send measurable time series of signal patterns which allow conclusions on universal structures and functional processes in networks. In general, a hierarchy of signal patterns can be characterized by their real-valued spectra. The variability of a signal with respect to its periodic components is measured by its spectrum. The spectrum of signals is approximately proportional to $1/f^b$ with exponent $b > 0$, i.e., it varies inversely proportional to the power of frequency. This phenomenon is called $1/f$ noise (Mandelbrot, 1999).

$1/f$ spectra provide the criteria to distinguish the different forms of signal patterns in complex networks. Signal patterns of subcellular gene and protein networks as well as the noise of electronic messages are analyzable in global logistics and communication systems. The signals can indicate self-organizing structures, long-term trends, and isles of order in a sea of real-valued random noise. Their time series analysis can be applied to all kinds of complex networks: signal series of gene and protein nets, ECG dates of the heart, EEG waves of the brain, signal series of energy grids or stock data of financial markets. Time series analysis is important for early warning systems to prepare clients, patients, and users for future events.

Extended Church's thesis of computability

During evolution, cyberphysical systems in nature at first emerged as subcellular logistics, control and information systems in complex gene and protein networks on the basis of genetic and protein codes. Neurons were developed as specialized cells of information systems on the basis of neurochemical codes. Their networks are models of brains as well as ant populations or human societies. During evolution, effective information and logistic procedures were developed without symbolic representation in computer models. Subcellular, cellular, and neural self-organization generated the appropriate digital and analog networks. They are equivalent to complex systems which are modeled by non-linear differential equations. Dynamical systems and their differential equations can be simulated, at least in principle, by analog and digital computer models. Examples are cellular automata or neural networks with deterministic and stochastic algorithms.

At this point, a deep equivalence of evolutionary, mathematical, and technical procedures become obvious, leading to an extension of Church's thesis of computability: not only digital and effective procedures with mathematical symbols can be simulated with computers in the sense of a universal Turing machine, but also atomic, molecular, and cellular processes of nature with appropriately extended information systems. If the extended Church's thesis holds, then new avenues of computational technologies are opened beyond Turing computability: all dynamic effective procedures can be modeled on a universal analog and digital computer (Mainzer and Chua, 2011; Mainzer, 2016a). The symbolic and mathematical codes of a computer are only a human kind of information processing representing atomic, molecular, cellular, and evolutionary processes (Mainzer, 2016b). According to complexity theory, there are different degrees of analog and digital computability. According to reverse mathematics, the different degrees correspond to subclasses of classical mathematics. The question arises how far these processes are away from constructive procedures. This is not only an epistemic and philosophical question, but also economically interesting: The computational amount of simulation and problem solving cost time and money.

Chapter 17

Philosophical Outlook

Classification of constructive and computational problem solving is deeply rooted in fundamental questions of logic and philosophy. In the Middle Ages, William of Ockham (Ockham 1287–Munich 1347), Franciscan friar and scholastic philosopher, was an early forerunner of this research program (Ockham, 1967–1988). He strongly argued that only individuals exist, rather than supraindividual universals, essences, or forms. He denied the real existence of metaphysical universals and demands a reduction of ontological assumptions. With respect to constructive positions, he argued for efficient reasoning with a principle of parsimony in explanation and theory building. In modern times, this principle became popular as Ockham's razor. In the interpretation of Bertrand Russell, Ockham's razor only allows an explanation in terms of the fewest possible causes, factors, or variables (Russell, 2000).

Actually, Ockham himself argued that universals are generated by abstraction from individuals by the human mind. Thus, for him, they do not exist outside the human mind, but they have a mental existence as internal representation. In epistemology, Ockham is in the middle between two extreme positions (Spade, 1994). On the one side, Platonic philosophers advocated a "real" existence of ideas independent of the human mind. On the other side, radical nominalist like the medieval philosopher Roscelin stated that universals even have no existence in the human mind, but they are only names and signs as used in sentences. Obviously, these early epistemic positions foreshadow modern positions in the philosophy of mathematics. There are still believers in a Platonic world of abstract mathematical sets and structures, but there are also radical formalists reducing mathematics to a game of symbols. Following Ockham, mathematics is a product of the human mind. Historically, he was sometimes called a conceptualist or "terminist" contrary to a "nominalist" (from Latin "nomen" for "name") (Turner, 1913).

Ockham's razor and plea of ontological reductionism did influence not only modern logic and science, but also literature. In Umberto Eco's novel *Il nome della rosa* (1980), the central character is a Franciscan friar named William of Baskerville (Eco, 1983). The fictional name alludes both to the philosopher William of Ockham and the detective Sherlock Holmes in Conan Doyle's famous novel *The Hound of Baskerville*. According to Ockham's razor, William of Baskerville only follows logical rigorousness and facts, and avoids speculative hypotheses in his detective work.

As philosopher, Eco's William of Baskerville is a nominalist in the medieval debate on universals. His novel ends with a reference to nominalism: *Stat rosa pristina nomine, nomina nuda tenemus* (The rose of old remains only in its name; we hold only naked names). Obviously, Brouwer's intuitionism is no radical nominalism because, for him, intuitionistic mathematics is generated by the human mind of a "creative subject". For Brouwer, Heyting's formalizing of intuitionistic logic was not considered to be essential for mathematics. Constructive problem solving should start with basic individuals and argue step by step without circular conclusions ("predicativism"). Constructivism also follows the principle of parsimony and tries to reduce abstract entities as far as possible. In mathematics, the basic individuals are the natural numbers. All sets, functions, and functionals are abstractions of the human mind.

The historical William of Ockham also considered the consequences of his philosophical position in theology and politics (Ockham, 1998). In these fields, the individuals are the people. State and church are only abstract universals invented by the human mind. In the center of Ockham's studies, there are individual people, citizens and believers, with their interests and needs. According to his principle of parsimony, state and church as supraindividual universals should only be accepted as much as necessary to organize social and religious lives of people. These were obviously radical positions with great impact to modern ideas in the reformation of church, state, and society. Eco's novel impressively illustrates how bloody and brutal the struggle on universals may be in the world of politics and religion. Even nowadays, the fight on state and religion as supraindividual power is still going on.

But we must not forget that, for Ockham, parsimony is only a methodological demand of efficient human reasoning (Spade, 1999). He never claimed that the universe itself is created by God with a kind of divine principle of parsimony. These questions are undecidable for him and beyond the intellectual abilities of humans. Contrary to the Thomistic doctrine, Ockham considered the world to be contingent, and only God is believed to be a truly necessary entity. In the age of early Enlightenment (18th century), the question of parsimony in nature came up again in a famous debate between Voltaire and Pierre-Louis Moreau de Maupertuis (1698–1759), physicist and president of the Prussian Academy of Science in Berlin. In classical mechanics, Maupertuis's law states that the path followed by a physical system is the one of least length. It is a special case of the more generally stated principle of least action. Maupertuis generalized his mechanical law ontologically as God's universal principle of parsimony in creation. But, actually, there are many examples in nature which do not follow this line of parsimony. Therefore, Voltaire made fun about Maupertuis's theological speculation.

In the early 20th century, David Hilbert still used the principle of least action to unify the general theory of relativity with a (physically false) version of restricted elementary particle physics. Obviously, Hilbert's unification law was mathematically elegant and simple. But, from a physical point of view, Einstein made fun of this "childish" theory of everything. According to Einstein, we should keep our theories simple, but not too simple (and false).

Coming back to mathematics, theorems and proofs should be reduced to constructive principles as far as possible. But ordinary mathematics cannot be totally reduced to constructive principles. In constructive reverse mathematics, we learnt that there are degrees of constructivity and computability. This insight is also supported by constructive proof mining and program extraction. Therefore, in mathematics, Ockham's principle of parsimony demands to reduce the number of abstract entities (e.g., sets and functionals of any type), axioms, and principles as far as possible. But, like in economy, a proof has its "price" of necessary and sufficient principles which may be more or less constructive. There is no free lunch — even in mathematics.

The crucial question arises how far abstractions are necessary. Without any doubt, in ordinary mathematics, real and complex numbers, continuous and analog concepts are extremely useful to find robust and reliable problem solving. From a constructive point of view, it is a question of "price" which we are ready to "pay" with the assumption of axioms and principles. But, Ockham's principle of parsimony is only a useful methodological demand of efficient reasoning. It cannot exclude Platonism in principle.

With respect to modern quantum physics, physical reality seems to be discrete. Quantum processes could be considered quantum algorithms — the universe as a quantum computer. Nevertheless, quantum field theories use continuous functionals and spaces, real and complex numbers, to compute and predict events with great precision. Are analog and continuous concepts only useful inventions of the human mind in order to solve problems in a discrete world with approximations? In the sense of constructive reverse mathematics, the question arises how far these abstractions of the human mind are away from constructive concepts. In previous chapters, we considered real numbers as natural constants of physics which do not change in space and time. They are defined by fundamental proportions and relations of physics which are highly confirmed by experiments. The fine-structure constant is an example of a dimensionless real number. Like mathematical relations, they are exactly defined. We remember the real number π which is exactly defined by the relation of circumference and diameter of a circle. But, as decimal expansion, there are only approximately finite values which can be computed or measured by physical instruments in space and time.

If successful explanations and predictions of physical theories actually depend on real numbers as axiomatically defined whole entities, then they could be interpreted as hints on a mathematical reality "behind" or "beyond" (better: independent of) physical space and time. In Chapter 15, we discussed the possibility that physical reality is only a part in the universe of mathematical structures. It is the old Platonic belief in symmetries and structures as mathematical ground of the world (Mainzer, 1988). From Ockham's point of view, this assumption is a waste of ontological entities, but it cannot be excluded in principle and is still attractive in science and philosophy since more than 2000 years.

Mathematics is not used only to model the world. Since antiquity, mathematics opens new avenues of technologies. In engineering sciences, real and complex numbers are used to compute secure and reliable technical machines. Analog neural networks with synaptic weights of real numbers promise neuromorphic computational structures of future technologies. Digital technologies and the analog world of life, humans and society should converge to master the increasing complexity of our civilization. We are living in a world with exponentially growing power of algorithms! Therefore, we end with a plea for more foundational and interdisciplinary research to get controllable and reliable tools. Since its very beginning, mathematical science is deeply rooted with the foundations of human existence.

References

Aczel, P. (1978): The type theoretic interpretation of constructive set theory. In: Macintyre, A.; Pacholski, L.; Paris, J. (Eds.), *Logic Colloquium '77*. North-Holland: Amsterdam-New York, 55–66.

Aigner, M.; Ziegler, G.M. (2001): *Proofs from The Book*. Berlin 2nd edition, 3.

Alber, G.; Beth, T.; Horodecki, M.; Horodecki, P.; Horodecki, R.; Rötteler, M; Weinfurter, H. Werner, R.; Zeilinger, A. (2001): *Quantum Information. An Introduction to Basic Theoretical Concepts and Experiments*. Springer: Berlin.

Albert, R.; Barabási, A.-L. (2002): Statistical mechanics of complex networks. *Reviews of Modern Physics* 74(1), 47–97.

Aoki, M.; Yoshikawa, H. (2007): *Reconstructing Macroeconomics — A Perspective from Statistical Physics and Combinatorical Stochastic Processes*. Cambridge University Press: Cambridge.

Arbib, M.A. (1987): *Brains, Machines, and Mathematics*. Springer: New York.

Arora, S.; Barak, B. (2009): *Computational Complexity. A Modern Approach*. Cambridge.

Artzner, P.; Delbaen, F.; Eber, J.-M.; Heath, D. (1999): Coherent measures of risk. In: *Mathematical Finance* 9(3), 203–228.

Audretsch, J.; Mainzer, K. (Eds.) (1994): *Philosophie und Physik der Raum-Zeit*. B.I. Wissenschaftsverlag: Mannheim 2nd edition.

Audretsch, J.; Mainzer, K. (Eds.) (1996): *Wieviele Leben hat Schrödingers Katze? Zur Physik und Philosophie der Quantenwelt*. Spektrum akademischer Verlag: Heidelberg 2nd edition.

Avigad, J.; Gerhardy, P.; Towsner, H. (2007): Logical stability of ergodic averages. In: *arXiv:0706.1512v1* (math.DS).

Awodey, S.; Warren, M.A. (January 2009): Homotopy theoretic models of identity type. In: *Mathematical Proceedings of the Cambridge Philosophical Society* 146(1), 45–55.

Awodey, S. (2012): Type theory and homotopy. In: Dybjer, P. *et al. Epistemology versus Ontology.* Logic, Epistemology, and the Unity of Science. Springer. pp. 183–201.

Awodey, S. (2014): Structuralism, Invariance, and Univalence. In: Philosophia Mathematica. Oxford University Press. 22(1), 1–11.

Bachelier, L. (1900): Théorie de la spéculation. Dissertation. In: *Annales Scientifiques de l'École Normale Supérieure* 17, 21–86.

Bachmann, H. (1967): Transfinite Ordinalzahlen. Springer: Berlin 2nd edition.

Baez, J.; May, P.J. (2010): *Towards Higher Category Theory.* Springer: Berlin.

Baier, C.; Katoen, J. (2008): *Principles of Model Checking.* MIT Press: Cambridge, MA.

Balch, T.; Parker, L. (eds.) (2002): *Robot Teams: From Diversity to Polymorphism.* Wellesley MA.

Barrow-Green, J. (1997): *Poincaré and the Three Body Problem.* American Mathematical Society: Providence, Rhode Island.

Beeson, M. (1985): *Foundations of Constructive Mathematics.* Springer: Berlin.

Bell, J.S. (1964): On the Einstein-Podolsky-Rosen-Paradoxon. In: *Physics* 1, 195–200.

Beltrami, E. (1999): *What is Random? Chance and Order in Mathematics and Life.* Copernicus: New York.

Bennett, C.H. (1995): Logical Depth and Physical Complexity. In: R. Herken (Hrsg.), *The Universal Turing Machine. A Half-Century Survey,* Springer: Wien 1995, 207–235.

Berger, J. (2005): The fan theorem and uniform continuity. In: Cooper, S.B.; Löwe, B.; Torenvliet, L. (Eds.): *New Computational Paradigms.* Springer: Berlin, 18–22.

Berger, J.; Ishahara, H. (2005): Brouwer's fan theorem and unique existence in constructive analysis. In: *MLQ Math. Logic Q.* 51, 360–364.

Berger, J.; Svindland, G. (2016a): Convexity and constructive infima. In: *Archive for Mathematical Logic* 55, 873–881.

Berger, J.; Svindland, G. (2016b): A separating hyperplane theorem. The fundamental theorem of asset pricing, and Markov's principle. In: *Annals of Pure and Applied Logic* 167, 1161–1170.

Berger, U.; Eberl, M.; Schwichtenberg, H. (2003): Term rewriting for normalization by evaluation. In: *Information and Computation* 183, 19–42.

Berners-Lee, T. (1999): *Weaving the Web: The Original Design and Ultimate Destiny of the World Wide Web by the Inventor.* San Francisco.

Bernstein, J. (1974): Spontaneous symmetry breaking, gauge theories, the Higgs mechanism and all that. In: *Revise Reports of Modern Physics* 46, 7–48.

Bertot, Y.; Castéran, P. (2004): *Interactive Theorem Proving and Program Development. Coq'Art: The Calculus of Inductive Constructions.* Springer: New York.

Bezem, M.; Coquand, T.; Huber, S. (2014): A model of type theory in cubical sets. In: R. Matthes, A. Schubert (eds.), *19th International Conference on Types for Proofs and Programs (TYPES 2013).* Dagstuhl, Germany: Schloss Dagstuhl–Leibniz-Zentrum fuer Informatik, 107–128.

Birkhoff, G.D. (1931): Proof of the ergodic theorem. In: Proc. Natl. Acad. Sci. USA 17 (12), 656–660.

Bishop, E. (1967): *Foundations of Constructive Analysis.* McGraw-Hill: New York.

Bishop, E.; Bridges, D. (1985): *Constructive Analysis.* Springer: New York.

Black, F.; Scholes, M. (1973): The pricing of options and corporate liabilities. In: *Journal of Political Economy* 81, 637–654.

Blum, L.; Shub, M.; Smale, S. (1989): On a theory of computation and complexity over the real numbers: NP-completeness, recursive functions and universal Machines. In: *Bulletin of the American Mathematical Society* 21(1), 1–46.

Blum, L.; Smale, S. (1993): The Gödel incompleteness theorem and decidability over a ring. In: Hirsch, M.; Marsden, Shub, M. (Eds.): *From Topology to computation: Proceedings of the Smalefest.* Springer: New York, 321–339.

Blum, L.; Cucker, F.; Shub, M.; Smale, S. (1998): *Complexity and Real Computation.* Springer: New York.

Boltzmann, L. (1896): *Vorlesungen über Gastheorie.* Barth: Leipzig.

Börger, E. (1989): *Computability, Complexity, Logic.* North-Holland: Amsterdam.

Bouwmeester, D.; Ekert, A.; Zeilinger, A. (Hrsg.) (2000): *The Physics of Quantum Information. Quantum Cryptography, Quantum Teleportation, Quantum Computation.* Springer: Berlin.

Braitenberg, V., Radermacher, F.J. (eds.) (2007): *Interdisciplinary Approaches to a New Understanding of Cognition and Consciousness.* Ulm.

Bridges, D.S.; Richman, F. (1987): *Varieties of Constructive Mathematics.* London Math. Soc. Lecture Notes 97. Cambridge University Press: Cambridge.

Bridges, D.S.; Vita, L.S. (2006): *Techniques of Constructive Analysis.* Universitext. Springer: New York.

Bromley, A.G. (1982): Charles Babbage's Analytical Engine 1938. In: *Annals of the History of Computing* 4, 196–219.

Brooks, R.A. (1999): *Cambrian Intelligence: The Early History of the New AI.* The MIT Press: Cambridge MA.

Brouwer, L.E.J. (1907): *Over de Grondslagen der Wiskunde* [On the foundations of Mathematics]. Maas & Van Suchtelen: Amsterdam.

Brouwer, L.E.J. (1927): Über Definitionsbereiche von Funktionen. In: *Mathematische Annalen* 97, 60–75.

Brouwer. L.E.J. (1975): *Collected Works I* (ed. A. Heyting). North-Holland: Amsterdam.

Brouwer, L.E.J. (1981): *Brouwer's Cambridge Lectures on Intuitionism* (ed. D. van Dalen). Cambridge university Press: Cambridge.

Carrington, G. (1994): *Basic Thermodynamics.* Oxford University Press: Oxford.

Chadzelek, T.; Hotz, G. (1997): *Analytic machines.* Technical Report 12/97, Sonderforschungsbereich 124 (VLSI-Entwurfsmethoden und Parallelität), Universität des Saarlandes: Saarbrücken.

Chadzelek, T. (1998): *Analytische Maschinen.* Diss. Universität des Saarlands: Saarbrücken.

Chaitin, G.J. (1969): On the length of programs for computing finite binary sequences: statistical considerations. In: *J. ACM* 16, 145.

Chaitin, G.J. (1998): *The Limits of Mathematics.* Springer: Singapore.

Chaitin, G.J. (2007): *Thinking about Gödel and Turing. Essays on Complexity 1970–2007,* World Scientific: Singapore.

Church, A.; Kleene, S.C. (1936): Formal definitions in the theory of ordinal numbers. In: *Fundamenta Mathematica* 28, 11–21.

Church, A. (1941): *The Calculi of Lambda-Conversion.* Princeton, repr. New York, 1965.

Churchland, P.S.; Sejnowski, T.J. (1992): *The Computational Brain.* MIT Press: Cambridge MA.

Cohen, E. (2000): Taps: A first-order verifier for cryptographic protocols. In: *Computer Security Foundations Workshop,* volume CSFW-13 Proceedings 13 IEEE 3–5 July, 144–158.

Colander, D.; Föllmer, H.; Haas, A.; Goldberg, M.; Juselius, K.; Kirman, A.; Lux, T.; Sloth, B. (2008): The financial crisis and the systemic failure of academic economics. In *Discussion Papers* 09-03, Department of Economics, University of Copenhagen.

Cont, R. (2006): Model uncertainty and its impact on the pricing of derivative instruments. *Math. Finance* 16, 519–542.

Conway, J. H.; Guy, R. K. (1996): Cardinal Numbers. In: *The Book of Numbers*. Springer: New York, 277–282.

Coquand, T.; Huet, G. (1988): The calculus of constructions. In: *Information and Computation* 76(2–3), 95–120.

Corry, L. (2004): *David Hilbert and the Axiomatization of Physics (1998–1918). From Grundlagen der Geometrie to Grundlagen der Physik*. Kluwer Academic Publisher: Dordrecht.

COTESYS 2012: DFG Research Cluster of Excellence on Humanoid Robots. Technische Universität München: Munich.

Cramér, H. (1969): Historical review of Filip Lundberg's works on risk theory. In: *Scandinavian Actuarial Journal*, 6–9.

Cucker, F.; Koiran, P. (1995): Computing over the reals with addition and order. Higher complexity classes. In: *J. of Complexity* 11, 358–376.

Cucker, F.; Matamala, M. (1996): On digital nondeterminism. In: *Mathematical Systems Theory* 29, 635–647.

Cucker, F.; Rosseló, F. (1993): Recursiveness over the complex numbers is time bounded. In: *13th Foundations of Software Technology and Theoretical Computer Science*. Springer: New York, 260–267.

Cyber-Physical Systems. Program Announcements & Information. *The National Science Foundation*, 4201 Wilson Bouleward, Arlington, Virginia 22230, USA, 2008-09-30.

Danielsson, J.; Embrects, P.; Goodhart, C.; Keating, C.; Muennich, F.; Renault, O.; Song Shin, H. (2001): An academic response to Basel II. *Special Paper Series* No. 130 May 2001, LSE Financial Markets Group, ESRC Research Centre.

Davis, M. (1958): Computability & Unsolvability. McGraw-Hill: New York.

Davis, M.; Putnam, H.; Robinson, J. (1961): The decision problem for exponential diophantine equations. In: *Ann. of Math.* II 74, 425–436.

Deco, G.; Schürmann, B. (2001): *Information Dynamics. Foundations and Applications*. Springer. Berlin.

Delbaen, F.; Schachermayer, W. (1994): A general version of the fundamental theorem of asset pricing. In: *Mathematische Annalen* 300(1), 463–520.

Detlefsen, K.; Scandolo, G. (2005): Conditional and dynamic convex risk measures. In: *Finance Stoch.* 9(4), 539–561.

Detlefsen, M. (1986): *Hilbert's Program*. Reidel/Kluwer Academic: Dordrecht.

Deuflhard, P.; Bornemann, F. (1994): *Integration gewöhnlicher Differentialgleichungen* II (*Numerische Mathematik*). De Gruyter: Berlin.

Deutsch, D. (1985): Quantum theory, the Church-Turing principle and the universal quantum computer. In: *Proceedings of the Royal Society of London* A 400, 97–117.

428   *References*

Deutsch, D.; Eckert, A. (2000): *Concepts of quantum computation.* In: Bouwmeester *et al.* (2000), Kap. 4.

Doncel, M.G.; Hermann, A.; Michel, L.; Pais, A. (Hrsg.) (1987): *Symmetries in Physics 1600–1980.* Seminari d'Història de les Ciències. Universitat Autònoma de Barcelona: Bellaterra (Barcelona).

Drake, F.R. (1974). *Set Theory: An Introduction to Large Cardinals.* North-Holland: Amsterdam.

Dubucs, J.; Bourdeau, M. (Eds.) (2014): *Constructivity and Computability in Historical and Philosophical Perspective.* Springer: Heidelberg.

Dummett, M. (1977): *Elements of intuitionism.* Clarendon Press: Oxford.

Dybjer, P.; Palmgren, E. (2016); Intuitionistic Type Theory. In: *The Stanford Encyclopedia of Philosophy,* The Metaphysics Research Lab (CSLI), Stanford University: Stanford (open access).

Ebbinghaus, H.-D.; Hermes, H.; Hirzebruch, F.; Koecher, M.; Mainzer, K.; Neukirch, J.; Prestel, A.; Remmert, R. (1991): *Numbers.* Springer: Berlin 3rd edition.

Ebeling, W.; Freund, F.; Schweitzer, F. (1998): *Komplexe Strukturen: Entropie und Information.* B.G. Teubner: Stuttgart.

Eco, U. (1983): *The Name of the Rose.* Harcourt: San Diego.

Eilenberg, S.; Cartan, H. (1956): *Homological Algebra.* Princeton University Press: Princeton.

Eilenberg, S.; MacLane, S. (1942): Group Extensions and Homology. In: *Annals of Mathematics* 43, 757–831.

Ekert, A.; Jozsa, R. (1996): Quantum computation and Shor's factoring algorithm. In: *Rev. mod. Phys.* 68, 733–753.

Ekert, A.; Gisin, N.; Huttner, B.; Inamori, H.; Weinfurter, H. (2000): Quantum cryptography. In: Boumeester *et al.* (2000), chapter 2.4.

Embrechts, P.; Klüppelberg, C.; Mikosch, T. (2003): *Modeling Extremal Events for insurance and Finance.* Springer: Berlin.

Enderton, H.; Luckham, D. (1964): Hierarchies over recursive well-orderings. In: *Journal of Symbolic Logic* 29, 183–190.

Euclid (1782): *The Elements of Euclid (with Dissertations).* J. Williamson (translator and commentator). Clarendon Press: Oxford.

Everett, H. (1957): "Relative state" formulation of quantum mechanics. In: *Review of Modern Physics* 29, 454–462.

Feferman, S. (1962): Transfinite progressions of axiomatic theories. In: *Journal of Symbolic Logic* 27, 259–316.

Feferman, S. (1964): Systems of predicative analysis. In: *Journal of Symbolic Logic* 29, 1–30.

Feferman, S. (1968a): Systems of predicative analysis II: Representations of ordinals. In: *Journal of Symbolic Logic* 53, 193–213.

Feferman, S. (1968b): Lectures on Proof Theory. In: *Proceedings of the Summer School in logic (Leeds)*. Springer: Berlin, 1–107.

Feferman, S. (1977): Categorical Foundations and Foundations of Category Theory. In: R. Butts (ed.), *Logic, Foundations of Mathematics and Computability*. Reidel: Dordrecht, 149–169.

Feferman, S. (1996): Kreisel's "unwinding" Program, in: P. Odifreddi (Ed): *Kreisleriana. About and Around Georg Kreisel. Review of Modern Logic*, 247–273.

Feferman, S. (2002): Predicativity. Lecture at a meeting of the American Philosophical Association March 28, 2002.

Feferman, S. (2006): Turing's Thesis, in: Notices of the American Mathematical Society 53(10), 1200–1206.

Feynman, R.P. (1967): *The Character of Physical Law*. The M.I.T Press: Cambridge Mass.

Feynman, R.P. (1982): Simulating physics with computers. In: *International Journal of Theoretical Physics* 21 6-7, 467–488.

Föllmer, H.; Schied, A. (2004): *Stochastic Finance. An Introduction into Discrete Time*. De Gruyter: Berlin 2nd edition.

Föllmer, H., Schied, A. (2008): Convex and coherent risk measures. *Working Paper*, Institute for Mathematics, Humboldt-University Berlin, October.

Fredkin, E. (1990): Digital Mechanics: An informational process based on reversible universal CA. In: *Physica* D, 254–270.

Friedman, H. (1975): Some systems of second order arithmetic and their use. In: *Proceedings of the International Congress of Mathematicians* (Vancouver, B.C., 1974), 1. Canad. Math. Congress. Montreal, 235–242.

Friedman, H.; Simpson, S.G. (2000): Issues and Problems in Reverse Mathematics. In: *Computability Theory and Its Applications. Contemporary Mathematics* 257, 127–144.

Frydman, R.; Goldberg, M.D. (2007): *Imperfect Knowledge Economics*. Princeton University Press: Princeton.

Frydman, R.; M.D. Goldberg, M.D. (2008): Macroeconomic Theory for a World of Imperfect Knowledge. In *Capitalism and Society* 3(3), 1–76.

Gentzen, G. (1934): Untersuchungen über das logische Schließen I. In: *Mathematische Zeitschrift* 39(2), 176–210.

Gentzen, G. (1936): Die Widerspruchsfreiheit der reinen Zahlentheorie. In: *Mathematische Annalen* 112, 493–565.

Gentzen, G. (1938): Die gegenwärtige Lage in der mathematischen Grundlagenforschung. In: *Deutsche Mathematik* 3, 260.

Gerhardy, P.; Kohlenbach, U. (2008): General logical metatheorems for functional analysis. In: *Trans. Amer. Math. Soc.* 360, 2615–2660.

Glendinning, P. (1994): *Stability, Instability, and Chaos. An Introduction to the Theory od of Nonlinear Differential Equations.* Cambridge University Press: Cambridge.

Gonthier, G. (2008): Formal Proof — The Four-Color Theorem. In: *Notices of the American Mathematical Society* 55(11), 1382–1393.

Goodstein, R.L. (1964): *Recursive Number Theory.* North-Holland: Amsterdam.

Gödel, K. (1931): Über formal unentscheidbare Sätze der Principia Mathematica und verwandter Systeme I. In: *Monatshefte für Mathematik und Physik* 38(1), 173–198.

Gödel, K. (1933): Zur intuitionistischen Arithmetik und Zahlentheorie. In: *Ergebnisse eines Mathematischen Kolloquiums* 4, 34–38.

Gödel, K. (1958): Über eine bisher noch nicht benützte Erweiterung des finiten standpunktes. In: *Dialectica* 12, 280–287.

Goldstine, H. (1977): *A History of Numerical Analysis from the 16th through the 19th Century.* Springer: New York.

Greenspan, A. (2008): We will never have a perfect model of risk. *Financial Time* 17 March.

Grötschel, M.; Lovács, L.; Schryver, A. (1988): Algorithms and Combinatorial Optimization. Springer: Berlin.

Grothendieck, A. *et al.* (1971–1977): *Séminaire de Géométrie Algébrique* Vol. 1–7, Berlin: Springer.

Haken, H. (1988): *Information and Self-Organization.* Springer: Berlin.

Haken, H. (1990): *Synergetk. Eine Einführung.* Springer: Berlin.

Hansmann, U. (2001): *Pervasive Computing Handbook.* Springer: Berlin.

Hawking, S. W. (2005, 2009): Information loss in black holes. In: *Physical Review D. American Physical Society.* 72(8) 2005: 4. Retrieved 8 October 2009.

Hayek, F.A. (1948): *Individualism and Economic Order.* The University of Chicago Press: Chicago.

Hayek, F.A. (1974): The pretence of knowledge, Nobel Lecture 1974. In: *New Studies in Philosophy, Politics, Economics, and History of Ideas.* The University of Chicago Press: Chicago.

Hazewinkel, M. (ed.) (2001): Homotopy. In: *Encyclopedia of Mathematics,* Springer: New York.

Herken, R. (Hrsg.) (1995): *The Universal Turing Machine. A Half-Century Survey.* Springer: Wien.

Herbrand, J. (1967): Investigations in proof theory: The properties of true propositions. In: van Heijenoort, J.: *From Frege to Gödel: A Source Book in Mathematical Logic, 1879–1931.* Harvard University Press 1967, 525–581.

Hermes, H. (1978): Aufzählbarkeit, Entscheidbarkeit, Berechenbarkeit. Einführung in die Theorie der rekursiven Funktionen. Springer: Berlin 3rd edition (1st edition 1961).

Heyting, A. (1934): *Mathematische Grundlagenforschung. Intuitionismus. Beweistheorie.* Springer: Berlin repr. 1974.

Hilbert, D. (1900): Mathematische Probleme. In: *Nachrichten der Königlichen Gesellschaft der Wissenschaften zu Göttingen, mathematisch-physikalische Klasse.* Heft 3, 253–297.

Hilbert, D. (1918): Axiomatisches Denken. In: *Mathematische Annalen 78,* 405–415.

Hilbert, D. (1926): Über das Unendliche. In: *Mathematische Annalen 95.* 161–190.

Hinman, P.G. (1978): *Recursion-Theoretic Hierarchies.* Springer: Berlin.

Hintikka, J.; Remes, U. (1974): *The Method of Analysis — Its Geometrical Origin and Its General Significance.* North-Holland: Dordrecht.

Hoffmann, D.W. (2013): *Grenzen der Mathematik. Eine Reise durch die Kerngebiete der mathematischen Logik.* Springer: Berlin 2nd edition.

Horowitz, G.T.; Maldacena, J. (2004): The black hole final state. In: *Journal of High Energy Physics 2,* 8.

Hotz, G.; Schieffer, B.; Vierke, G. (1995): *Analytical Machines.* Technical Report TR95-025. Electronic Colloquium on computational complexity. Universität des Saarlands: Saarbrücken.

Howard, W.A. (1968): Functional interpretation of bar induction by bar recursion. In: *Compos. Math.* 20, 107–124.

Howard, W.A. (1969): The formulae-as-types notion of construction. In: Seldin, J.P.; Hindley, J.R. (Eds.), *To H.B. Curry: Essays on Combinatory Logic, Lambda Calculus and Formalism,* Academic Press: Boston, MA, 479–490.

Howard, W.A. (1973): Hereditarily majorizable functionals of finite type. In: Troelstra, A.S. (ed.): *Metamathematical Investigation of Intuitionistic Arithmetic and Analysis.* Springer: Berlin 1973, 454–461.

International Monetary Fund, *Global Financial Stability Report* (2009): *Responding to the Financial Crisis and Measuring Systemic Risk.* Washington, April 2009.

Ishahara, H. (1990): An omniscience principle, the König's lemma, and the Hahn-Banach theorem. In: *Z. Math. Logik u. Grundlagen Math.* 36, 237–240.

Ishahara, H. (1992): Continuity properties in constructive mathematics. In: *J. Symbolic Logic* 57, 557–565.

Ishahara, H. (1993): Markov's Principle, Church's thesis, and Lindelöf's theorem. In: *Indag. Math.* 4, 321–325.

Ishahara, H.; Schuster, P. (2004): Compactness under constructive scrutiny. In: *MLQ Math. Log. Q.* 50, 540–550.

Ishihara, H. (2005): *Constructive Reverse Mathematics. Compactness Properties.* Oxford University Press. Oxford Logic Guides 48, 245–267.

Ishahara, H. (2006): Reverse mathematics in Bishop's constructive mathematics. In: *Philosophia Scientiae* 6, 43–59.

Joyal, A. (2002): Quasi-categories and Kan complexes. In: *Journal of Pure and Applied Algebra* 175, 207–222.

Kant, I. 1787): *Werke.* Akademie Textausgabe. De Gruyter: Berlin 1968. Band 3: *Kritik der reinen Vernunft.* (Nachdruck der 2. Auflage 1787).

Kaufmann, M.; and J.S. Moore, J.S. (2004): Some key research problems in automated theorem proving for hardware and software verification. In: *RACSAM, Rev. R.Acad. Cien. Serie. A. Mat.*, 98(1), 181–195.

Keyl, M. (2002): Fundamentals of quantum information theory. In: *Physics Reports. A Review Section of Physics Letters* 369, 431–548.

Kleeberg, J.M.; Schlenger, C. (2000): Value-at-Risk im Asset Management. In: L. Johannig, B. Rudolph (eds.): *Handbuch Risikomanagement.* Uhlenbruch-Verlag, 973–1014.

Kleene, S.C. (1938): On notation for ordinal numbers. In: *Journal of Symbolic Logic* 3, 150–155.

Kleene, S.C. (1945): On the interpretation of intuitionistic number theory. In: *Journal of Symbolic Logic* 10, 109–124.

Kleene, S.C. (1955a): Hierarchies of number theoretic predicates. In: *Bull. of the American Math. Soc.* 61, 193–213.

Kleene, S.C. (1955b): On the forms of the predicates in the theory of constructive ordinals (second paper). In: *American Journal of mathematics* 77, 405–428.

Kleene, S.C. (1956): Representation of events in nerve nets and finite automata. In: Shannon, C.E.; McCarthy, J. (Eds.). *Automata Studies.* Princeton University Press: Princeton, 3–41.

Kleene, S.C. (1958): Arithmetical predicates and function quantifiers. In: *Transactions of the American Mathem. Society*, 312–340.

Kleene, S.C. (1959): Quantification of number-theoretic functions. In: *Compositio Math.* 14, 23–40.

Kleene, S.C. (1974): *Introduction to Metamathematics.* North Holland: Amsterdam 7th edition.

Knights, F. (1921): *Risk, Uncertainty, and Profit.* Ph.D. Yale.

Kobayashi, K. (1981): *On compressibility of infinite sequences.* Technical Report C-34. Department of information Sciences. Tokyo Institute of Technology: Tokyo.

Kohlenbach, U. (1993): Effective moduli of uniqueness from ineffective uniqueness proofs. An unwinding of de La Vallée Poussin's proof for Chebycheff approximation. In: *Ann. Pure Appl.Log.*, 27–94.

Kohlenbach, U. (1999): The use of a logical principle of uniform boundedness in analysis. In: Cantini, A.; Casari, E.; Minari, P. (Eds.). *Logic and Foundations of Mathematics*. Kluwer Academic: Dordrecht, 93–106.

Kohlenbach, U. (2005): Some logical metatheorems with application to functional analysis. In: *Trans. Amer. Math. Soc.* 357(1), 89–128.

Kohlenbach, U. (2008): Applied Proof Theory: Proof Interpretations and Their Use in Mathematics. Springer: Berlin.

Kolmogorov, A.N. (1932): Zur Deutung der intuitionistischen Logik. In: *Math. Z.* 35, 58–65.

Kolmogorov, A.N. (1965): Three approaches for defining the concept of information quantity. In: *Probl. Information Transmission* 1, 1–7.

Kreisel, G. (1959): An interpretation of analysis by means of constructive functionals of finite types. In: Heyting, A. (ed.): Constructivity in Mathematics. North Holland. Amsterdam, 101–128.

Kreisel, G. (1960): La predicativité. In: *Bull. de la Soc. Math. de France* 88, 371–391.

Kreisel, G. (1967): Informal rigour and completeness proofs. In: Lakatos, I. (ed.): *Problems in the Philosophy of Mathematics. Proceedings of the International Colloquium in the Philosophy of Science* 1965. North-Holland: Amsterdam, 138–186.

Lambek, J. (1968): Deductive Systems and Categories I. Syntactic Calculus and Residuated Categories. In: *Mathematical Systems Theory* 2, 287–318.

Lambek, J. (1969): Deductive Systems and Categories II. Standard Constructions and Closed Categories. In: *Category Theory, Homology Theory and their Applications I*. Springer: Berlin, 76–122.

Lambek, J. (1972): Deductive Systems and Categories III. Cartesian Closed Categories, Intuitionistic Propositional Calculus, and Combinatory Logic. In: *Toposes, Algebraic Geometry and Logic* (Lecture Notes in Mathematics, Volume 274). Springer: Berlin, 57–82.

Ledrappier, F.; Young, L.-S. (1985): The metric entropy of diffeomorphisms. Part I: Characterization of measures satisfying Pesin's entropy formula. In: *Annals of Mathematics* 122, 509; Part II: Relations between entropy, exponents, and dimension. In: *Annals of Mathematics* 122, 540.

Leibniz, G.W. (1998): *Philosophical Texts*. Edited and translated by R.S. Woolhouse and R. Francks. Oxford University Press: Oxford.

Li, M.; Vitányi, P.M.B. (1993): *An Introduction to Kolmogorov Complexity and Its Applications*. Texts and monographs in Computer Science Series. Springer: New York.

Loeb, I. (2005): Equivalents oft he (Weak) Fan theorem. In: *Ann. Pure Appl. Logic* 132, 51–66.

Lord Turner (2009): *The Turner Review*. A regulatory response to the global banking crisis. March 2009. The Financial Services Authority, 25 The North Colonnade, Canary Wharf, London E14 5HS.

Lorenzen, P. (1955): *Einführung in die operative Logik und Mathematik*. Springer: Berlin.

Lorenzen, P. (1965): *Differential und Integral. Eine konstruktive Einführung in die klassische Analysis*. Akademische Verlagsgesellschaft. Frankfurt.

Lorenzen, P. (1980): *Metamathematik*. B.I. Wissenschaftsverlag: Mannheim 2nd edition.

Luckhardt, H. (1973*): Extensional Gödel Functional Interpretation*. Springer: Berlin.

Ludwig, G. (1978): *Die Grundstrukturen einer physikalischen Theorie*. Springer: Berlin.

Lux, T.; Westerhoff, F. (2009): Economics crisis. *Nature Physics* 5, 2–3.

Maccheroni, F.; Marinaci, M.; Rustichini, A. (2006): Ambiguity aversion, robustness, and the variational representation of preferences. *Econometrica* 74, 1447–1498.

Macintyre, A; McKenna, K.; van den Dries, L. (1983): Elimination of quantifiers in algebraic structures. In: *Adv. In Math.* 47, 74–87.

Mackensen, L. von (1974): Leibniz als Ahnherr der Kybernetik — ein bisher unbekannter Vorschlag einer "Machina arithmetica dyadicae". In: Akten des II Internationalen Leibniz-Kongresses 1972 Bd. 2. Steiner: Wiesbaden, 255–268.

Mainzer, K. (1970): Der Konstruktionsbegriff in der Mathematik. In: *Philosophia Naturalis* 12, 367–412.

Mainzer, K. (1973): Mathematischer Konstruktivismus, Ph.D. U. Münster.

Mainzer, K. (1977): Is the Intuitionistic Bar-Induction a Constructive Principle? In: *Notre Dame Journal of Formal Logic* 18(4), 583–588.

Mainzer, K. (1980): *Geschichte der Geometrie*. B.I. Wissenschaftsverlag: Mannheim.

Mainzer, K. (1981): *Grundlagen und Geschichte der exakten Wissenschaften*. Universitätsverlag: Konstanz.

Mainzer, K. (1988): *Symmetrien der Natur*. De Gruyter: Berlin, New York (English translation: *Symmetries of Nature*. De Gruyter: New York 1996).

Mainzer, K. (1994): *Computer — Neue Flügel des Geistes?* De Gruyter: Berlin, New York.

Mainzer, K. (1997): *Gehirn, Computer, Komplexität*. Springer: Berlin.

Mainzer, K. (2000): *Hawking*. Herder: Freiburg.

Mainzer, K. (2003): *KI — Künstliche Intelligenz. Grundlagen intelligenter Systeme*. Wissenschaftliche Buchgesellschaft: Darmstadt.

Mainzer, K. (2004): Dynamical Systems. In: A. Scott (Hrsg.): *Encyclopedia of Nonlinear Science*, Fitzroy Dearborn: London, 240–241.

Mainzer, K. (2005): *Symmetry and Complexity. The Spirit and Beauty of Nonlinear Science*. World Scientific: Singapore.

Mainzer, K. (2007a): *Thinking in Complexity. The Computational Dynamics of Matter, Mind, and Mankind*. Springer: Berlin 5th edition.

Mainzer, K. (2007b): *Der kreative Zufall. Wie das Neue in die Welt kommt*. C.H. Beck: München.

Mainzer, K. (2008): Organic computing and complex dynamical systems. Conceptual foundations and interdisciplinary perspectives. In: R.P. Würtz (ed.): *Organic Computing*. Springer: Berlin, 105–122.

Mainzer, K. (ed.) (2009a): Complexity. *European Review* 17. Cambridge University Press: Cambridge.

Mainzer, K. (2009b): Challenges of complexity in economics. In: *Evolutionary and Institutional Economics Review*. Japan Association for Evolutionary Economics 6(1), 1–22.

Mainzer, K. (2010): *Leben als Maschine? Von der Systembiologie zur Robotik und Künstlichen Intelligenz*. Mentis: Paderborn.

Mainzer, K.; Chua, L. O. (2011): *The Universe as Automaton. From Simplicity and Symmetry to Complexity*. Springer: Berlin.

Mainzer, K.; Chua, L. (2013): *Local Activity Principle*. Imperial College Press: London.

Mainzer, K. (2014a): *Die Berechnung der Welt. Von der Weltformel zu Big Data*. C.H. Beck: Munich.

Mainzer, K. (2014b): The new role of mathematical risk modeling and its importance for society. In: Klüppelberg, C.; Straub, D.; Welpe, I.M. (eds.): *Risk — A Multidisciplinary Introduction*. Springer: Berlin.

Mainzer, K. (2016a): *Künstliche Intelligenz. Wann übernehmen die Maschinen?* Springer: Berlin.

Mainzer, K. (2016b): *Information. Algorithmus-Wahrscheinlichkeit-Komplexität-Quantenwelt-Leben-Gehirn-Gesellschaft*. Berlin University Press: Berlin.

Mandelbrot, B.B. (1990): *The Fractal Geometry of Nature*. Spektrum Akademischer Verlag: Heidelberg.

Mandelbrot, B.B. (1997): *Fractals and Scaling in Finance. Discontinuity, Concentration, Risk*. Springer: New York.

Mandelbrot, B.B. (1999): *Multifractals and 1/f Noise. Wild-self-Affinity in Physics*. Springer: New York.

Mandelbrot, B.B.; Hudson, R.L. (2004): *The (mis)Behaviour of Markest. A Fractal View of Risk, Ruin, and Reward.* Basic Books: New York.

Mandelkern, M. (1988a): Limited Omniscience and the Bolzano-Weierstraß Principle. In: *Bull. London Math. Soc.* 20, 319–320.

Mandelkern, M. (1988b): Constructive complete finite sets. In: *Z. Math. Logik u. Grundlagen Math.* 34, 97–103.

Mantegna, R.N.; Stanley, H.E. (2000): *An Introduction to Econophysics. Correlations and Complexity in Finance.* Cambridge University Press: Cambridge.

Markowitz, H.M. (1959): *Portfolio Selection: Efficient Diversification of Investments.* John Wiley & Sons Inc.: New York.

Martin-Löf, P. (1975): An intuitionistic theory of types: Predicative part. In: H.E. Rose, J. Shepherdson (Eds.), *Logic Colloquium '73.* Amsterdam: North-Holland, 73–118.

Martin-Löf, P. (1982): Constructive mathematics and computer programming. In: L.J. Cohen, J. Los, H. Pfeiffer, K.-P. Podewski (Eds.), *Logic, Methodology and Philosophy of Science VI, Proceedings of the 1979 International Congress at Hannover, Germany.* North-Holland Publishing Company: Amsterdam, 153–175.

Martin-Löf, P. (1998): An intuitionistic theory of types. Twenty-five years of constructive type theory (Venice, 1995). In: *Oxford Logic Guides* 36, Oxford University Press: New York, 127–172.

Martin-Löf, P. (2009): 100 years of Zermelo's Axiom of Choice: What was the problem with it? In: S. Lindström, E. Palmgren, K. Segerberg, V. Stoltenberg-Hansen (Eds.), *Logicism, Intuitionism, and Formalism: What has become of them?* Springer: Dordrecht: 209–219.

Mataric, M.; Sukhatme, G.; Ostergaard, E. (2003): Multi-robot task allocation in uncertain environments. In: *Autonomous Robots* 14(2–3), 253–261.

Matiyasevich, Y.V. (1970): Enumerable sets are diophantic. In: Soviet Math. Dokl. 11, 345–357.

Matiyasevich, Y.V. (1993): *Hilbert's Tenth Problem.* The MIT Press: Cambridge MA.

Maxwell, J.C. (1871): *Theory of Heat.* Longmans: London.

Mayr, E.; Prömel, H.; Steger, A. (Eds.) (1998): *Lectures on Proof Verification and Approximation Algorithms.* Lecture Notes in Computer Science vol. 1967. Springer: Berlin.

McCauley, J.L. (2004): *Dynamics of Markets. Econophysics and Finance.* Cambridge University Press: Cambridge.

Minsky, M.L. (1961): Recursive unsolvability of Post's problem of "tag" and other topics in the theory of turing machines. In: *Annals of Math.* 74, 437–454.

Minsky, M.L. (1967): *Finite and Infinite Machines*. Prentice Hall: Engelwood Cliffs.

Moschovakis, Y. (1995): The logic of functional recursion. In: *Logic and Scientific Methods. 10th Intern. Congr. Logic, Methodology and Philosophy of Science*. Kluwer Academic Publishers: Dordrecht, 179–208.

Myers, G. W.; Casiado, R. (2009): *Algorithms in Bioinformatics*. 5th International Workshop WABI. Springer: Berlin.

Ng, Y.J. (2001): From computation to black holes and space-time foam. In: *Physical Review Letters* 86, 2946–2949.

Neumann, J. von (1932): *Physical Applications of the Ergodic Hypothesis*. In: *Proc. Natl. Acad. Sci. (USA)* 18(3), 263–266.

Neumann, J. von (1948): The general and logical theory of automata. In: Jeffress, L.A. (Ed.). *Cerebral Mechanisms in Behavior — The Hixon Symposium*. California Institute of Technology: Pasadena, 1–31.

Neumann, J. von (1963): *Collected Works* (ed. A.H. Taub). Volume A. Pergamon Press: Oxford.

Nicolis, G.; Prigogine, I. (1987): *Die Erforschung des Komplexen*. Piper: München.

Nicolis, G.; Nicolis, C. (2012): *Foundations of Complex Systems. Emergence, Information, and Prediction*. World Scientific: Singapore.

Nielsen, M.A.; Chuang, I.L. (2001): *Quantum Computation and Quantum Information*. Cambridge University Press: Cambridge.

Nipkow, T.; Paulson, L.C.; Wenzel, M. (2002): *Isabelle/HOL. A Proof Assistant for Higher-Order Logic*. Springer: Heidelberg.

Nolfi, S.; Floreano, D. (2001): *Evolutionary Robotics. The Biology, Intelligence, and Technology of Self-Organizing Machines*. The MIT Press: Cambridge M.A. 2nd edition.

Ockham, W. of (1967–1988): *Opera philosophica et theologica*. Ed. G. Gál *et al.* 17 vols. Franciscan Institute St. Bonaventure: New York.

Ockham, W. of (1998): *On the Power of Emperors and Popes*. Translated by A.S. Brett. University of Durham.

Ord, T. (2002): *Hypercomputation: Computing more than the Turing machine*. Report at the Department of Computer Science. The University of Melbourne.

Ord, T. (2006): The many forms of hypercomputation. In: *Applied Mathematics and Computation* 178, 143–153.

Palmgren, E. (1998): On universes in type theory. In: G. Sambin, J.M. Smith (eds.), *Twenty-five years of constructive type theory*, Clarendon Press: Oxford, 191–204.

Pappi Alexandrini *collectionis quae supersunt* 3 vols. (ed. F.O. Hultsch). Berlin 1876–1878.

Pascucci, A. (2011): *PDE and Martingale Methods in Option Pricing.* Springer: Berlin.

Penrose, R. (1966): *An Analysis of the Structure of Space-time.* Adams Prize Essay. Cambridge University Press: Cambridge.

Penrose, R. (1991): *The Emperor's New Mind.* Penguin: London.

Pfeifer, R.; Scheier, C. (2001): *Understanding Intelligence.* The MIT Press: Cambridge MA.

Pinasco, J.P. (2009): New Proofs of Euclid's and Euler's theorems. In: *American Mathematical Monthly* 116(2), 172–173.

Plotkin, G.D. (1977): LCF considered as a programming language. In: *Theoretical Computer Science* 5, 223–255.

Pohlers, W. (1989): *Proof Theory.* Springer: Berlin.

Post, E.L. (1936): Finite combinatory processes — Formulation I. In: *Symbolic Logic* I, 103–105.

Rathjen, M.; Edward R.; Griffor, E.R.; Palmgren, E. (1998): Inaccessibility in constructive set theory and type theory. In: *Annals of Pure and Applied Logic* 94, 181–200.

Richard, J. (1905): Les principes des mathématiques et le problème des ensembles. In: *Revue générales des sciences pures et appliquées* 16, 541.

Riedel. F. (2006); Dynamic convex risk measures: time consistency, robustness, and the variational representation of preferences. In: *Econometrica* 74, 1447–1498.

Robinson, J. (1949): Definability and decision problems in arithmetic. In: *Journal of Symbolic Logic* 14, 98–114.

Rodin, A. (2014): *Axiomatic Method and Category Theory.* Springer: Heidelberg.

Rössler, O.E. (1976): An Equation for continuous chaos. In: *Physics Letters* 57A(5), 397–398.

Rogers, H. (1967): *Theory of Recursive Functions and Effective Computability.* McGraw-Hill.

Russell, B. (1908): Mathematical logic as based on the theory of types. In: *Amer. Journal of Mathematics* 30, 222–262.

Russell, B. (1936): The limits of empiricism. In: *Proceedings of the Aristotelian Society* 36, 131–150.

Russell, B. (2000): History of Western Philosophy. Allen & Unwin: London, 462–463.

Savage, J.E. (1976): *The Complexity of Computing.* Wiley: New York.

Scheinkman, J.A.; M. Woodford, M. (2001): Self-organized criticality and economic fluctuations. In: *American Economic Review,* 417–421.

Scheinkman, J.A. (1990): Nonlinearities in economic dynamics. In: *Economic Journal,* Supplement, 100(400), 33–47.

Schmidhuber, J. (2002): Hierarchies of generalized Kolmogorov complexities and non-enumerable universal measures computable in the limit. In: *International Journal of Foundations of Computer Science* 13(4), 587–612.

Scholz, H. (1961): *Mathesis Universalis. Abhandlungen zur Philosophie als strenger Wissenschaft*, hrsg. von H. Hermes, F. Kambartel, J. Ritter. Benno Schwabe & Co: Basel.

Schumacher, B. (1995): Quantum coding. In: *Physical Review A.* 51(4), 2738–2747.

Schütte, K. (1977): *Proof Theory.* Springer: Berlin.

Schütte, K. (1965): Predicative well-orderings. In: *Formal Systems and Recursive Functions.* North-Holland: Amsterdam, 280–303.

Schwichtenberg, H. (2006): Minlog. In: F. Wiedijk (ed.): *The Seventeen Provers of the World.* Lecture Notes in Artificial Intelligence vol. 3600. Springer: Berlin, 151–157.

Schwichtenberg, H.; Wainer, S. S. (2012): *Proofs and Computations.* Cambridge.

Scott, D. (1982): Domains for denotational semantics. In: Nielsen, E.; Schmidt, E.M. (eds.): *Automata, Languages, and Programming.* Lecture Notes in Computer Science 140. Springer: Berlin, 577–613.

Setzer, A. (2000): Extending Martin-Löf type theory by one Mahlo-universe. In: *Archive for Mathematical Logic* 39, 155–181.

Sharpe, W.F. (1964): Capital asset prices: A theory of market equilibrium under conditions of risk. In: *Journal of Finance* 19, 425–442.

Sheperdson, J.C.; Sturgis, H.E. (1963): Computability of recursive functions. In: *J. Assoc. Comp. Mach.* 10, 217–255.

Shoenfield, J.R. (1967): *Mathematical Logic.* Reading (Mass.).

Shor, P.W. (1996): Polynomial-time algorithms for prime factorization and discrete logarithms on a quantum computer. In: *SIAM J. Computing* 26, 1484–1509.

Shub, M. (1993): Some rmarks on Bezout's theorem and complexity theory. In: Hirsch, M.; Marsden, J.; Shub, M. (Eds.): *From Topology to Computation.* Proceedings of the Smalefest. Springer: New York, 443–455.

Sieg, W. (2013): *Hilbert's Programs and Beyond.* Oxford University Press: New York.

Siegelmann, H.T.; Sontag, E. D. (1994): Analog computation via neural networks. In: *Theoretical Computer Science* 131, 331–360.

Siegelmann, H.T.; Sontag, E.D. (1995): On the computational power of neural nets. In: *Journal of Computer and Systems Science* 50, 132–150.

Siegelmann, H.T. (1995): Computation beyond Turing limit. In: *Science* 268, 545–548.

Siegelmann, H.T. (1999): *Neural Networks and Analog Computation. Beyond the Turing Limit.* Springer-Sience + Business Media: New York.

Simpson, S.G. (1999): *Subsystems of Second Order Arithmetic. Perspectives in Mathematical Logic.* Springer: Berlin.

Simpson, S.G. (2005): *Reverse Mathematics.* Lecture Notes in Logic 21. The Association of Symbolic Logic 2005.

Sinn, H.W. (2010): *Kasino-Kapitalismus.* Ullstein-Verlag: Berlin.

Skolem, T. (1967): Logico-combinatorial investigations in the satisfiability or provability of mathematical propositions: A simplified proof of a theorem by L. Löwenheim and generalizations of the theorem. In: van Heijenoort, J.: *From Frege to Gödel: A Source Book in Mathematical Logic, 1879–1931.* Harvard University Press 1967, 252–263.

Smale, S. (1974): A mathematical model of two cells via Turing's equation. In: *Lectures in Applied Mathematics*, Vol. 6 (American Mathematical Society) 15–26.

Small, M. (2005): *Applied Nonlinear Time series Analysis: Applications in Physics, Physiology, and Finance.* World Scientific: Singapore.

Soare, R.I. (1987): *Recursively Enumerable Sets and Degrees.* Perspectives in Mathematical Logic. Springer: Berlin.

Soare, R.I. (2016): *Turing-Computability. Theory and Applications.* Springer. New York.

Solomonoff, R. (1964): A formal theory of inductive inference Part I. In: *Information and Control Part* I–II 7(1–2), 224–254.

Spade, P.V. (1994): *Five Texts on the Medieval Problem of Universals: Porphyry, Boethius, Abelard, Duns Scotus*, Ockham. Hackett: Indianapolis IN.

Spade, P.V. (1999): *The Cambridge Companion to Ockham.* Cambridge University Press: Cambridge.

Spanier, E.H. (1966): *Algebraic Topology.* Springer: Berlin.

Specker, E. (1949): Nicht konstruktiv beweisbare Sätze der Analysis. In: *Journal of Symbolic Logic* 14, 145–158.

Spector, C. (1955): Recursive well-orderings. In: *Journal of Symbolic Logic* 20, 151–163.

Spector, C. (1962): Provably recursive functionals of analysis: a consistency proof of analysis by an extension of principles formulated in current intuitionistic mathematics. In: Dekker, J.C.E. (ed.): *Recursive Function Theory.* Proceedings of Symposia in Pure Mathematics 5. American Mathematical Society: Providence, 1–27.

Stoer, J.; Bulirsch, R. (2002): Introduction to Numerical Analysis. Springer: Berlin 3rd edition.

Stutz, R.M. (2009): Was Risikomanager falsch machen. In: *Havard Business Manager*, April, 67–75.

Sundholm, G. (2014): Constructive recursive functions, Church's thesis, and Brouwer's theory of the creative subject: Afterthoughts on a Parisian Joint session. In: *Dubucs Bourdeau 2014*, 1–35.

Takayasu, M.; Fukuda, K.; Takayasu, H. (1999): Application of statistical physics to the Internet traffic. In: *Physica A* 274 (1), 140–148.

Takayasu, M.; Takayasu, H.; Sato, T. (1996): Critical behaviors and 1/f noise in information traffic. In: *Physica A* 233, 824–834.

Tarski, A. (1951): *A Decision Method for Elementary Algebra and Geometry*. University of California Press: Berkeley.

Tegmark, M. (2014): *Our Mathematical Universe*. Alfred A. Knopf: New York.

The Univalent Foundations Program (2013): *Homotopy Type Theory: Univalent Foundations of Mathematics*. Institute for Advanced Study: Princeton, NJ.

Toffoli, T. (1977): Computation and construction universality of reversible cellular automata. In: *Journal of Computcr and Syatcm Scicnccs* 15, 213–231.

Troelstra, A.S. (1968): The theory of choice sequences. In: van Rootselaar, B.; Staal, J.F. (Eds.). *Logic, Methodology, and Philosophy of Science* 3. North-Holland: Amsterdam, 201–223.

Troelstra, A.S. (ed.) (1973): *Metamathematical Investigation of Intuitionistic Arithmetic and Analysis*. Springer: Berlin.

Troelstra, A.S. (1974): Note on the fan theorem. In: *Journal of Symbolic Logic* 39, 584–596.

Troelstra, A.S. (1977): Some models for intuitionistic finite type arithmetic with fan functional. In: *Journal of Symbolic Logic* 42, 194–202.

Troelstra, A.S.; van Dalen, D. (1988): *Constructivism in Mathematics*. An Introduction. 2 vols. North-Holland: Amsterdam.

Troelstra, A.S.; Schwichtenberg, H. (2000): *Basic Proof Theory*. Cambridge University Press: Cambridge 2nd edition.

Turing, A.M. (1936–1937): On computable numbers, with an application to the Entscheidungsproblem. In: *Proc. London Math. Soc.* Ser. 2 (42), 230–265.

Turing, A.M. (1939): Systems of logic based on ordinals. In: *Proc. London Math. Soc.* 2, 161–228.

Turing, A.M. (1950): Computing machinery and intelligence. In: A.M. Turing: *Intelligence Service. Schriften.* Berlin 1987, 147–182.

Turing, A.M. (1952): The chemical basis of morphogenesis. In: *Philos. Trans. Roy. Soc. London.* Series B 237, 37–72.

Turner, W. (1913): William of Ockham. In: Herbermann, C.: *Catholic Encyclopedia*. Robert Appleton Company: New York.

Veblen, O. (1908): Continuous Increasing Functions of Finite and Transfinite Ordinals. In: *Transactions of the American Mathematical Society* 9(3), 280–292.

Wätjen, D. (2004): *Kryptographie. Grundlagen, Algorithmen, Protokolle.* Spektrum akademischer Verlag: Heidelberg.

Voevodsky, V. (2012): A universe polymorphic type system. `http:// uf-ias-2012.wikispaces.com/file/view/Universe+polymorphic+ type+sytem.pdf`

Weber, H. (1893): Leopold Kronecker, in: *Jahresbericht der Deutschen Mathematiker-Vereinigung* 2, 1893, 19.

Wedde, H.J.; Lehnhoff, S.; Rehtanz, C.; Krause, O. (2008): Von eingebetteten Systemen zu Cyber-Physical Systems. Eine neue Forschungsdimension für verteilte eingebettete Realzeitsysteme. *Pearl 2008 — Informatik Aktuell.* Aktuelle Anwendungen in Technik und Wirtschaft 2007 12.

Weiser, M. (1991): The computer for the 21st century. In: *Scientific American* 9, 66–75.

Weizsäcker, C.F von (1985): *Aufbau der Physik.* Carl Hanser: München.

Weizsäcker, C.F. von (1992): *Zeit und Wissen.* Carl Hanser: München.

Weyl, H. (1918): *Das Kontinuum. Kritische Untersuchungen über die Grundlagen der Analysis.* De Gruyter: Leipzig.

Weyl, H. (1921): Über die neue Grundlagenkrise der Mathematik. In: *Mathematische Zeitschrift* 10 1921, 39–79.

Weyl, H. (1923): *Raum, Zeit, Materie. Vorlesungen über allgemeine Relativitätstheorie.* Berlin 1923, repr. Wissenschaftliche Buchgesellschaft: Darmstadt 1961.

Weyl, H. (1931): *Gruppentheorie und Quantenmechanik.* Hirzel: Leipzig.

Weyl, H. (1968): *Gesammelte Abhandlungen* 4 Bde. Springer: Berlin.

Weyl, H. (1927): *Philosophie der Mathematik und Naturwissenschaft.* R. Oldenburg: Munich 5th edition 1982.

Wheeler, J.A. (1990): Information, physics, quantum: The search for links. In: Zurek, W. H. (ed.): *Complexity, Entropy, and the Physics of Information.* Addison-Wesley: Redwood City, California.

Wigner, E.P. (1960): The unreasonable effectiveness of mathematics in the natural sciences. In: *Communications on Pure and Applied Mathematics* 13(1), 1–14.

Williams, D. (1991): *Probability with Martingales. Cambridge.* University Press: Cambridge.

Williams, M.R. (1985): A History of Computing Technology. Prentice-Hall: Englewoo Cliffs.

Wilson, E.O. (1971): *The Insect Society.* MIT Press: Cambridge MA.

Wilson, E.O. (2000): *Sociobiology: The New Synthesis.* 25th Anniversary Edition. MIT Press Cambridge MA.

Wolfram, S. (1986): *Theory and Applications of Cellular Automata.* World Scientific: Singapore.

Wolfram, S. (1994): *Cellular Automata and Complexity.* Addison–Wesley: Reading MA.

Wolfram, S. (2002): *A New Kind of Science.* Wolfram Media, Inc.: Champaign Il.

Wolfram, S. (2005): *The Mathematica Book.* Cambridge University Press: Cambridge 5th edition.

Wootters, K.; Zurek, W.H. (1982): A single quantum cannot be cloned. In: *Nature* 299, 802.

Yasugi, M. (1963): Intuitionistic analysis and Gödel's interpretation. In: *Journal of Mathematical Society of Japan* 15, 101–112.

York, M. (ed.) (2008): *Aspects of Mathematical Finance.* Springer: Berlin.

Zurek, W. (ed.) (1989): *Complexity, Entropy, and the Physics of Information.* Addison-Wesley: Redwood City.

Zuse, K. (1969): *Rechnender Raum.* Friedrich Vieweg & Sohn: Braunschweig.

Author Index

Subject Index

The Digital and the Real World

Computational Foundations of Mathematics, Science, Technology, and Philosophy

Klaus Mainzer

Technical University of Munich, Germany

Chapter 2:

p. 27, l. 16: correct (Hermes, 1961, 189)

p. 30, l. −7: add (Chaitin, 1998, 10)

p. 32, l. 11: add (Matiysevich, 1970, 1993)

Chapter 3:

p. 42, l. 2: replace "(Shoenfield, 1967)" by "The results on pages 42–47 are taken from Shoenfield, 1967, Chapters 7.6–7.8."

p. 44, l. −6: add "For the definition of sequence numbers of tupels see Shoenfield, 1967, 6.4."

p. 45, l. −4: replace "with…" by "for all y with $3 \cdot 5^y \in \sigma$"

Chapter 4:

p. 59, l. 1: add "(Troelstra and van Dalen, 1988, vol. 1, 43)"

p. 65, l. −4: add "The following treatment refers to Troelstra and Schwichtenberg, 2000, 261–264."

p. 75, l. −3: extend "(Feferman, 1964, 1968a; Mainzer, 1973; Pohlers, 1989)"

Chapter 5:

p. 81, l. 20: add "The following axioms are minor modifications of Scott's axioms by Schwichtenberg and Wainer, 2012, 251 (due to K.G. Larsen and G. Winskel, Using information systems to solve recursive domain equations. In: Information and Computation 91, 232–258 (1991)."

p. 84, l. 6: add "The following results are given in Schwichtenberg and Wainer, 2012, Chapters 6.1.3, 6.1.5."

Chapter 6:

p. 92, l. 2: add "(Troelstra and van Dalen, 1988, 187–188)"

p. 93, l. 13: add "(Troelstra and van Dalen, 1988, 208–2010)"

p. 94, l. 12: add "(Troelstra and van Dalen, 1988, 211–212)"

p. 95, l. 7: add "(Troelstra and van Dalen, 1988, 217–218)"

p. 96, l. 11: add "… Fig. 7 (Troelstra and van Dalen, 1988, 222)"

p. 98, l. 1: add "(Troelstra and van Dalen, 1988, 221)"

p. 99, l. 10: add "(Troelstra and van Dalen, 1988, 229)"

Chapter 7:

p. 101, l. −6: This chapter relies crucially on the book Kohlenbach, 2008 from which the author borrowed freely and extensively. Naturally, the errata listed below (and all remaining ones) are all due to the present author and not to U. Kohlenbach.

p. 102: cf. Kohlenbach, 2008, p.15

pp. 103–104: cf. Kohlenbach, 2008, pp. 13–14

pp. 104–106: cf. Kohlenbach, 2008, pp. 22–25

p. 105, l. 4: add "The converse implication holds in the logic $PL_{-=}$ without equality. But, for the formal extension $PL^2_{-=}$ of $PL_{-=}$ by ..."

p. 106, l. 7, 11, 12: delete $\forall x$

pp. 107–108: cf. Kohlenbach, 2008, pp. 26–27

p. 108, l. 4: "realize" instead of "realizes"

p. 108, l. 14: "on the sentence" instead of "of the sentence"

p. 108, l. 16: "The no-counterexample..." instead of "No-counterexample ..."

p. 108, l. 17–19: delete "For example ... recursive functionals."

p. 109, l. 12: "of the no-counterexample" instead of "of no-counterexample"

p. 109, l. 15: replace "sufficient" by "necessary"

pp. 109–112: cf. Kohlenbach, 2008, pp. 28, 31–35

p. 110, l. −13: "modus ponens rule" instead of "modus ponens conclusion"

p. 110, l. −8: "In the language" instead of "In language"

p. 111, l. −10: delete φ_3

p. 112, l. 7: delete $\exists x, g, v$

pp. 113–115: cf. Kohlenbach, 2008, pp. 47–49

p. 115, l. 10: replace yz by $0\ \underline{y}\ \underline{z}$

p. 115, l. 11: replace z by $\underline{z}$

p. 115, l. −7: delete "Before generalizing his concept," and start "Let us …"

p. 116, l. 14: delete "generalized"

pp. 116 (bottom)–120 (top): cf. Kohlenbach, 2008, pp. 97–101

p. 117, l. −4: delete "under the condition of $E - HA^\omega + AC + IP^\omega_{ef} + \Delta_{ef} \vdash A \vee A \to A$"

p. 118, point (5): add "Verification is easy."

p. 120, l. 1: add "For the full set-theoretic structure $\mathcal{S}^\omega$, it follows …"

p. 120, l. 2: replace $\mathcal{L}^\omega$ by $\mathcal{S}^\omega$

p. 120, l. 6: delete line 6

p. 123, l. 5: replace "functional" by "uniform"

p. 123, l. 13: add "The bound is uniform, because it does not depend on y. The extraction needs a variant of modified realizability mr (modified realizability mrt with truth) (see Kohlenbach, 2008, 98)."

p. 123, l. −7: correct "terms $\underline{t}$ such that"

p. 123, l. −5: correct "closed terms $\underline{t}^*$"

p. 123, l. −1: replace "program extraction" by "uniform bound extraction"

p. 123: cf. Kohlenbach, 2008, pp. 111, 115

p. 124, l. 19: add "(The proof of this extraction theorem needs mrt.)"

p. 124: cf. Kohlenbach, 2008, pp. 116, 123, 125

p. 125, l. 3: delete M^ω and add "… with A_0 quantifier-free and $\underline{x}$ of type 0."

p. 125, l. 4: delete "of M^ω"

p. 125, l. 14: add "The Markov Principle in all finite types is defined by $M^\omega := \bigcup_{\underline{\rho} \in T}\{M^{\underline{\rho}}\}$, where $M^{\underline{\rho}}: \neg\neg\exists \underline{x}^{\underline{\rho}}\, A_0(\underline{x}^{\underline{\rho}}) \to \exists \underline{x}^{\underline{\rho}} A_0(\underline{x}^{\underline{\rho}})$ with A_0 arbitrary quantifier-free formula of WE-HA$^\omega$ and $\underline{x}$ of arbitrary types $\underline{\rho}$. The formula $A_0(\underline{x}^{\underline{\rho}})$ may contain further free variables in addition to $\underline{x}$."

p. 125: cf. Kohlenbach, 2008, p. 126

p. 126: cf. Kohlenbach, 2008, pp. 129, 130

p. 127, l. 6: delete "under condition …"

p. 127, l. 11: replace $A_D\left(\underline{x}, \underline{Y}'\, z\underline{x}\,\underline{x}'\,\underline{y}'', \underline{a}\right)$ by $A_D\left(\underline{x}', \underline{Y}'\, z\underline{x}\,\underline{x}'\,\underline{y}'', \underline{a}\right)$

pp. 127–131 (top): cf. Kohlenbach, 2008, pp. 130–138

p. 128, l. 12: add "… induction rule (from which, in this case, the induction axiom can be derived)"

p. 129, l. 4: replace $(\neg\neg A \to A)^D$ by $\neg\neg A \to A$

p. 129, l. 10: add "… E $-$ HA$^\omega$ (see Kohlenbach, 2008, 126–127)."

pp. 134 (bottom)–135: cf. Kohlenbach, 2008, pp. 141, 142

p. 136, l. 3: add "Theorem of uniform bound…"

p. 136: cf. Kohlenbach, 2008, p. 145

p. 137, l. 17: replace "an infinitely" by "a finitely"

p. 137, l. −9, −10: correct "KL can be derived from WKL if we assume the axiom of choice."

pp. 137–138 (top): cf. Kohlenbach, pp. 149–150

pp. 138 (below)–139 (top): cf. Kohlenbach, 2008, pp. 163, 165–166, 168

p. 138, l. −9: insert "QF (quantifier-free)"

p. 139, l. −4: correct $\underline{t}^*$

pp. 139 (below)–140 (middle): cf. Kohlenbach, 2008 pp. 174–175

p. 140, l. 1: insert "Theorem of uniform bound …"

p. 140, l. 14: insert "… König's lemma (UWKL). The following results stem back to U. Kohlenbach, On uniform weak König's lemma. In: Pure Appl. Log. 114, 103–116 (2002)."

pp. 141–142 (top): cf. Kohlenbach, 2008, pp. 180, 182–184

p. 142, l. −2; p. 143, l. 6: replace QF-AC^{1,0} by QF-AC^{0,0}

pp. 142 (bottom)–147 (top): cf. Kohlenbach, 2008, pp. 199–211

p. 143, l. −1: replace $A_D(y, uy, v))^D$ by $A_D(y, \underline{u}y, \underline{v}))^D$

p. 144, l. 1: replace $B(AxB)$ by $\underline{B}(\underline{A}x\underline{B})$

p. 144, l. 12: replace $\underline{A}(\underline{t_U}, \underline{t_B})$ by $\underline{A}(t_x, \underline{t_B})$

p. 145, l. 21: WE-PA$^\omega$ + QF-AC +DC $\vdash A(\underline{a})$

p. 145, l. −8: replace by $\underline{t}, \underline{a}$ by $\underline{ta}$

p. 147, l. 6: replace BR$_{0,1}$ by B$_{0,1}$

p. 147, l. 18: replace "Kohlenbach (1999) … classical inconsistencies" by "The following uniform Σ_1^0-boundedness principle Σ_1^0-UB is also classically inconsistent, but consistent in, e.g., system E-PA$^\omega$. Therefore, in the following theorem, Σ_1^0-UB can be applied to get a classically correct bound (Kohlenbach, 2008, 225)."

p. 147, l. $-4, -3$: delete "The uniform … given proof."

pp. 147 (bottom)–148 (middle): cf. Kohlenbach, 2008, pp. 225, 228

p. 148, l. 9: replace "Bolzano-Weierstraß" by "Dini's theorem"

p. 149, l. 9: correct $j_1 n_1, j_1 n_2$

pp. 149–152 (middle): cf. Kohlenbach, 2008 pp. 78–83

p. 152, l. -1, p. 153, l. 13: correct $\mathrm{WA} \overset{\frown}{-} \mathrm{HA}^\omega \upharpoonright$

pp. 152 (middle)–153(middle): cf. Kohlenbach, pp. 269, 271

p. 153, l. 6: The following application refers to a hierarchy of systems $G_n A^\omega$ corresponding to the Grzegorczyk hierarchy (Kohlenbach, 2008, 55).

p. 153, l. -13: replace "(Gerhardy and Kohlenbach, 2008)" by "(Kohlenbach, 1993)"

pp. 154–155 (middle): cf. Kohlenbach, 2008, pp. 279–280

p. 155, l. 3: correct $=_X$

p. 155, l. -2: add $\forall k \in X$

p. 155, l. -1: correct 2^{-k}

pp. 155 (middle)–156 (middle): cf. Kohlenbach, 2008, pp. 284–285

p. 156, l. 3: correct $d_K(y_1, y_2) <_{\mathbb{R}} 2^{-k}$

p. 156, l.9: replace WE-PA$^\omega$ by WE-HA$^\omega$

p. 156, l. 24:
$$\forall x \in X \; \forall y \in K \; \forall k \in \mathbb{N} \; \exists n \leq \delta(x,k)(F(x,y,n)) <_{\mathbb{R}} 2^{-k})$$

p. 156, l. 25:
$$\forall x \in X \; \forall y \in K \; \forall k \in \mathbb{N} \; \forall n \geq \delta(x,k)(F(x,y,n)) <_{\mathbb{R}} 2^{-k})$$

p. 156 (2nd half): cf. Kohlenbach, 2008, pp. 291, 388, 389, 392, 436, 437, 442, 443, 446

p. 157, l. 18: replace $st(x)$ by $S(x)$

p. 157, l. 19: correct T^X

pp. 157–158: cf. Kohlenbach, 2008, pp. 382, 388, 389, 392

p. 158, l. 8: add "A $\forall$-formula (resp. $\exists$-formula) A has the form $A \equiv \forall \underline{a}^{\underline{\sigma}} A_{qf}(\underline{a})$ (resp. $A \equiv \exists \underline{a}^{\underline{\sigma}} A_{qf}(\underline{a})$) with A_{qf} quantifier-free and types $\underline{\sigma}$ of degree 1^* or $(1, X)$."

p. 158, l. 13–17:
$$\forall x^\sigma \; \forall y \leq_\rho s(x) \; \forall z^\tau \big(\forall u^0 B_\forall(x,y,z,u) \to \exists v^0 \, C_\exists(x,y,z.v) \big)$$
provable in $\mathcal{A}^\omega[X,d] \Longrightarrow \forall y \leq_\rho s(x) \forall z^\tau (\forall u \leq \Phi(x,b) B_\forall(x,y,z,u) \to \exists v \leq \Phi(x,b) \, C_\exists(x,y,z,v)$ holds …

p. 158, line 18: add "From a logical point of view, the metatheorems on abstract spaces are so-called semi-formal rules: The validity of the extracted bound follows from the provability of the premise. A sentence of $\mathcal{L}(\mathcal{A}^\omega[X,d])$ holds in a (nonempty) bounded metric space (X,d) if it holds in the full set-theoretic model of $\mathcal{A}^\omega[X,d]$ (in the sense of Kohlenbach, 2008, 392)."

pp. 158 (bottom)–159: cf. Kohlenbach, 2008, pp. 436, 438

p. 159, l. 13–17:
$$\forall x^\sigma \, \forall y \leq_\rho s(x) \, \forall z^\tau \big(\forall u^0 \, B_\forall(x,y,z,u) \to \exists v^0 \, C_\exists(x,y,z,v) \big)$$
provable in $\mathcal{A}^\omega[X,d] + \exists\text{-UB}^X \Rightarrow \forall y \leq_\rho s(x) \, \forall z^\tau (\forall u \leq \Phi(x,b)$
$B_\forall(x,y,z,u) \to \exists v \leq \Phi(x,b) \, C_\exists(x,y,z,v))$ holds …

p. 159, l. 20–21: replace "Applications … of metric spaces." by "$\exists$-UB^X, like Σ_1^0-UB, is also classically false, but the principle can be used to get true propositions of the special form in the bound extraction theorem. The principle can be understood as an effective version of a non-effective procedure in non-standard analysis: Classically false uniform propositions are true in structures which are generated by transition to ultrapowers X^U. For propositions of a special logical form, their validity in the given structure X follows from their validity in the ultrapowers X^U."

p. 159, l. −5: correct "incompleteness"

pp. 159 (bottom)–160: cf. Kohlenbach, 2008, pp. 442, 443, 446

pp. 161–162: cf. Kohlenbach, 2008, pp. 499, 500

p. 162, l. 3: add "This bound has been improved and generalized to uniformly convex Banach spaces in U. Kohlenbach, L. Leustean: A quantitative Mean Ergodic Theorem for uniformly convex Banach spaces. In: Ergodic Theory and Dynamical Systems Vol. 29, pp. 1907–1915 (2009)."

Chapter 8:

p. 165, l. −6: correct "… RCA$_0$ refers to the computable sets."

p. 166, l. −1: add "… arithmetical hierarchy (Simpson 2005; C. Fan, Reverse Mathematics 2010, 9 (http://www.math.uchicago.edu/~may/VIGRE/VIGRE2010/REUPapers/Fan.pdf)."

<u>p. 170, 1.14:</u> For the following results, consider Friedman and Simpson, 2000 and Simpson, 2005.

<u>p. 174. l. 13:</u> correct "... $\beta(n) = 0)) \to \cdots$"

Chapter 9:

<u>p. 183, l. 4:</u> add "(see Dybjer and Palmgreen, 2016, Chapter 3.4)".

<u>p. 184, l. 2:</u> add "(Dybjer and Palmgreen, 2016, Chapter 3.5)"

<u>p. 186, l. 5:</u> add "(see, e.g., E. Palmgren, The Grothendieck construction and models for dependent types. In: Notes March 2013, minor editing May 28, 2016 (http://staff.math.su.se/palmgren/ iterated_presheaves_and_dependent_types_v8.pdf); F. Lamarche, Modeling Martin-Löf type theories in categories. In: Journal of Applied Logic 1 2014, 28–44)"

<u>p. 186, l. 9:</u> add "(Dybjer and Palmgren, 2016, Chapter 3.8)"

<u>p. 187, l. −6:</u> add "(Dybjer and Palmgren, 2016, Chapter 4.3)"

<u>p. 190, l. −9:</u> add "... (CiC) (C. Paulin-Mohring, Introduction to the calculus of inductive constructions. In: HAL archives-ouvertes.fr (https://hal.inria.fr/hal-01094195/file/CIC.pdf); Calculus of constructions, Wikipedia https://en.wikipedia.org/wiki/Calculus_ of_constructions)"

<u>p. 192, l. −4:</u> add "(G. Huet, G. Kahn, C. Paulin-Mohring, The Coq Proof Assistant. A Tutorial, April 4 2013, TypiCal Project (formerly LogiCal) http://flint.cs.yale.edu/cs430/coq/pdf/Tutorial.pdf; The Coq Proof assistant. A short introduction to Coq, https://coq.inria.fr/ a-short-introduction-to-coq)"

p. 195, l. −12: add "(The Univalent Foundations Program 2013, 4, 77f; Homotopy type theory, Wikipedia https://en.wikipedia.org/wiki/Homotopy_type_theory)"

p. 198, l. 12: extend "(The Univalent Foundations Program 2013, 15)"

Chapter 10:

p. 202, l. −13: add: "The results of Chapters 10 and 11 are essentially taken from the excellent textbook on real computing of Blum, Cucker, Shub and Smale, 1998 and other papers of a research group of Steve Smale on these issues."

p. 203, l. −1: add "… (Fig. 9) (Blum *et al.*, 1998, 7)."

p. 204, l. −8: add "… (Fig. 10) (Blum *et al.*, 1998, 7)."

p. 205, l. −7: add "… (Fig. 12) (Blum *et al.*, 1998, 11)."

p. 206, l. −10: add "… (Fig. 13) (Blum *et al.*, 1998, 40)."

p. 207, l. 2: extend "(Blum *et al.*, 1998, 40–41)"

p. 207, l. −7: add "… halts (Blum *et al.*, 1998, 45)"

p. 208, l. 9: add "… Fig. 14 (L. Blum, F. Cucker, M. Shub, and S. Smale, Complexity and Real Computation: A Manifesto. In: International Journal of Bifurcation and Chaos 6, 11 (1996))."

p. 208, l. −1: extend "… (Blum *et al.*, 1989; Blum *et al.*, 1998, 71–72)"

p. 211, l. 12: add "… quasi-algebraic) sets (Blum *et al.*, 1998, 50)."

p. 211, l. −4: add "… over a ring (Blum *et al.*, 1998, 52)."

p. 212, l. 8: add "Newton's method (Blum *et al.*, 1998, 55–56)."

p. 213, l. 11: add "… for dynamical systems (L. Blum, F. Cucker, M. Shub, and S. Smale, Complexity and Real Computation: A Manifesto. In: International Journal of Bifurcation and Chaos 6, 6–7 (1996))."

p. 215, l. 9: extend "… Hotz, 1997):"

p. 216, l. 2: add "… (see Fig. 15) (Chadzelek, 1998, 10)"

p. 216, l. −5: add "… called δ-$\mathbb{Q}$-machine (Chadzelek, 1998, 13)"

p. 217, l. 9: add "… real functions (Chadzelek, 1998, 14):"

p. 218, l. 10: add "… halting problem (Chadzelek, 1998, 31):"

p. 221, l. 4: add "… generalized (Chadzelek, 1998, 53 f):"

p. 221, l. −2: extend "(Chadzelek, 1998, 58)"

p. 222, l. 5: add "… without division) (Chadzelek, 1998, 61–62)"

p. 224, l. −14: add "… proven (Chadzelek, 1998, 65):"

p. 225, l. −12: add "… chosen $t_0 = 0$ (Chadzelek, 1998, 66):"

p. 225, l. −6: "… on real computing (Blum *et al.*, 1998, 97–98)."

p. 226, l. 10: correct "sets"

p. 226, l. −1: correct "… contradiction to (a) and (b))."

p. 227, l. 5: replace "undecidability" by "decidability"

p. 227, l. 6: add "Proof by program (Blum *et al.*, 1998, 98)."

p. 227, l. −9: correct "c_S" and "χ_S"

Chapter 11:

p. 229, l. −9: add "… computing cost (Blum *et al.*, 1998, 86 f)."

p. 230, l. −10: add "(Blum *et al.*, 1998, 87)."

p. 232, l. 10: add "… In Table 1 (Blum *et al.*, 1998, 111)."

p. 233, l. −16: correct "P = NP"

p. 233, l. −15: correct "…, HN is decidable and …"

p. 234, l. 7: add "… permit (Blum *et al.*, 1998, 394):"

p. 235, l. 3–4: correct "… in a field (Blum *et al.*, 1998, 398). For a ring R, we define:"

p. 235, l. 9–10: replace $\mathbb{R}$ by R

p. 235, l. −1: correct "example, we consider $\Sigma_k^{\mathbb{R}}$-completeness of 4-FEAS (Blum *et al.*, 1998, 399)."

p. 237, l. 4: add "… machine concept (Blum *et al.*, 1998, Chapters 23.1–2)."

Chapter 12:

p. 243, l. 8: extend "… mathematical sense. This chapter is essentially taken from results and proofs of Hava T. Siegelmann *et al.* on these issues."

p. 244, l. 4: extend "(Siegelmann und Sontag, 1995, 134–35)"

p. 246, l. 17: add "… of regular languages (Siegelmann, 1999, 25–27)"

p. 250, l. 9: add "... response time $r(\omega) = T(\omega) + O(|\omega|)$ (Siegelmann und Sontag, 1995, 136)"

p. 252, l. −3: add "... encode ist input and output (Siegelmann und Sontag, 1994, 337 f)."

p. 253, l. −1: add "... output node (Siegelmann und Sontag, 1994, 338–339)."

p. 254, l. −10: extend "... (Siegelmann und Sontag, 1994, 339)."

p. 254, l. −5: sentence in italics

p. 255, l. −8: extend "...following steps (Siegelmann und Sontag, 1994, 340–341)."

p. 257. l. −3: extend "... (see Fig. 19) (Siegelmann und Sontag, 1994, 341)."

p. 258, l. 2: add "simulation network $\mathcal{N}_S(\mathcal{C})$ (Siegelmann und Sontag, 1994, 342 f):"

p. 259, l. 2: extend "*et al.* (1994, 343) suggested)"

p. 259, l. −5: add "... (Figs. 20 and 21) (Siegelmann und Sontag, 1994, 344)."

p. 261, l. −14: add "... in two steps (Siegelmann und Sontag, 1994, 345)."

p. 262, l. 12: extend "... Theorem (2) (Siegelmann und Sontag, 1994, 346 f)."

p. 263, l. −3: add "... following notations (Siegelmann und Sontag, 1994, 347–348)."

p. 265, l. −8: add "... the following way ... (Siegelmann und Sontag, 1994, 348–350)."

p. 267, l. −5: add "… threshold circuits (Siegelmann und Sontag, 1994, 350 f):"

p. 268, l. −5: add "polynomial time (Siegelmann und Sontag, 1994, 353):"

p. 268, l. −1: replace ":=" by "="

p. 270, l. 6: add "… for analog computing (Siegelmann und Sontag, 1994, 355)"

p. 271, l. −16: add "… total bit size n. With that in mind, we consider the simulation of generalized networks by neural networks (Siegelmann und Sontag, 1994, 358):"

p. 271, l. −5: extend "Siegelmann (1995a, 546)"

p. 272, l. −6: add "… by table (Siegelmann, 1995a, 547)"

p. 274, l. 12: add "… the binary alphabet $E = \{0, 1\}$ (Siegelmann, 1995a, 547):"

p. 276, l. −6: extend "… Fig. 22 (Siegelmann, 1999, 79)."

p. 278, l. −12: add "for $i > 1$. With that in mind, the following theorem holds (Siegelmann, 1999, 88):"

p. 280, l. 6: add "… closed under $O(\cdot)$ (Siegelmann, 1999, 89)."

Chapter 13:

p. 285, l. 14: extend "… 0 and 1 (Hoffmann, 2013, 343; Mainzer, 2016b, 30)?"

p. 286, l. 16: add "… (Fig. 23) (Hoffmann, 2013, 351)"

p. 289, l. −9: add "2013, 352):"

p. 291, l. 2: add "… following proposition (Hoffmann, 2013, 351–352) implies …"

p. 292, l. 15: "$n = |P_S|$ (Chaitin, 2007; Hoffmann, 2013, 359)"

Chapter 15:

p. 332, correct

A	B
0	1
1	0

Chapter 16:

p. 374, l. 1: replace "We set" by "J. Berger and G. Svindland use the following sets to find criteria for the existence of an arbitrage strategy (Berger and Svindland, 2016b,1162):"

p. 374, l. −12: add "With that in mind, Berger and Svindland prove the following lemma (Berger and Svindland, 2016b, 1167):"

p. 375, l. 5: replace "y_n" by "$\mathcal{Y}_n$"

p. 375, l. 10: add "… with $C \cdot p = \pi$. The following proof was given in Berger and Svindland, 2016b, 1168–1169:"

p. 375, l. −9: replace "x_n" by "$\mathcal{X}_n$"

p. 375, l. −3: correct "2016b" (delete "a")

p. 378, l. 9: add "… located subsets of Hilbert spaces (Berger and Svindland, 2016b, 1166)."